T0323932

Made in Africa

Made in Africa

Hominin Explorations and the Australian
Skeletal Evidence

Steve Webb
Professor of Australian Studies,
Bond University and Adjunct Professor,
Australian Centre for Human Evolution,
Griffith University, Brisbane, Australia

ACADEMIC PRESS

An imprint of Elsevier

Academic Press is an imprint of Elsevier
125 London Wall, London EC2Y 5AS, United Kingdom
525 B Street, Suite 1800, San Diego, CA 92101-4495, United States
50 Hampshire Street, 5th Floor, Cambridge, MA 02139, United States
The Boulevard, Langford Lane, Kidlington, Oxford OX5 1GB, United Kingdom

Notices
Knowledge and best practice in this field are constantly changing. As new research and experience
broaden our understanding, changes in research methods, professional practices, or medical treatment may become necessary.

Practitioners and researchers must always rely on their own experience and knowledge in evaluating
and using any information, methods, compounds, or experiments described herein. In using such
information or methods they should be mindful of their own safety and the safety of others, including parties for whom they have a professional responsibility.

To the fullest extent of the law, neither the Publisher nor the authors, contributors, or editors, assume
any liability for any injury and/or damage to persons or property as a matter of products liability,
negligence or otherwise, or from any use or operation of any methods, products, instructions, or
ideas contained in the material herein.

Library of Congress Cataloging-in-Publication Data
A catalog record for this book is available from the Library of Congress

British Library Cataloguing-in-Publication Data
A catalogue record for this book is available from the British Library

ISBN: 978-0-12-814798-6

For information on all Academic Press publications visit our website
at https://www.elsevier.com/books-and-journals

Working together
to grow libraries in
developing countries

www.elsevier.com • www.bookaid.org

Publisher: Andre Wolff
Acquisition Editor: Anna Valutkevich
Editorial Project Manager: Katerina Zaliva
Production Project Manager: Mohana Natarajan
Designer: Greg Harris

Typeset by Thomson Digital

Dedication

*This book is dedicated to all my fellow scientists
who work on many continents in many disciplines,
some I have had the privilege to meet, others not,
and who seek to understand the human story
as much as I do. I also want to include my children,
Ali, Sam and Charlotte, who are just brilliant,
funny, clever and happy, that is all you can ask for,
and none of whom work in human evolution.*

Contents

Part III
The Willandra Lake Collection: A Record

10. Willandra Lakes Skeletal Collection: A Photographic and Descriptive Catalogue

Introduction

Over the last 30 years the story of modern humans has been shown to be very complex and it is likely it will become increasingly so. We now know that the ancestors of everybody on the planet originally came from Africa but the journey to becoming who we are today is the complex part. Our knowledge of that process is growing through fossil and genomic research that are helping retrace those early footsteps now faded from our memory. Resurrecting memories is demanding, challenging and sometimes contains surprising revelations. The evidence for our journey is diverse, often strange and much is missing. Not everybody is an expert in the diverse research skills that combined make our investigations possible so we all must learn as we go along. The job is not simple nor do we know everything about our past. Sometimes what is discovered does not make sense; at other times those findings are so surprising they can easily be dismissed as error in our methods, mistakes, or poor judgement on the part of the researchers involved. Those are some of the reasons that the journey of the first people to Australia is not well understood. We are ignorant of specifics and specifics are what would allow us to understand modern human dispersals out of Africa and we do not have as much knowledge of those as we would like. Ironically, the evidence we in Australia have is now under threat through fossil reburials.

This book, therefore, is about *us*. All of us, wherever we live, whoever we are, whatever our colour, creed, shape or size. It is a story of our long collective history. It is about who we are and how we moved around the world, surviving Ice Ages and adapting to the many environments we found along the way and that allowed us to live almost anywhere on the planet. Why the topic of *us*, surely it's been done before? It has to some degree but I wanted to write it because I am passionate about us; our energy, enthusiasm, bravery, determination, many failures and, importantly, the way we reached Australia. In other words this book looks more at modern human travellers as people rather than just hominins on a chess board or objects robotically carrying out mechanical dispersals across vast landscapes. The story becomes even more special when you view it from Australia. That is because, rather than just trudging across thousands of kilometres of land, the people who arrived here needed to design and use a good watercraft capable of making an open ocean crossing and that was a significant first in human achievement.

I try to piece together the evidence we have so far for how people arrived in Australia, one of the distant end-points of modern human dispersal. I am not interested in being pedantic about detail or in all the components that might be associated with that journey; indeed, there is not enough evidence to be pedantic. Instead I want to put together the most likely story on the available evidence with a touch of poetic licence. To do that, I look at the background to the long and complicated trans-world movement itself. The complexity has not only made the story difficult to relate, there are also blanks in it. Moreover, the story is constantly changing because of expanding realm of fieldwork and the science and technology which is an integral part of archaeological method and which drives it along. There are also unexpected discoveries exposing diffuse strands of evidence that does not always make sense but instead raises many new questions and uncertainties.

Writing about Australia's first arrivals has never been undertaken against a background of what we know about the first Australians. Secondly, almost all those first Australians have never been described before on the international stage. There is a third reason, however, which is the most important one for writing this book. We may no longer be able to see, research or access those earliest humans in future. They are being returned to the Aboriginal communities of western New South Wales who claim them as their ancestors. At present, we are not sure whether they will be buried; locked away from research or will be made available for study sometime in the future. After many years experience listening to Aboriginal people's commentary regarding the study of Aboriginal skeletal remains, I believe the first option will prevail.

Modern humans had ancestors and they had travelled the world blazing a trail for later people. I begin the book by looking at their travels and achievements while recording who reached where and when. That story also goes further back: how long ago did those early hominins first leave Africa? We know far less about those events than the later ones, of course, but we do know those ancient hominin excursions across the planet occurred over a much longer period of time, they were successful and long enough for evolutionary change to occur among those early travellers.

The Out of Africa story, therefore, spans at least 2 million years probably much longer and it culminated in people eventually reaching Australia. Some of those people are the ancestors of the people found in the Willandra. That valuable collection is the largest modern human sample from the region and may contain the only evidence for modern human–archaic mixing in this part of the world. It is all we have that represents the outstanding achievements, abilities and adventures of modern humans who journeyed from Africa to this continent.

The origin of the first Australians has been debated for decades. I have been involved in that story for nearly 35 years and have been involved in prominent changes of thinking about those origins. How we saw the Willandran story 30 years ago was very different from what it is today. Of course, we have hindsight and a large archive of research from around the world. But it is also due to the way the story of modern humans has unfolded and the different types of evidence that emerged over that time. Exceptional strides in technology that developed during that time together with the growth in genomics and geomorphological dating have enhanced that story. These have opened our eyes and minds to the possibilities for understanding human evolution in a way not possible before and those developments have highlighted how little we really understood about the human story. The result is that the story of people coming to Australia is seen now in a very different light from that of three decades ago. And I suppose that is why I am writing this book.

But the oldest human remains in Australia need to be put into World context against the backdrop of recent modern human research and not as parochially focussed as it was in the past. The international significance of the Willandrans was obvious then but is even more so now. But what has become glaringly important is what they mean in a much wider sense: as important players in the Out of Africa story and modern human dispersals around the world. In other words, it is now clear the Willandra collection is a vital part of the history of us all. That part marks the end of arguably the longest and greatest journey of modern human before the last Ice Age.

If the Willandra collection is buried or scientific enquiry about them is denied, it will be a great tragedy not only for understanding our origins but also Australia's Aboriginal community. Future generations will question why scientific research into that story and their heritage could not continue; moreover it will be the destruction of an important stage in Australia's history that can never be replaced. Aboriginal people themselves will not have the opportunity that I have had to study the collection and that is why that data needs to be put on the record as evidence of the final stages of exploration and travel of our ancestors and who crossed the planet as far as their ingenuity could take them. This is that record.

The book uses some technical terms and uses data gathered by many scientists in many disciplines working collaboratively and sharing a passion for understanding our past. Without their work this book could not have been written. The book is fully referenced but the speed of progress in the field makes it impossible to include all relevant references in the field. Moreover, I wanted to present some ideas, arguments and interpretations of my own on various issues. The references included will lead the reader to salient researchers, their papers and other workers not mentioned directly.

LATE AFTERNOON IN AN ICE AGE

With our inability to time travel, we can only imagine what we would have seen on those ancient Ice Age shorelines of northern Australia as we watched the rafts arrive. What would you have seen and rafts appeared and people jumped into the sea from them and waded ashore…?

If we could time travel back to the last Ice Age, we might sit on a beach that today lies 40 m below sea level. Time is not important because nobody on the planet now knows what year it is or even thinks in those terms. Compared with today the world is virtually empty of people. For those that are around there is no such thing as years, months and days of the week. The sun comes up and it goes down, they recognise seasons, the phases of the Moon and that stars change their position in the sky but that is all there is for the concept of time among humans during the last Ice Age.

The sea before us is calm and slowly undulating in a shallow swell. It might be an Ice Age but there is no ice. Instead, it is warm and humid. We are the only people on the beach which is on a continent that nobody knows exists. Our beach stretches north to a place that one day will be called West Papua. The shore line is broken only by a few creeks that flow across it and a very wide estuary half way along that drains overflow from a great lake to the east, Lake Carpentaria, that is half the size of the Caspian Sea.

We are not alone. Occasionally a large animal walks onto the beach. Rarely, you might see a giant lizard patrol the shore. Giant salt water crocodiles 8 m long sometimes drift through the sea in front of us and meander along the rivers and creeks flowing from inland. Swamps along the rivers serve as places to nest. A distant thunder cloud punches a giant flattened out mushroom into the sky. A tropical storm is brewing on the western horizon. Staring towards the horizon the brilliant eye-stinging glints off the sea reflect the sinking afternoon sun. There is nothing to see but the glittering ocean. Then something occurs for the first time. The first continental discovery by humans in 2 million years is about to happen. To make that discovery, humans must complete the first ocean crossing. Then a small craft appears way out on the horizon marking the completion of a 20,000 km journey begun far away in Africa. The first Australians are approaching....

On reaching Australia, they were at the end of an amazing journey, landing on a beach 20,000 km away from their original homeland. But sitting on those ancient shorelines of northern Australia watching the rafts or boats arrive: what would you have seen and heard as people stepped ashore...? All we know is that they were originally MADE IN AFRICA.

Part I

The Longest Walk

Chapter 1

A View from Kakadu

Sets of 20,000 year old footprints in the Willandra, western New South Wales

TIME PAST

The truth is, of course, that my own people, the Riratjingu, are descended from the great Djankawu who came from the island Baralku far across the sea. Our spirits return to Baralku when we die. Djankawu came in his canoe with his two sisters, following the morning star which guided them to the shores of Yelangbara [Port Bradshaw] on the eastern coast of Arnhem Land. They walked far across the country following the rain clouds. When they wanted water they plunged their digging stick into the ground and fresh water flowed. From them we learnt the names of all the creatures on the land and they taught us all our law. (the words of the Son of Malawan from Yirrkylla, Northern Territory, in Isaacs [ed.] 1980:5, taken from Jacob, 1991:330) Namarrgon, the lightning man, like so many of the 'First People' entered the land on the northern coast. He was accompanied by his wife, Barrginj, and their children. They came with the rising sea levels, increasing rainfall and tropical storm activity. The very first place Namarrgon left some of his destructive essence was at Argalargal (Black Rock) on the Cobourg Peninsula. From there the family members made their way down the peninsula and then moved inland, looking for a good place to make their home. (after Chaloupka, 1993:56)

The northern boundary of Kakadu National Park fronts the Van Diemen Gulf on the southern edge of the Arafura Sea. The first people to land in Australia came across that sea and there are many Aboriginal Dreaming stories that relate it as those above do. Further east along the coast from Kakadu another Aboriginal Dreaming story describes those events that occurred in the 'Dreaming' time long ago. It describes when the ancestral Djanggawul sisters arrived from the island of Branko. That is the first of the two stories above. But if they were following the Morning Star, they were travelling west to east across the

Made in Africa. http://dx.doi.org/10.1016/B978-0-12-814798-6.00001-7

3

Arafura Sea looking at the early morning rising star. The story goes on to tell how ancestors landed on the East Arnhem Land coast near Yirrkala. Taking the form of Goanna lizards, they journeyed inland and jabbing their digging sticks into the ground they released fresh water. On they went, as they moved they created the land, then the people. Dreaming stories often relate how ancestral beings travelled the land and as they did so they gave the people their language, lore, dances, songs, ceremonies and rituals and taught them how to treat the land. The land then taught them what they needed to know to survive. If one reads between the lines it is not far off what could have been the experience of the first people as they floated from islands in the west and landed on this continue then continue their journey across it, long, long ago.

If we could time travel back at least 65,000 years ago, we could sit on a beach linking Australia to Papua New Guinea when the world was on average 10° cooler than it is today. Today that beach is covered by 40 m of sea. The exact date is not really important and nobody on the planet at that time knows what year it is or even thinks in those terms. There is no such thing as years months and days of the week in the minds of the world's human inhabitants. The sun comes up and goes down, they can count those events and recognise regular changes in the environment that we call 'seasons', but that is all there is as a concept of time among humans living in the Ice Age before last.

The sea before us is calm and slowly undulating in a shallow swell. It might be the onset of an Ice Age but there is no ice. Instead, it is warm and humid. We are the only people on the beach and on this 10 million km^2 continent and nobody knows it even exists. Our beach stretches north to a place that one day will be called West Papua. A few creeks flow across the beach breaking the shoreline and a wide estuary half way along it drains overflow from a great lake half the size of the Caspian Sea to the east, Lake Carpentaria.

We are not completely alone; hours ago we saw a large animal way up the beach. They are a rarity. A giant lizard far bigger and longer than a Komodo Dragon is known to patrol the shore but they are not around today. There are giant salt water crocodiles, the oldest well over 100 years, that grow to more than 7 m. One could be patrolling the water in front of us. A distant thunderhead cloud punches into the sky signalling a tropical storm brewing over the western horizon. We stare at the sparkling sea that reflects the lowering afternoon sun and gaze towards the horizon. There is nothing to see but an eye-stinging glitter of the Sun on the water. It is a pivotal time in the story of Australia. We can now see a small craft away in the distance. Australia is about to receive its first humans that have made the first ocean crossing and complete a journey that began 20,000 km away in Africa.

TIME PRESENT

I visited Kakadu National Park just before Christmas 2012. I had visited many times before but this time was special. Kakadu is Australia's largest national park situated in the middle of a square-shaped block of land that pushes into the Arafura Sea. The area is known to Australians as the 'Top End' of the Northern Territory. I was there just before Christmas at the height of the Monsoon or 'wet' season, but 'the wet', as it is known up there, had not arrived. The climatic extremes at this time of the year are vitally important for rejuvenation, replenishment and nourishment of the special environment of the north Australian tropics. Aboriginal people know that *Namargon* the ancestral lightning spirit man is the one they have to thank for the 'wet'. He brings it to the region providing renewal and growth for Arnhem Land's bush food that will be used in the coming 'dry' or wintertime. The Monsoon develops over Southeast Asia and travels southeast on the south-east trade winds to the north Australian coast. It brings storms, torrential rains, thunder and lightning, the occasional vicious cyclone and the essence of life. The time before the monsoon arrives is called the 'build-up' when it was said that among the first white settlers in the region, as well as many that followed, suicides increased as the humidity and heat gradually built. Suicide was a response to the oppressive, thick cloak of damp heat that builds for weeks before the first big storms build and explode through blue-black mountains of cloud bringing drenching downpours and hopefully some respite from the heat, reducing the heavy atmosphere to almost bearable.

Those were the conditions on my visit. The monsoon was late that year. It was oppressively hot and humid with each movement drawing gallons of sweat out of my rapidly dehydrating body. Nevertheless, the place was buzzing, literally, with an urgency of life of all kinds. Insects, animals, birds and other wildlife were waiting for the monsoon to 'break' and bring relief from the hot, humid blanket laying across the country. Billabongs, waterways and flood plains were low on water or parched and cracked and the sky shimmered with a feint grey opacity that watered down its usual blue. Kakadu is part of Arnhem Land where temperatures in the 'stone country' up on the escarpment stagger between 50°C and 60 °C from reflective heat that bounces off rock surfaces and 90% humidity. At this time of year climbing rocks in Kakadu is not for the fainthearted or those who do not like being friendly with flies.

I was there to view an art site, one of the estimated 5000 or more scattered across the 34,000 km^2 of the Arnhem Land Plateau. Some suggest there are twice that number or even more and 'more' seems the most likely figure. It is almost certain that no one will ever know exactly or see them all. The reason is that it would take several lifetimes to walk the difficult and deeply incised Arnhem Land sandstone country and inspect every likely nook and cranny that could be a potential site of hidden artistic treasure. One would need to explore a myriad of deep gorges; traverse the difficult, undulating escarpment surface strewn with piles of tower-block-sized boulders and cliffs that form dark caves and overhangs. Steep and often vertical cliff faces would need to be climbed to explore high, perched ledges. Drops would have to be made into deep shadowy canyons of hidden vegetation to investigate all the caves, chasms, rock shelters and expansive vertical gallery walls that are profuse across this Archaean block that really did form in the *Dreamtime* over 3.5 billion years ago.

The Arnhem Land Block is one of many massive foundation blocks of granitic rock that forms the Australian continent. Some go 16–20 km deep into the Earth and act as continental cornerstones. The vast net of rain-hewn fissures that split the ancient Arnhem Land block house a wide variety of special ecosystems and habitats, some almost impossible to reach, that lie deep in narrow canyons. Down there jungle pockets grow, in gorges and thickets of vegetation crowd the giant crags and cracks that split the rugged stone country across the top of the Arnhem Land escarpment. All these places lock away the unique plants and animals as well as some we are almost sure we have yet to discover. One can stand on the escarpment and look around knowing that there are probably cultural and/or biological treasures perhaps only 50 m or less away and you have no idea they exist. They may be near or far but it is certain they are there. The mystery of the Arnhem Land plateau is wrapped in the pristine and unknown nature of the region and in the certain knowledge that you will not find answers to that mystery. I was once reassured by someone who had worked there for years as an environmental researcher, often with the assistance of a helicopter, that they were convinced deep in the plateau there were canyons where 'black fellas' had not set foot possibly for hundreds of years and 'white fellas' had never been.

Kakadu is an ecological gem set in Arnhem Land's crown of environmental splendour. It is unique in its natural and biological composition and because of the variety of creatures and plants it contains. It is festooned with caves and rock shelters once waiting for the first people to arrive and make camps in them. Today, the monsoon forms vast wetlands in enclosed basins flooding rivers and creeks and infusing them nutrients, rejuvenating the wildlife, triggering another breeding season and sparking life into the region. Massive stacks of gigantic stone blocks and steep cliff faces and pinnacles reach up to mysterious wide shelters and entrances where people painted, camped, carried out ceremonies, sang, told stories of the land and ancestral beings that created the country and provided places to bury their dead for tens of thousands of years. The area is almost impossible to survey thoroughly. There are no roads or tracks across the escarpment, so arduous walking or helicopter is the only way to see the country. Even those two methods have their limitations because helicopter time is expensive and landing is difficult in many places and the exhaustion of hard exhausting trekking across such country for days, knowing where to look and physically accessing many areas is time consuming and sometimes impossible, making the vastness of the region very difficult for site surveys. It would take an extremely long time to thoroughly look for archaeological sites, rock shelters and living areas of those past inhabitants, let alone excavate, examine and properly research the findings from such places. That alone emphasises the imperative to look after, maintain and preserve the country as thousands of generations of Indigenous people have. It is their story that is contained in the region, defines it and brings the country to life. One look at Arnhem Land from the air makes the outsider feel an overwhelming awe. Most will immediately feel the impossibility of travelling across and seeing all the rocky places likely to have been occupied during the last 60 millennia at least. It also presents a rather sad feeling. Besides the beautiful and intriguing country, the magnificent art and natural heritage of the region, the feeling of never knowing even part of the story that is held in the country and knowing it will probably forever remain unknown, diminishes the viewer. That story also contains the travels and exploration of the first people to arrive here and the trepidations and triumphs of their journey across this country on their way to explore a vast unknown and uninhabited continent. They were oblivious to the fact that this was the end of a long road that had begun in Africa. Once they crossed the continent before them they could go no further.

THE 'WRITING' ON THE WALL!

The above picture of Arnhem Land can also be used to describe the Kimberly 600 km to the west. These two ancient granite bastions and the land between contain the earliest cultural record of the initial human settlement of Australia. Thus, it seemed so inconsequential to visit one particular art site in Kakadu. Aboriginal art is not done to create a pretty picture or to particularly please the viewer. It is done to tell a story, connect with country, depict an event, indicate a special place or record the Dreaming and bring into vision the associated spirits and creatures that are responsible for those things or lived at that time. Also it is not necessarily permanent or preserved. A lot of rock art often placed in natural galleries has for generations been over painted with other figures and symbols. Aboriginal art is, therefore, utilitarian; it has a purpose beyond mere creativity, pleasure, aesthetics and illusion.

There is always a feeling of privilege visiting an art site and, with comparatively few open to the public, viewing one rarely seen is a special privilege. The one I was visiting was particularly special because it made a direct link between the viewer and some of the earliest people of the area. It interested me because I believed it depicted something that the first people to arrive in this part of Australia could have been confronted with and probably puzzled by and I will never see. On the grey vertical rock face was the image of a very big and very different type of kangaroo from modern varieties and it showed some special features. It had an unusually thick tail, a thick wiry fur coat and very fluffy ears. It also sported a very long and large single claw on a very long and very big foot. Its primary feature of interest however was its short face unlike the deer-like muzzle of all modern kangaroos. It was a short-faced kangaroo. Such kangaroos were usually very big animals and part of a suite of animals called megafauna that went extinct during the last Ice Age. The people's fright or puzzlement would have been because of its size, the fact the arrivals had not seen kangaroos before, let alone giant ones, and this very big animal had two heads: the second one poking out from where its stomach should be! The artist had sent me a 'photograph' of a kangaroo that had been extinct for tens of thousands of years (Fig. 1.1).

FIGURE 1.1   Spot the extinct kangaroo!

I was with a ranger and two local Indigenous people; traditional owners from the nearby community. They showed no particular effects from the oppressive December conditions except profuse sweating as we all were. But they were as enthusiastic as I was to see close up what I had told them about. I first noticed the painting though my telephoto lens from some way off several months previously. The painting was in a section of the park that was off-limits to the public so I was visually trespassing. I was aware that to get to the painting I needed permission to enter the area. So, I arranged with the Northern Territory Parks and Wildlife Service and local clan groups to go with them and a park ranger to view the site close up. The painting was placed high on a rock wall with a handy but very narrow ledge in front of it. I climbed up and over some large boulders at the base of the gallery and then over more narrow, cracked and unsafe-looking rocks to reach the ledge that was even narrower than I had first thought. Keeping my body close to the rock face for balance on the narrow ledge, I reached the large fresco that had intrigued me since I first saw it. This close it seemed like a jumble of painted lines going in all directions. That impression came from the fact that the fresco was over-painted with other images, symbols, creatures and possibly spirit beings. They were superimposed across the broad, high and smooth rock face that made a natural canvas for the artist or artists. I now came face to face with what I believed was Australia's equivalent of the famous European cave art that has for so long been upheld as the yardstick of the cultural coming-of-age of modern humans. Such paintings in the tunnels and caves of the Pyrenees and surrounding regions have been marked as signposts that we were emerging from being cave men to cave artists and 'proper' people. The paintings mark the development of art and emergence of artists with imagination able to reflect and recreate their world in the same way we can. It joins us together over many millennia: although we may not fully understand it, we can relate to it. So, what were we now looking at in Kakadu?

It has been suggested that some rock art images in Arnhem Land and the Kimberley represent some of the largest marsupial animals that ever lived (Murray and Chaloupka, 1984; Akerman, 1998, 2009; Akerman and Willing, 2009). Although not all those interested in Australia's ancient art agree. The images have been claimed as the megafauna species that went extinct well before the peak of the last Ice Age and that, importantly, could give them a basic date. While partially masked by more recent art, this image showed the outline of an extremely large and well-muscled kangaroo measuring 235-cm long by 185-cm high. The large size of the creature as painted may or may not indicate that the artist wanted to present to the viewer a very large animal. But other details of the painting are important in depicting a very different animal. For example, short paint strokes were made in the outline and within the body area, perhaps depicting coarse fur while also showing long, rounded ears with equally long hair on them. The tail had a thick base and although slightly tapering remained stout along its entire length with a stubby end. The design is quite unlike a modern kangaroo tail and was indicative of the type of robust tail required to support a large, heavy body when the animal sat upright. The testes placed just below the tail base indicated a male animal and they are situated just where we would expect to find them, a detail the artist felt necessary to include and something that underpins the artist's need to accurately depict the animal's form. Again, it reminds me of the accurate depictions found in the many animals found in European caves. The legs are heavily muscled and suggest great strength with the limb ending in a partially lifted foot with circular motifs emphasising ankle and toe joints. At the end of the long foot is a single, large pointed claw. The most distinctive feature of the kangaroo, however, is its short face with rounded muzzle. The face is somewhat faded around the distal nasal region but close inspection and faded lines extending from the brow show the rounded and shortened facial shape.

There were 23 short-face species of kangaroo grouped in six genera living in the mid-late Quaternary (Webb, 2013). Although presumed extinct by the late Quaternary a recent reassessment of megafauna extinction timing proposes their demise was staggered with seven short-faced species still around possibly 40,000 years after humans had arrived on the continent (Webb, 2013; Wroe et al., 2013; Clarkson et al., 2017; Tacon and Webb 2017). They include the largest species *Procoptodon goliah* (~250 kg), the 150 kg *'Simosthenurus' pales*, the 120 kg *Simosthenurus occidentalis*, *'Procoptodon' oreas* (100 kg), all larger than any modern kangaroos and the smaller *Sthenurus andersoni* (75 kg) and *'Procoptodon' gilli* and *browneorum* (50–55 kg). Short-faced kangaroos lived widely across the southern continent and in southeast Queensland but some may have lived quite well in the north particularly during glacial conditions. At that time the Kakadu environment was savannah-like, contrasting with the tropical forest there today. Any of the three largest species listed above could be candidates for the painted image although the presence of such a long prominent claw suggests it might be an image of the largest of the short-faced varieties, *Procoptodon goliah*, a creature well known for this feature or even an unknown variety of large macropod.

The image I was looking at was an Australian equivalent of a European cave bear, mammoth, cave lion, spotted horse or woolly rhino. It was a megafauna species like them. Like European artists the artists here painted easily identified animal forms reflecting species known to them, accurately drawn with an amazing skill. The Kakadu example also has spiritual or mythical beings and creatures painted over the kangaroo, by later artists. In Aboriginal art spirit beings are often identified by exaggerated or abstract forms as stick-figures and have features or adornment having little resemblance to everyday creatures. The most spiritually powerful images often have a composite form made up of body parts of various animals. Examples of such figures overlay this painting. Realism is exemplified not only in paintings of everyday terrestrial and aquatic wildlife but also in recent times non-Aboriginal historic objects such as European ships, Macassan praus, military aircraft, firearms and other introduced objects all of which are visible in Arnhem Land galleries (e.g. Chaloupka, 1993; May, Tacon, Wesley, & Travers, 2010). They are easily recognisable and can often be identified to particular types and makes. Moreover, Aboriginal artists paint excellent and detailed likenesses of extant animals they live with that can often be identified to species level (Chaloupka, 1993).

If the last megafauna lasted to between 40 and 25 ka (Miller et al., 1999; Webb, 2013) that could place this painting equal with the oldest painted art anywhere in the world. Alternatively, if the image is not that old then the image would mean that some megafauna species still inhabited Arnhem Land perhaps as late as 35–30 ka, as the very latest date accepted for the existence of large short-face kangaroos (Ibid; Wroe et al., 2013). New dates of rock art sites in Sulawesi, situated to Australia's immediate north, are vital in important to this story. Uranium-series dating of small rocky growths (coralloid speleothems) that have grown over and under 12 hand stencils and two naturalistic animal depictions in infilled outline from seven sites in the Maros area of Sulawesi, have revealed the oldest ages for these forms of rock art in the world. The earliest minimum age for a hand stencil is nearly 41 ka at Leang Timpuseng and the oldest animal painting, of a babirusa 'pig-deer' (*Babyrousa* sp.) at the same site, is 37 ka. A second animal painting (probably pig) at another site has a minimum age of 52–36 ka (Aubert et al., 2014). All of them are painted in the same style of the earliest rock art of the Kimberley and Kakadu but no depictions of extinct or 'mega' animals have been discovered so far in Sulawesi although I can't think of what megafauna lived there at that time. The art there and northern Australia may have resulted from a shared practice undertaken by modern humans as they spread through the region. One important question emerges out of these discoveries and that is when did artistic creativity or the ability to reflect objects in the world around us arise? Perhaps even more intriguing is the question, why did it seemingly emerge between 40 and 50,000 years ago? Is it a co-incidence that by the time modern humans arrived in Europe they began arctitic expression about the same time they reached Sulawesi and then Australia although these places are 20,000 km apart? In other words, did the ability to draw fairly accurate images emerge among modern humans as a factor of their dispersal? And if it was, did it arise back near Africa, along the way or indiginously in separate destinations? If it was the latter then modern human artistic ability developed independently among groups separated by great distance. It has been suggested that why teenagers do and act the way they do is because many of their cerebral neurons have not yet fully 'joined' up. Then around 25–28 years, those neurons complete the final connection and they begin to suddenly see what their parents were getting at.… Perhaps something similar happened among modern humans. If art had to begin somewhere and at some time, why then? The human brain was not evolving in size any longer but it may have been evolving in its neural connections and networks and other ways with the result that art emerged as another of our abilities, almost overnight, possibly driven by the new experiences brought about by and during our dispersal experiences. I return to this neural enrichment in later chapters.

It seems that while the earliest European Palaeolithic rock art has been in the spotlight as a cognitive stage in human evolution, that of Arnhem Land and elsewhere, has been in comparative shade. The Spanish and French cave paintings of Altamira and Lascaux were painted after the peak of the last Ice Age less than 20,000 ago, but the Chauvet Cave paintings are older, around 35 ka, long before the Ice Age maximum. An Italian site at Fumane has revealed even earlier paintings between 36 and 41 ka. Discovery of simple art claimed to be Neanderthal has been found in El Castillo Cave, Spain (~40 ka) and Gorham's Cave in Gibraltar (~39 ka). Some reject it is art, however, because of a long-standing view that Neanderthals were incapable of such skills and the dates may reflect an overlap between them and the arrival of modern humans and the art. It has been suggested there was some adoption of modern human tool making techniques by Neanderthals. If true, a similar copying of simple designs by them may not be fanciful. After all, we were both very close at some point, so much so that we were not averse to having sex with each other and no doubt we copied one another in various ways.

I would not be the first to suggest that the Late Pleistocene cave paintings of Europe probably marked a half-way point in human cognition between wanting to express a concept or describe something orally and being able to record and present it in writing. Art then must have been drawn or painted as a lasting record of the urge to explain, depict, demonstrate something, make a point, reflect images of the mind or the secular or spirit world, tell a story that would last beyond the artists life time or perhaps it was a combination of all these with a dash of art, being useful for ceremony and ritual, thrown in. It was likely that paintings like the one I now stood so close to were telling me something in the only way the artists knew how and that was to show me something they had seen or were aware of rather than just a figment of their imagination. They and we were effectively talking across a vast amount of time as we might do standing in the Altamira, Lascaux and Chauvet Caves. Sadly, I had lost the code or lacked the cultural heritage or wit of the time to explain the whole story or perhaps any of it. Yet I knew what the subject was because through the skill of the artists I could see what they were seeing as surely as if I were looking at the fighting rhino's or charging lions in Chauvet. One thing seems certain, this painting in Kakadu was probably as old as those of Lascaux and Chauvet or even El Castillo. Why I know that is because we do know when the giant short-faced kangaroos went extinct and at the very latest that was 25,000 years ago unless Arnhem Land was one of their last refuges and they lasted a lot longer there than anywhere else. Before, we had no other evidence of art of that age in the region and without this species being painted, to claim the region's art was that old could have been easily dismissed. We now know people were painting next door to Australia 10,000 years earlier and people had landed in Arhem Land at least 65,000 years ago (Clarkson et al., 2017). It is a sad fact, however, that we are at present unable to accurately date the kangaroo painting or any of the other examples claimed to be megafauna images among Australian rock art in the same way they have been able to do in Sulawesi because we lack both a method to do so and the corolloid speleothems overlying the image.

There is little doubt in my mind that this painting is of an extinct species of megafauna kangaroo and is at least contemporary with Ice Age European art. With that possibility it seems that two groups of modern humans living on either side of the world tens of thousands of years ago were linked by their art. They drew and painted together on different continents, in different environments and in a different world from the present but they had common themes and no doubt common goals to their work even though they had never had any contact with each another. It was a time when the world had a different geography, different environments across continents and some of those continents, now separated, were joined. Ancient Australian art shines a beacon on and stamps the arrival not just of people here, such as archaeological debris in a living site might do, it marks a cognitive linkage between humans across vast distances. For the first time, the mark of modern humanity on the planet reached beyond stone tool making and cave dwellings into a new cerebral world of thought processes that say: *we will never go back from here.* These were the first human messages to reach Australia laid down by people talking to us across time as well as recording and showing us their world. It does not matter what their motive was, or what exactly they were trying to say, we will never know that, it is enough that they were able to speak to us across tens of thousands of years. To some extent, I can understand their silent message that needs no knowledge of their spoken language (whatever that sounded like) as I stood on the same ledge as they did, as close as they ever stood to the wall where they left that painted message so long ago. The ledge was perhaps wider then but I was almost sharing their breath that bounced back off the stone as they painted.

The painting was doing something else. It was showing a vibrant flag of presence of the people who had crossed the world and had the capability to send images and knowledge into the future. They had the ability to transmit their culture forwards. It was a cultural base they had brought with them and which was being added to every day they lived on this new continent. The first Australians and their neighbours were doing similar things to their European counterparts; they were drawing animals. Should that be surprising? They were drawing animals they knew, everyday species for them, that for one reason or another, practical, spiritual, ceremonial or secular seemed important enough to replicate on rocky canvases using ochres of red and yellow, white clay and charcoal, for paint that has lasted an incredibly long time. They were painting beasts known to us today as well as those that fascinate us because they have gone: the Australian megafauna. The question put earlier repeats itself: as they walked the world, did they always have art as part of their cerebral baggage or did they develop it on their journey? The latter would imply art as a mental and manipulative skill must then have developed at least twice on either side of the world. Perhaps these people took with them the embryonic skills required for drawing and artistic expression when they left Africa, as ochre stained Abalone shell palettes and engraved ochre pieces found in Blombos Cave (100–75 ka) on the coast of southern Africa, might suggest. Tentative support for that comes from ochre found in a 65-ka archaeological site not far from the short-faced kangaroo painting but more of that in Chapter 5. Those artists were also making their art, whatever that was, at a time when at least four, possibly more, other human species roamed the world.

A LANDING, BUT WHERE?

As I stood literally face to face with the giant kangaroo on the narrow ledge I cautiously turned my head to look across the nearby country. I tried to imagine what the region would have looked like when not only giant kangaroos lived there but also when the first people arrived thousands of years earlier. There was one thing I was sure about; my precarious position told me that this ledge must have been wider when the kangaroo was painted!

Looking north across Kakadu's Magella Plain wet lands from Ubirr Rock, not far from the kangaroo site, the landscape now looks very different from that the first people saw. They shared that kangaroo's Ice Age landscape when the present coastline, now 60 km away, was over 200 km away. What was Kakadu like 65,000 years ago or earlier when these modern human travellers emerged out of island South East Asia and walked through the 'stone country'? They could have moved inland through open woodland and found a suitable rock shelter to camp in. The coast then was far beyond the modern horizon, as far away as it had been for the previous 75,000 years. It was a coast unfamiliar to us where they landed, but it was one that had existed longer than the present one. Australia's northern continental shelf is shallow and gently sloping which meant that falling sea levels rapidly exposed it, particularly during the first 50 m drop. Lowered sea levels were present for over 80% of the Pleistocene so our present levels are unusual. At those times, the extended coast effectively put the Arnhem Land Escarpment 200 km further inland and out of sight of anyone landing on the exposed shelf during maximum low stand. To us, there would have looked very different with a different environment in the Kakadu region from that of today. The monsoon was weaker during glacials exacerbating the process of continental aridification that allowed desert to creep closer to coasts while hyper-aridity was taking place in the centre of the continent.

It is easy to say people arrived here or landed there but in the phraseology, we lose the environmental complexity and variability that accompanied these events. The simple view is that people arrived somewhere along our 4000 km northern coastline that today stretches from the southwestern Kimberley on the Indian Ocean, near the town of Broome, east to the Gulf of Carpentaria and Cape York whose east coast fringes the Coral Sea. It is reasonable to suggest people did not land at either end of that ancient coast. It has always been assumed that they took the shortest crossing from the islands to the northwest placing them somewhere in the centre between the central Kimberley and the Bathurst and Melville Islands. And with a differently shaped coastline and a blocked off Carpentarian region, preventing Cape York landings, that idea is enhanced. However, an overland dispersal through Irian Jaya-Papua New Guinea from landings in the west of that island could have resulted in them being funnelled to the east through the New Guinea Highlands. That could see them exiting north of Cape York and make a dry-foot crossing into Australia. An entry from Papua New Guinea is entirely possible. An alternative possibility, when considering that the difficult mountainous terrain may have discouraged an inland move-ment, a coastal movement would have brought them to the centre of northern Australia via the palaeocoastline. Another alternative is any movement from the west along the northern coast of Papua New Guinea would take them down that coast far enough to make a crossing to Cape York especially if they had avoided the Central Highland regions of the island.

A direct crossing would be more successful, safer and quicker. But without knowing the quickest way, or how to plot the shortest route, we cannot assume that is the route they took and must assume a more random approach. On the other hand, their natural abilities, experience and knowledge of the sea, probably helped them in ways we may not recognise. The fact they arrived is, in itself, testament to those abilities. So perhaps we can assume one of the most likely landing spots was somewhere in the middle one third of the northern Australian coastline (whatever it looked like at the time) opposite the nearest point of origin on the islands. The geographical position of arrival also depended on what the northern Australian coast looked like which depended, in turn, on sea levels at the time, and they dras-tically changed its shape and length which fluctuated throughout the most likely time of the arrival. Sometimes the coast reached out to meet them at the places they were likely to leave from as glacial sea levels lowered.

A LANDING: IN REALITY

It is easy to believe that as the first sailors approached the coast, by tasting the sea water they may have picked up on it freshening as fresher water from Lake Carpentaria and southern PNG entered and mixed with the sea. Spotting different vegetation flotsam and watching different varieties of bird life they could also judge how close they were to land. But from 90 ka–70 ka (the landing time), the coastline was very different from today and the shortest crossing occurred around 65 ka when sea levels were at their lowest since 140 ka. Before that, higher sea levels required longer and subsequently more difficult voyages to Australia. To take this possibility a little farther, it might be suggested that any people living in the Indonesian island group between 133 and 114 ka may not have had the opportunity to cross to Australia because sea levels were even higher than today. Lowered sea levels modified ocean currents that dictated sailing directions and presented opportunities to land in or reach some places more easily than others, thus favouring some crossings over others. During the lowest sea levels (120–130 m) the coast formed a square-shaped embayment that fronted the east Timor Sea and linked Australia to Papua New Guinea. The Timor Sea and Arafura Sea shrank and the Gulf of Carpentaria became the Arafura Plain with the giant Lake Carpentaria isolated in the middle. The lake varied in size from 28 to 165,000 km^2 depending on sea levels. It received water from a 1250,000 km^2 catchment covering mainly northern Australia but also part of southern New Guinea. A large drainage channel drained west from Lake Carpentaria flowing across the exposed shelf to the Timor Sea. Lowered sea levels also changed the shape of coastlines throughout the region including that of western Papua. One salient change saw southern Papua extend westward over 500 km to engulf the Aru Islands making them distant hills on the new plain. The same happened at Yos Sudarsa on the southwestern tip of Papua. The coast linking Australia and Papua became a catch-all for any adventurous souls pushing east across the shrunken Timor Sea. They would run into a dead end formed by the new coastline stretching north from about Melville and Bathurst Islands, that were hills, and that joined up with the Aru Hills in the northwest. Changing sea levels brought change to coastal environments that began to colonise the exposed shelf. In effect, this process shifted environments back and forward following sea levels. The process changed vegetation structures and plant compositions as well as drew animal populations out onto the emergent shelf. At glacial maximum low sea levels expanded the Australian continent from 7500,000 km^2 to 10,000,000 km^2: it is the largest continental shelf in the world.

Lowered sea levels reduced landing options because the coastline was shorter than today's. Therefore, although the north Australian Ice Age coast was shorter, finding the camps of the first arrivals and where they lived is next to impossible. That is because we know the most likely landing place, such as the area between the Kimberley and Darwin and between Arnhem Land and Papua, was the widest part of the shelf, in some places extending 200–400 km from the present Australian coast. So, we have to know what the sea levels were doing at the time people first landed to assemble a likely landing geography but that is difficult. While we have a good idea of what the sea level was like at any one time over the last 100,000 years, we do not yet know exactly when people first arrived and thus cannot match one to the other to establish a likely landing place.

As well as low sea levels, glacials changed the interior of the Australian continent. Except for two rather small glaciers in the Snowy Mountains and central Tasmania we missed out on the kind of massive ice development seen in the Northern Hemisphere. Our Ice Age geography was, nevertheless, very different from today especially in the north and centre of the continent. The major change was aridity that spread to the coast in many areas. Central deserts expanded into regions now covered in tropical grasslands and open forests while in some areas they met up with the coast. Vegetation moved further north with expanding desert close behind. A transitional region between the two contained less tropical and florally more open conditions across the southern portions of the Top End. During glacials the desert lapped at Arnhem Land's southern boundary. It's now a verdant splendour of wet sclerophyll forest, jungle pockets and rainforest thickets that was largely replaced by an open, drier and more arid environment with savannah vegetation. Denser, tropical vegetation only continued in the shaded, deep, damp and remote gorges of the plateau where pockets of the original vegetation mosaic changed little. The broad change to open woodlands made the region generally easier to traverse on foot than it might be today, although the arrivals were tropically adapted having spent generations living in and moving through Southeast Asia. The difficulties of the escarpment country would have made the journey ten times as long as crossing even ground. However, people need not have moved into the escarpment country but searched for an easier although longer route around that difficult country perhaps following rivers that over millennia had cut down through the ancient granite blocks. It is not difficult to imagine that surface fresh water was not that plentiful inland and indeed got rapidly rarer the farther people moved south away from the coast.

In the mid-1970s the archaeologist Sandra Bowdler proposed the *Coastal Colonisation Theory* (Bowdler, 1977) to explain the initial movement of the first people into Australia. It was a general strategy explaining how the first people were most likely to have explored Australia given the way they had been living for many generations before their arrival. That strategy was one based on the sea; their movement through islands and following coastlines and was based on the premise that they would have had a *marine economy*. As the name implies the boat people depended on the sea as their main economic and nutritional resource because they were essentially sea farers or perhaps a better name would be *marine nomads*. They would know the sea and its workings; they would have boats and know how to use them as well as knowledge of currents and tides and the vagaries of the ocean and what it could do. Such people could fish, catch sea mammals and turtles and of course gather shellfish and crustaceans while taking advantage of terrestrial mammals when the opportunity arose. Perhaps they had become *marine super nomads* after originally being *terrestrial super nomads*. As a coastal or sea going people their skills tied them to the coasts. That was an environment they knew and it would provide fresh water as they discovered rivers and streams flowing to the sea. They could then explore and move inland. Those same rivers would provide the same kinds of food resources as the coast.

Freshwater shellfish and crustaceans, fish as well as animals attracted to fresh water were all available on and in rivers. Those food resources formed a good nutritious diet and one that they would not have to tramp miles to obtain, if you were lucky, or have to chase for hours, probably expending as much energy as they would receive from the protein they caught.

Coastal colonisation made sense in 1977 and it still does today. Recent discovery of a 50-ka site on Barrow Island, that was a hill on the exposed continental shelf during low sea levels, might go some way to reinforcing that idea although those people moved there at least 15,000 years after the initial colonisation and so could have just been a local group exploiting exposed continental shelf (Veth et al., 2017). Knowing inland was arid or semi-arid and something to avoid if possible also supports the idea. Who would want to stay Inland where game was probably rare and when it was around it was difficult to catch because you were unfamiliar with the terrain and its behaviour. Vegetable foods were also scarce and new people were also unfamiliar with them, but fish, shellfish, sea mammals and turtles were very familiar. These people were not arid adapted and had not experienced desert conditions perhaps for generations or many thousands of years. The nearest arid environments or deserts were the Taklamakan and Gobi Deserts of northwest China and Mongolia and the Thar Desert of western Rajasthan in India and these people had not experienced any of these places for many generations.

MOVING ON

While the coastal colonisation model is a logical idea for the movement of people around Australia, the coast was a different shape and in most places not where we find it today. Strict adherence to it would in most cases now place the first camp sites well under water and lost forever. Therefore, looking at the culture of the people will never be possible unless inland sites are found. We can only hope they did eventually wander inland and human nature would suggest they did just that when it was safe to do so. Small lakes and puddles from occasional downpours were available and cooler temperatures and lower evaporation rates aided in maintaining surface water, any rainfall that filled streams and ponds would last longer. During glacials major north Australian rivers such as the Adelaide, Daly, Victoria, Ord and Drysdale Rivers stretched out across the emergent continental shelf but may not have received the catchment rains and the consequent flow strength they have today. That would mean while they gained length across the shelf their headwater regions were probably a lot drier. Therefore, while some exploration of these major systems might have taken place after landfall their effective flow was now on the continental shelf because they did not always extend as far inland as they do today. If that were the case people would have been less likely to explore too far inland during glacials. Landings along the shelf that in some places were hundreds of kilometres from the present coast would have been areas more likely favoured for exploration as the recent 50-ka date on Barrow Island, may suggest.

The earliest people here met animals that were totally new to them and they were walking into and using new and very different environments and vegetation then suddenly they were confronted by desert. They would eventually meet cool temperate and even Alpine environments with snow and ice. It was certainly a very different place from their last home on the tropical islands they had moved through. They had to survive and they did just that, but how fast they moved across Australia is anyone's guess. But some did camp near Kakadu not far from where their descendants would paint a giant short-faced kangaroo on a rock face. Whichever route was taken as they began their transcontinental journey they were patently successful. They continued journeying through and/or round the continent eventually ending up at the tip of southwestern and inland southeastern Australia by at least 50 ka. That would suggest some traversed the length of the west coast and its continental shelf in which case they would have had to cope with arid regions inland that kept them on the coastal fringe wherever that was. Some may have penetrated south-central areas of the continent, eventually finding the northern Flinders and perhaps exploring large rivers that ran inland, one of which could have brought them to the Willandra Lakes.

But who were those people and what was their history?

REFERENCES

Akerman, K., 1998. A rock painting, possibly of the now extinct marsupial *Thylacoleo* (marsupial lion), from the north Kimberley, Western Australia. The Beagle, Records of the Museums and Art Galleries of the Northern Territory 14, 117–121.

Akerman, K., 2009. Interaction between humans and megafauna depicted in Australian rock art? Antiquity 322, 323–328.

Akerman, K., Willing, T., 2009. An ancient rock painting of a marsupial lion, *Thylacoleo carnifex*, from the Kimberley Western Australia. Antiquity 319, 105–110.

Aubert, M., Brumm, A., Ramil, M., Sutikna, T., Saptomo, E.W., Hakim, B., et al., 2014. Pleistocene cave art from Sulawesi Indonesia. Nature 514, 223–228.

Bowdler, S., 1977. The coastal colonisation of Australia. In: Allen, J., Golson, J., Jones, R. (Eds.), Sunda and Sahul, prehistoric studies in Southeast Asia, Melanesia and Australia. Academic Press, New York, NY, pp. 205–246.

Chaloupka, G., 1993. Journey in time. Reed Books, Sydney, 100–101.

Clarkson, C., Jacobs, Z., Marwick, B., Fullager, R., Wallis, L., Smith, M., et al., 2017. Human occupation of northern Australia by 65,000 years ago. Nature 547, 306–310.

Jacob, K.T., 1991. In the beginning. Ministry of Education, Western Australia.

May, S., Tacon, P., Wesley, D., Travers, M., 2010. Indigenous observations and depictions of the 'other' in northwestern Arnhem Land Australia. Australian Archaeology 71, 57–65.

Miller, G.H., Magee, J.W., Johnson, B.J., Fogel, M.L., Spooner, N.A., McCulloch, M.T., et al., 1999. Pleistocene extinction of *Genyornis newtoni* human impact on Australian megafauna. Science 283, 205–208.

Murray, P., Chaloupka, G., 1984. The Dreamtime animals, extinct megafauna in Arnhem Land rock art. Archaeology in Oceania 19, 105–116.

Tacon, P., Webb, S.G., 2017. Art and megafauna in the Top End of the Northern Territory, Australia: Illusion or reality. In: David, B., Tacon, P., Delannoy, J-J., Geneste, M. (Eds.), The Archaeology of Rock Art in Western Arnhem Land, Australia, *Terra Australia 47*ANU Press, Canberra, pp. 145–164.

Veth, P., Ward, I., Ulm, S., Ditchfield, K., Dortch, J., Hook, F., et al., 2017. Early occupation of a maritime desert, Barrow Island North-West Australia. Quaternary Science Reviews 168, 19–29.

Webb, S.G., 2013. Corridors to extinction and the Australian megafauna. Elsevier, New York, NY.

Wroe, S., Field, J.H., Archer, M., Grayson, D.K., Price, G.J., Louys, J., et al., 2013. Climate change frames debate over the extinction of megafauna in Sahul Pleistocene Australia—New Guinea. PNAS 110 (22), 8777–8781.

Chapter 2

Ancestors of the Ancestors

The next three chapters deal with many early hominin and modern human players in the world's pageant of human evolution. It presents not only a survey of salient specimens representing *Homo*'s earliest ancestors who played an important part in the story of human evolution but also formed the backdrop to the development of modern humans. Accordingly, they are the ancestors of our ancestors. Some readers might say certain fossils should have been included but the book is not intended to be a complete catalogue of hominin specimens. The ones included are those I feel are important one for our story. It is worth noting also, there are fossil hominins we cannot peg accurately in the scheme of things although ideas about them abound and I will mention some of those. Other specimens are not well dated and it is difficult to know how they relate to those that are. Some look different to others of the same age and from the same place, while some that are widely separated look very similar, it's confusing. Therefore, not all fit into the story in both time and space and I am not able to remedy that here. Although confusing, it nevertheless shows the story of our evolution is not straightforward, nowhere near complete and leaves it looking a little illogical and makes it difficult to follow at times. It used to be very straightforward but it has become an increasingly complex and convoluted story the more we learn about it, so much so that sometimes it calls for a little leap of faith here and there to tie ends together or sunder previously assumed links depending on new evidence. Indeed, as we try to move our understanding of human evolution forward, it often seems like we are not moving it anywhere let alone forward. As I have said, not all fossil hominin specimens have been

Made in Africa. http://dx.doi.org/10.1016/B978-0-12-814798-6.00002-9

included and neither are all the arguments that surround them. I describe only those important to the emergence of modern humans and representative of lineages that seem to lead to us, modern humans. That is particularly relevant to the story of humans coming to Australia but we should begin that story a long way back, way before the recent discovery of our emergence 300,000 years ago. Hublin et al. (2017); Richter et al. (2017).

EVOLUTION'S FORCES

Humans evolved according to processes that apply to all living things and these should always be kept in mind during discussions of human evolution and human movements over great distances and long periods of time. Those processes are the forces of evolutionary biology that drive evolutionary change in all organisms and they usually take place in various strengths and combinations across wide spatial and temporal gulfs. Such changes produce physical or morphological change and force speciation. They alter appearances in one direction or another and include the major forces of adaptation and selection that operate to allow organisms to dwell successfully in many strange and difficult environments around the planet. The rules or mechanisms are well understood and their effects have been studied in animal groups the world over. They must have operated among hominins as they moved into different regions and occupied different and sometimes highly contrasting environments for various lengths of time. Individually and collectively those mechanisms changed their skeletal morphology and provided them with different physical and biological abilities to cope with the changes and challenges they faced as they dispersed across the planet. While prominent, adaptation and selection easily slip off the tongue as primary evolutionary mechanisms but they are joined by others that are also important to and for the efficacy of the whole evolutionary pathway. I will mention them often under the collective term *9EM* (the nine basic evolutionary mechanisms). They include: *founder effect, random genetic drift, random mutation, isolation, extinction, genetic assortment, adaptive or natural selection, population size* and *inbreeding*. All these played their part, acting in different degrees, as forces tugging one way and another in different combinations at different times to change all hominins and enforce evolutionary change. Their effect and speed of change depended on living circumstances determined by enviro-climatic factors, population size, parental and population genetics, hybridisation between different groups, group isolation and group/population success or growth. They affected hominins as they travelled the planet in different ways and in different proportions including separation from their ancestors and as those travellers experienced different surroundings and life histories. I will invoke *9EM* selective forces many times during the book as a gentle reminder of what is going on 'stealthily behind the scenes' as people dispersed across and settled various areas of the world especially as they divided from parental groups and met others who were not directly related to them or with whom they had shared a common ancestor, hundreds of thousands of years before.

How did the *9EM* act within human evolution? The answer requires a brief explanation of how each mechanism works. The main reason for invoking these is that we need to consider them as they operated as hominin groups separated on their journeys and became small and/or isolated groups/bands and as further divisions occurred in those bands:

Founder effect. When a small population or group moves away from its parent population it takes with it only a sample of the alleles or genes of the parent group. The new group will not, therefore, have all the genes of the parent group or its characteristics and so will present an altered genetic expression of its own. That could eliminate certain characteristics among future generations and form a unique and characteristic genetic expression of its own. Future mixing with other unrelated groups could alter that expression removing it further from its parental origins.

Random genetic drift. This is a random undirected change in allele frequency that, by chance, occurs in all populations, but particularly in small ones in whom they have their greatest effect. Drift would have been prominent among small populations that took part in widely spread and separated dispersals across the globe.

Random mutation. Mutation is the ultimate source of genetic variation and so is vital in the evolutionary process. The process occurs all the time and may act more rapid or slower in different populations. Most random mutations are harmful and are eliminated from a given population but others can become incorporated in the genome over time.

Isolation. Isolation of a population prevents the introduction of fresh alleles and the benefit that can come from the vigour that process can confer. Fresh genetic input is a natural way to being variety and thus genetic toughness into a population and isolation prevents that. Isolation can also mean a drift towards extinction of the group genome or characteristics carried by that group alone.

Extinction. This is the ultimate elimination of genes and characteristics carried by a population or groups. Elimination of genes can also mean the end of certain characteristics that confer identity on that population which are lost forever.

Genetic assortment. Natural genetic assortment among small groups does not work as effectively as it does among larger groups with larger numbers of people and the opportunity for mixing between individuals. The process will have less expression as group size is reduced.

Adaptive or natural selection. Adaptation is the progressing modification that improves the chances of survival and reproduction in a given environment. This has a better chance of taking place in larger populations and groups possessing a broader range of genomic variation.

Population size and inbreeding. The common theme in all nine mechanisms is population size, its degree of isolation, growth and reduction. Hominin migration was severely tested by these factors and the evolutionary mechanisms mentioned would have impacted on all such migrations over the last 2 million years particularly on small groups who had moved far from their parent population. Small groups/populations could also be subject to inbreeding between relatives/siblings, reducing the number of heterozygous offspring and raising the possibility of expressing harmful alleles in the local genome more often. These conditions reduce overall population vigour damaging its propensity to produce viable offspring. Moreover, the mal-expression of alleles can cause morphological alterations and pathological change over time, particularly in the cranial skeleton.

WHAT'S IN A NAME?

Having surveyed the mechanisms of change operating in our evolution, we can see how we encounter hominins with different morphologies even though some differences may be small or considered inconsequential. Nevertheless, some palaeoanthropologists take these small differences to be reason enough to place hominins in different baskets, separating them and relieving some of their previously accepted relationships to others. That can result in changes in how lineages are seen and who is related to whom. Knowing how hominins changed over time in accordance with the effects of the 9EM, it is appropriate to briefly visit the world of taxonomy that uses differences between hominins to reach conclusions about how they are associated or lie on different lineages as they travelled the world during the last 2 million years.

It is worth defining a few points of terminology used in the following chapters. Some believe the term 'modern humans' is imprecise but I have chosen to use it. Other definitions may also be more correct such as 'anatomically modern humans' (AMH) and, of course, *Homo sapiens* is the correct scientific term that often includes a sub-species extension, *Homo sapiens sapiens*. I will use 'modern humans'. 'Archaic' *H. sapiens* is used to denote the immediate stage before fully modern humans emerged and sits between *Homo heidelbergensis* and fully *H. sapiens* people although in effect this is just a stage of a clinal evolutionary process from one to the other. A few individuals lie morphologically between us and more ruggedly built archaic *sapiens* people making it difficult to discern the two and raising arguments where the dividing line should be. It is not important in my view, because it is difficult to put specific boundaries between certain hominins and us and accurately define that boundary. Although some would hate me for saying it, there are areas of palaeoanthropology that are not precise science, and the transition between species is one of them, tending to cause differences of opinion and rendering exactitude defunct. But taxonomists have struggled with those problems for years. In these terms, the range of temporal and spatial variation among hominins throughout human history has been wide and it's a problem that should be dealt with and accepted, but I discuss that later.

The term 'Archaic' used here defines hominins who look modern in some respects but still retain ancient skeletal traits, such as a robust build, particularly on and around the cranium but they still might be our close ancestors. I also use the term 'archaic' to mean hominins not directly on the modern human line. Going further back, I use a combination of terms for that widespread and famous hominin group *Homo erectus* that includes 'erectines' or *erectus* that came before *H. heidelbergensis*. An issue to make clear is *time* or segments of it, more particularly the geological timescale during which the history of *Homo* took place. I am not necessarily concerned with what went on before that, because I am only interested in *Homo*: our direct family that extends back to a possible combination of *Homo habilis*, and *Homo rudolfensis* that I often call 'habilines'. Numerical ages are indicated by the standard abbreviation of 'ka' for thousands of years and 'Ma' for millions of years. The time of *Homo* extends through the Pleistocene Epoch and possibly back to the later part of the Pliocene Epoch as far as 5.33 Ma, far older than the time dealt with here. The boundary between these two Epochs is termed the *Plio-Pleistocene boundary* at 2.58 Ma. The Pleistocene finished when the Holocene Epoch began at 11.7 ka (Ogg, Ogg, & Gradstein, 2016). Going backwards, the Pleistocene is usually sub-divided into Upper (130–11.7 ka), Middle (770–130 ka) and Lower (2.58–0.77 Ma) stages. Technically, the Lower Pleistocene is sub-divided further into the Gelasian (earliest half) and Calabrian (last half) separated at 1.8 Ma but those divisions are not used here because it is standard in palaeoanthropology to stick to Upper, Middle and Lower stages that are occasionally and more precisely divided further using *early*, *mid* and *late* divisions. The Pleistocene is generally regarded as the *human Epoch* because it brackets the major time slice that covers the development of humans although it now seems as though *Homo* is not playing fair and extending itself back into the Late Pliocene.

I have outlined these terms because the general reader may not be familiar with them and even if they are they may not know the exact dates that these temporal divisions signify that have changed from time to time due to alterations in the accepted sedimentary boundaries defining them. After those points of nomenclature, I want to look at hominin taxonomy broadly and how and why we name various hominin species.

SPLITTING, LUMPING AND NAMING OURSELVES

We have yet to isolate genetic material from Lower Pleistocene hominins. The oldest so far is mtDNA extracted from a 350-ka *H. heidelbergensis* femur discovered at one of the Sierra de Atapeurca sites in Spain (Meyer et al., 2014). The findings suggest we should rethink the model of Middle Pleistocene human evolution particularly the status of *H. heidelbergensis*, pre-Neanderthals and Denisovans and I will discuss that later. Primarily, palaeo-DNA research can overturn previous ideas based on skeletal morphology alone and by doing so can radically change accepted models, and following the previous discussion regarding the *9EM* is very understandable. It will be interesting to see what might happen to such models when DNA is eventually extracted from even earlier hominins that that is only a matter of time. When achieved it will have the potential to overturn the old methods of hominin taxonomy completely. Past assessment of ancient fossil hominins using metrical and non-metrical methods separated one from another and placed them in different genera and species, their results were then used to construct evolutionary trees. Two groups of palaeoanthropologists do this, they are unofficially known as 'splitters' and 'lumpers'. Splitters 'discover' intermediates or denote slightly different hominins by observing their morphological differences, some quite small and often difficult to quantify accurately but that requires a balance of features that can be different due to natural variation between hominins. The usual process can then result in species once thought to be the same being separated. Taxonomic 'lumpers' tend to do the opposite, accepting small differences between species usually as within a range of expected variation within a species. Thus, they group fossils together usually arguing against the reasoning of splitters and rejecting minor skeletal differences, but it is hard to know who is right in these arguments.

Dividing species relies on a perceived degree of morphological variation and every species has a natural range of morphological variation, so it is vital to understand that before a division of types can take place. It usually requires a large sample of fossils of the same type to establish a range. But in the world of fossil hominins a large sample is often illusive if not impossible to attain. Moreover, sexual dimorphism and age cause variation among hominins of the same species and those ranges also should be established which again relies on sample size and knowledge of expected ranges. The sex of a fossil hominin is not always obvious particularly if it consists of a few fragments or there are only one or two other specimens for comparison, if you are lucky. All this applies not only to cranial but post-cranial remains and while one or other of these might be available for one specimen they might not for another. Palaeoanthropologists deal with large slices of time that affects morphology but so does a hominin's demographic history. Moreover, there may only be a few specimens scattered over hundreds of thousands of years and/or across continents. Hopefully, the drawbacks of species allocation using morphology alone will be enlightened by improved methods of DNA extraction from fossil bone, particularly on earlier specimens.

The evolutionary process leading to us has been long convoluted taking at least 7 million years, the suspected time when we departed from the anthropoid line. As with other animal families our evolutionary lineage now has a series of genera and species names. They help separate/define perceived stages of development and inter-relationships along a continuum of becoming what is assumed to be modern. Other branches of our family tree have also been given names for their different stages. Fossil discoveries often change the appearance of the human family tree, and, on occasion, call for a rethink about where the position of its branches should be or if they should be there at all and we will see examples of this in the following chapters. Then there are the twigs, where should they be and on what branches and what fossils make up twigs. It is for these reasons there have been many tree shapes in the garden of human evolution over the years and no doubt there will be a few more to come. That is not a fault with the research it merely says we are learning all the time and models change because of it and that's how science works.

Sometimes, the stages along our evolutionary pathway seem real enough while at other times they seem unlikely. Taxonomic terms or names are useful but, at times, they do not seem so or reflect what they are trying to represent, so some taxonomic divisions may not be justified. But scientists like to name things and they continue to do so. We invent names for ourselves, and the ancestors of our ancestors all the time. We also argue about what constitutes divisions between them, how appropriate a species name is, where exactly to put the divisions and what terms might be more appropriate. The different stages through which we have passed have increased so much that we now have a rather confusing list of ancestors as taxonomy tries to keep up with them (Table 2.1).

Names depend on whether they stand the test of time and future evidence as we discover new hominins. That will inevitably bring new family branches and re-evaluation of what named stages mean in the light of those discoveries. Arguments over our taxonomic correctness and, in some cases, its value, occur all the time, but then palaeoanthropologists like a good argument. Occasionally, arguments over taxonomic terms cause enmity among researchers forming divisions and 'sides', dividing former colleagues and lumping others into various 'camps'. Such arguments often take up more of the palaeoanthropological literature than they should and more than their fair share of discussion, justification and the writing time of researchers. But, is it worth it?

Taxonomy provides an understanding of the biology of our planet. As humans, we require compartmentalisation of data to handle it. Compartmentalisation makes lots of different data easier to understand and we can pigeonhole and manage it when it all becomes too large to look at holistically. Then, he separate parts can eventually be combined to make up a more understandable whole. It is something we do with everything we study in the natural world. It demarcates stages in human evolution so we can cut a long convoluted and complex story into manageable pieces. And that makes the complex world of human evolution a little easier. We must have cognitive pictures and taxonomy helps us paint them. But it may not always be correct where it places a demarcation at least when it comes to humans.

Looking at the building blocks of all living things, and how those blocks relate, clarifies our understanding of the world and the organisms in it. But palaeoanthropology is now focussing on modern humans and our evolution in a way never before attempted. One thing to remember is that for many years palaeoanthropologists tried to put fossil hominins on a single string like a row of conkers from a chestnut tree. It was done to assemble a single pathway of modern human descent from our ancient ancestors and thus make that pathway both logical and easy to understand. The trouble is that our biological evolution was never uniform nor did it follow precise pathways; the *9EM* took care of that. The visual uniformity that taxonomy sometimes presents cannot be taken

TABLE 2.1 Major Hominin Genera, Species and Named Specimens since Our Separation from Anthropoids

Genus	Species	Origin	Age
Sahelanthropus	tchadensis	Chad (Sahel desert)	~7.0 Ma
Orrorin	tugenensis	Kenya	6.1–5.7 Ma
Ardipithecus	kadabba	Ethiopia	5.8–5.2 Ma
	ramidus	Ethiopia	4.1 Ma
Australopithecus	anamensis	Ethiopia, Kenya	4.2–3.9 Ma
	afarensis	Ethiopia, Kenya, Tanzania	3.85–2.95
	bahrelghazali	Chad	3.5–3.0 Ma
	aethiopicus	Ethiopia, Kenya	2.7–2.3 Ma
	garhi	Ethiopia	2.5 Ma
	africanus	South Africa	3.3–2.1 Ma
	sediba	South Africa	1.98 Ma
Kenyanthropus	platyops	Kenya	3.5–3.2
Paranthropus	robustus	South Africa	1.8–1.2 Ma
	boisei	Ethiopia, Kenya, Tanzania, Malawi	2.3–1.2 Ma
Homo	habilis	East/South Africa	2.4–1.4 Ma
	rudolfensis	East Africa	2.5–1.8 Ma
	georgicus	Georgian Republic	1.8
	erectus	Africa, Asia	1.9Ma–350 ka
(Homo erectus)	ergaster	Africa	1.9Ma–350 ka
	antecessor	Europe	1.2Ma–800 ka
	naledi	South Africa	300–200 ka?
	heidelbergensis	Europe, Asia	800–200ka
	rhodesiensis	Africa	400–125 ka
	neanderthalensis + Denisovans	Europe, Israel, Asia	300–40 ka
(East Asian Archaics)[a]	Jinniushan, Dali, Mapa, Penghu, Hexian	China	300–200 ka
Homo	floresiensis	Flores, Indonesia	195–17 ka
	sapiens	Africa	<250 ka

[a]These hominins are all Homo but as yet lack species names and are known more widely by their individual site origin.

to be accurate as it stands and at best uses very broad and often imprecise labels to do its job, at least with hominins who are such a variable lot.

The aim of this chapter is to look at the difficult but intriguing issue of what we know about our emergence through our very ancient ancestors and trace the path that gave rise to us and, thus, those people who first came to Australia. Australia should not be forgotten in the juggernaut of world human evolution because it was here one group of modern humans eventually settled, at one of the farthest points from Africa. So, we need to look at the people from whom we descended (or we think we descended); where *they* came from; their endeavours, and try to compose a picture of how our emergence took shape. That includes the dispersal process of people leaving Africa over the last 2 million years and how eventually it moulded those first Australian arrivals.

First, we must look at the ancestors of our ancestors, however, the more we discover the less logical things seem. The old way of stringing fossils in neat lines, using a mantra of this begat that and that begat that, has become severely battered over the years. The fossil record now shows us we knew nothing about hominin relationships when it came to matching fossil evidence but that does not prevent us from still doing it because those mental pictures are easier to deal with. The trouble was that the more lineages were constructed the less likely they seemed to fit, particularly our lineage. (It is for these reasons that I will, in the following chapters, leave unanswered question marks scattered throughout the text.) But once again, palaeogenetics or the ability to research the molecular biology of hominins and other animals came on the scene but more about that later. Therefore, to understand where the first people to enter Australia came from we must start back at the beginning of *Homo*, over 2 million years ago and we should expect some degree of confusion in the process.

OUR EARLY ANCESTORS AND THE ADVENT OF ICE AGES

Generically, *Homo* was here to stay. The term *Homo* defines a genus that defined our earliest boundary where we changed into a recognisable human form, albeit an archaic form. That boundary defined where we recognise we had changed enough from an ape-like ancestor (*Australopithecus*) to a hominin recognisable as something closely resembling a human and just that little bit more likely to look like one of us even though such a division is not easy to demarcate. That change occurred across the Plio-Pleistocene boundary between 3 Ma and 2 Ma when *H. habilis* emerged. The changeover time between *habilis* and australopithecines is a grey area, much argued about and, unfortunately, with few fossils to help. Those we have appear to be contradictory and confusing. Two examples within that time frame are fragmentary mandibles showing transitional hominin characteristics that morphologically lie between *Australopithecus* and *habilis*. The first was found in Ethiopia's Hadar region in 1994 (AL 666-1) (2.3 Ma); the second was also from the Hadar region, found later at Ledi-Gururu (LD 350-1) and dated to 2.8 Ma (Villamoare et al., 2015). Many anatomical and behavioural factors mark hominin transitions but a consensus of definition is lacking over which one is correct. That prompts arguments about what signifies a species and what does not, which I have already touched on. The *9EM* that push us from one species to another are sometimes subtle and require time but they operate through different catalysts and they are a little easier to pinpoint.

Our evolution has been difficult to not only explain but also quantify as a phenomenon. Like all organisms, we are moulded and affected in many ways by the world we live in, and the world reacts most radically when our climate changes particularly if that change is radical and remains for some time. The world's climate began to act very radically between 3 and 2 Ma. It has always been difficult to define exactly what drove our development towards an increasingly cleverer creature that evolved a big brain, a good upright balance and free manipulative hands that gave us an increasingly widening edge over other animals. However, major evolutionary changes in our behaviour and morphology appear to coincide with global climatic changes associated with onset of cyclic glaciations particularly as they became stronger and regular in the late Pliocene. Climatic change brought alteration to the environment, coasts and landscapes and changes to water and food availability, particularly the spread of grasslands that resulted in faunal demographic change. As primitive scavengers and gatherers, and possibly occasional hunters, 2–3 million years ago hominins were as affected by climatic change as any other species but because they lacked the speed and rapid deployment capabilities of most animal populations they had to work harder to survive. Therefore, changing environments and landscapes were catalysts for evolutionary mechanisms that drew out their adaptive strategies and presented them with survival challenges. Climatic change also made hominins move. One interesting aspect of the Plio-Pleistocene glaciations is that they continued to slowly intensify bringing increasing challenges (Fig. 2.1). Regular worldwide climatic changes, although slow to take place at first, are difficult to dismiss as one of the fundamental mechanisms of hominin development. It was also slow but laced with subtle changes.

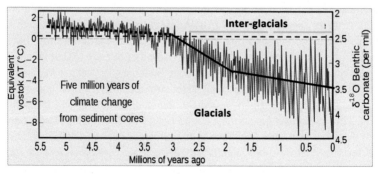

FIGURE 2.1 **Plio-Pleistocene glacial cycling.** *(After Lisiecki & Raymo, 2005)*

Some early hominin changes occurred around 5.5 Ma about the time when glacial–interglacial cycling began although in a minor way. Cycling continued largely unchanged in strength and amplitude for nearly two million years to around 3.5 Ma. The change marked by *Sahelanthropus chadensis* (~7.0 Ma), the first recognised departure from the hominoid line, into something close to an *Ardipithecus* (an early type of australopithecine) coincided with the start of such cycles around 5.5 Ma. The first 2 million years involved about 50 cycles, although they were moderate in both variability and intensity and had smaller amplitude swings compared to Middle and Late Pleistocene glacials. Different types of ardipithecines and australopithecines came and almost went, or perhaps these two genera just represent variation within a single genus (Table 2.1). We now know that australopithecines had migrated south by then to South Africa and that has been proven by the discovery of an almost complete skeleton of 'Little Foot' a young australopithecine near Johannesburg. Hominins continued stealthily progressing through the Pliocene as average world temperatures slowly reduced from 2°C above today's average to that of today by 3.0 Ma (Fig. 2.1, *red dotted line*). That, however, does not reflect present temperatures, because glacial minimums brought temperatures to an equivalent of the present. Between 3.3 and 3.0 Ma cycling patterns changed in two ways. First, the world cooling trend increased (Fig. 2.1, *first solid line*), and, second, Ice Ages themselves became more severe increasing the amplitude of temperature swings. From 3.0 Ma, average world temperatures dropped below those of today 80–90% of the cycle time, for the first time in hominin history and any time since Snowball Earth. Their severity markedly increased with the depth of cooling and maximum interglacial temperatures were, at best, 80–90% of the cycle time now on a par with those of today. This trend lasted from ~3.0 to ~1.8 Ma. In terms of hominins, this cooling pattern began with late australopithecines, continued during *H. habilis* and *H. rudolfensis*, and ended with early *H. erectus*. It also marked the time of the first hominin explorations out of Africa that we know of. A cooling world brought reduced sea levels helping them reach places like Indonesia. Could it have been then when a group of the northern *Australopithecus afarensis* broke away and moved to southern Africa, giving rise to the southern australopithecines such as *A. africanus* and *A. sediba*?

It may be no coincidence humanity evolved exponentially over the last three and a half million years in concert with planetary climate changes that were intense and that were frequently extremely changeable, factors that were ideal to drive the *9EM* evolutionary mechanisms. Initially changes were slow but later they increased, seemingly in step with an enlarging brain or was it the reverse? The advent of increasingly stronger glacial–interglacial cycling changed landscapes in a way never experienced by hominins or any other fauna. The relationship between the two became mixed with drastic environmental change and commensurate alteration of vegetation pushing animals to adjust to changing patterns of vegetation and move into new areas, and hominins followed.

Dietary changes for animals and hominins responded to ecological changes as vegetation mosaics shifted together with altering patterns of savannah and woodland and fluctuating desertification. These processes imposed stresses and changes to the necessities for survival that included quicker decision making; the ability to change habitat and movement to better places; improved survival techniques as well as a growing ability to find, adapt to and survive in new and different surroundings. These would have made hominins flexible generalists with flexible behaviours. Those that could not change, such as the robust *Paranthropus* hung on for as long as they could but eventually went extinct as their preferred ecosystems deteriorated and changed but they did not. In evolutionary terms, there could have been nothing better than this long-term climatic phenomenon to install in our early ancestors heads a mechanism for strategic thinking. It also imposed a range of selective pressures that more than anything else would increase their ability and tenacity to survive with the help of cerebral enlargement and rearranging neural networks.

From 2 million years ago, the cooling trend eased but average world temperatures continued trending down in a sharper trajectory. Glacials became increasingly colder, increasing their depth further, deeper and consequently imposing greater environmental swings around the globe. Just prior to 2.0 Ma interglacials that normally gave some respite from glacials also became much cooler and present interglacial temperatures were not reached; that pattern lasted until around 1.5 Ma. From that time forward, interglacial temperature patterns became more variable but generally cooler than our present one. It was in this phase that we think early hominins first left Africa and travelled far. After 1.5 Ma, glacials cooled further with deeper cyclical patterns. It was only during interglacials that average world temperatures reached and occasionally exceeded those of today, although that was not consistent during the last 1 million years.

To summarise, the trend in average world temperature from 5.5 Ma to the present decreased in a stepwise pattern with a steepening of the trend at around 3.3 Ma, the time of the emergence of the earliest stone artefacts. Australopithecines disappear and habilines (*habilis* and *rudolfensis*) emerged between 3–2 Ma followed by *H. erectus* towards the end. Is it a coincidence that our earliest *Homo* ancestors emerge during the onset and transition of the world's climate and and an accelerating cooling trend that followed? These conditions of drastic climatic change must have acted as forcing mechanisms affecting hominin evolution in many ways as Ice Ages intensified and world temperatures continued to decline imposing radical changes on environments everywhere. Climatic conditions between 2.0 and 1.5 Ma heralded the first intense cold phase of the continuing glacial pattern with cool interglacials bringing only minor relief from glacials. Drastic environmental changes that accompanied glacials, including desertification, could have precipitated hominin movement away from North and sub-Saharan Africa particularly. It was at this time that small hominin groups pushed further north possibly into better conditions in places like the southern Caucasus, East Asia and Indonesia.

ERECTUS EMPIRES

Our genus contains several species representing various developmental types and divergent stages over the last 2 million years and I have already mentioned some of those. Until now, the stalwart transitory species link between *Australopithecus* and *H. erectus*, has been widely accepted as *H. habilis* or *H. rudolfensis* that existed between 2.5 and 1.4 Ma and all hominins since then are regarded as *Homo* (Table 2.1) except for a few unnamed Middle Pleistocene species from China and Denisovans. Originally, *H. erectus* was categorised as a single species composed of several adult and sub-adult fossils found on several continents. The first fossil discoveries were made in Java and China, respectively, African erectines were discovered later. There is a wide range of cranial variation among erectines but the few post-cranial remains there are suggest they generally differed little from each other and from modern people in their morphology although they were generally shorter than modern humans. Cranial variation, however, was a great attraction for those who felt observed differences meant something about the status and position of various hominins on the evolutionary ladder and in some cases, they probably represented different species. The logic for those beliefs was that variation must relate to ancestry and geographic division where *9ME* work their selective powers in a sparsely populated world, particularly through isolation and founder effect. Isolation of a small group or population in the Lower Pleistocene virtually meant death, or at the very least held the promise of eventual extinction unless more individuals could be added to the gene pool. The *9EM* alone produced different cranial forms among *erectus* groups, even if only slightly different in some cases. One famous example of an observed regional difference occurs between Chinese and Javan erectines in the form of their respective supra-orbital sulci, a gutter located above and behind the eyes that divides the orbits from the frontal bone. This feature is deep in Chinese *erectus* but in Javan *erectus* it was shallow forming a much more even transition from brow to forehead. Thus, the presence or absence of a supra-orbital sulcus became famous for separating a Javan from a Chinese *erectus*. Originally, these two groups were given different generic names based partly on their geographic homeland but also on their respective morphologies such as their different shaped sulcus. The man who worked on the initial *erectus* excavation in China, Canadian anatomist Davidson Black, called the first Chinese *erectus* fossils, found at Zhoukoutien in the late 1920s and 1930s, *Sinanthropus erectus*. Eugene Dubois following his discovery of the first *erectus* at Trinil in 1891 gave the generic name of *Pithecanthropus erectus*. This and other morphological differences between Javan and Chinese erectines was later ignored when they were lumped into one genus and species, *H. erectus* (upright man), in 1950 by the eminent evolutionary biologist and taxonomist Ernst Mayer.

Forty years ago, the species name *erectus* was revised dividing African *erectus* from those found outside that continent (Groves & Mazak, 1975). The latter retained the name *H. erectus* but those living inside Africa became *H. ergaster* (working man), although not everybody recognises this division and I am one. I also have never understood why at that time Javan and Chinese erectines were also not given different species names considering their extensive separation from each other as well as their different cranial morphologies sufficient to tell them apart quite easily. Those grounds alone might warrant a separate species name that reflected their original taxonomic identification of *P. erectus* and *Sinanthropus pekinensis* that, for me, remains a puzzle.

Three *erectus* species names are in use at present but they do not really suggest anything unique, peculiar or special about any of them that, in my view, *really* separates them. The three names, *erectus* (upright man), *ergaster* (working man) and *antecessor* (explorer man) which is the name given to European erectines, are also slightly confusing because each can be applied to the other and some researchers often do that. However, they were all *upright*, they *worked* equally as much as each other at surviving and making stone artefacts, and each *explored* extensively. Morphological differences between erectines probably emerged from geographic isolation and undergoing different regional selection processes associated with environmental variation and lifestyle as they do in other animal species. *H. erectus* is, as I have said, a very morphologically variable hominin and there is great variety among fossils attributed to it, as is the case for the later *H. heidelbergensis* that I talk about below. However, *H. erectus* probably requires a revision that reflects its temporal variation more thoroughly and that may not necessarily be a factor of its origin.

The modern human story does not start 200,000 years ago; it is a continuum spanning at least 2 million years. At its most basic, the generally accepted story of human evolution goes this way: it began with an early *H. habilis* that evolved into *H. erectus* that evolved into *H. heidelbergensis* in Europe and *H. rhodesiensis* in Africa and the latter evolved into an early *H. sapiens*. At least that is the most supported and easily understood story. But there are doubts about every stage and even the origins of the hominin species involved. For example, there have been calls for *H. heidelbergensis* to be revised because it has become a species catch-all for any hominin that occurs later than *erectus*, but only in Europe. It has more advanced features than *erectus*, such as a bigger cranial capacity, it is usually somewhat larger in its body so it should be classified as something different than *erectus*. It could just as easily be referred to under a *late erectus* banner. I have more to say about this in later chapters but this is a good point to visit the long story of our emergence and recall the basic evolutionary stepping-stones in a story that spans the Old World. It is not as easy or straightforward as the list above would suggest. It is an increasingly complex story that involves different phases or categorisations of 'us' and plenty of unknowns.

In the last couple of decades, our idea of the time when the first hominins left Africa has doubled from 1 to 2 million years. Discoveries in China, Java and Georgia now firmly indicate the first African exits must have occurred at or before 2.0 Ma. That means it was not necessarily *H. erectus* that left it could have been *H. habilis*. Mode 1 Olduvai-like non-biface core-chopper tools found in China and the Middle East also suggest that the hominin making the journey could have been *H. habilis* (Fig. 2.2). It has been suggested an early exit occurred before 2 million years ago in the late Pliocene either by *H. habilis* or *H. rudolfensis* or both or even an *Australopithecus*. Another view is that early East Asian inhabitants may have originated from an australopithecine-like hominin that left Africa 3.0–2.5 Ma and migrated to the region and that *H. erectus* originated in East Asia (Kramer, 1994; Raghavan et al., 2003). At present, the likelihood that this is the case seems doubtful, especially because there is a total lack of supporting evidence, although palaeoanthropological discoveries being what they are, that evidence may turn up tomorrow! It is suggested that an australopithecine (among many ideas) could have also been the ancestor of Indonesia's *Homo floresiensis* ('Hobbit') but I discuss that later also.

An early hominin African exit emerged between 1991 and 2005 when archaeological excavations in Dmanisi in the Georgian Republic revealed five very archaic looking hominin skulls (Gabunia et al., 2000; Lordkipanidze et al., 2013). However, that exit has been challenged by the suggestion that the Dmanisi hominins may not have left Africa, instead they may have come from the east (Dennell, 2011). Therefore, right at the beginning of our story there is controversy. This book, however, is not here to unravel all the controversies that surround palaeoanthropology, it is about looking at the evidence for the story of modern humans and their eventual arrival in Australia although it does not shy away from controversial subjects, a few ideas, and a little arm waving. We must begin with a foundation from which to work. So, while being aware of the counter arguments at several stages of the standard story I have chosen to continue with the most widely accepted story at present although counter arguments will be noted where they occur and it may turn out that they are right or not far off the mark.

The Dmanisi hominins have been dated to between 1.9 and 1.75 Ma placing them contemporaneous with hominins in East Asia but earlier than the 1.5 Ma Ubeidiya site in Israel, that has been known for some time but its age was

FIGURE 2.2 Left, Olduwan Mode 1 basic chopper (1.8 Ma) and (right) flakes and choppers (2.5 Ma). *(Steve Webb)*

doubted as too early (Tchernov, 1988). The Dmanisi artefact assemblage is of Mode I and does not hint at the later Acheulean Mode 2 type hand axes associated with *H. erectus* that appears around 1.6 Ma (Fig. 2.3). The Dmanisi finds tends to support the Ubeidiya findings in Israel of an early exit but Dmanisi revealed a great deal of unique palaeoanthropological data. First, the number and exceptional preservation of the crania found there was unusual for hominins of their age and because a cluster of fossil hominins of that age is not a regular occurrence in archaeological digs. The Atapuerca cave site revealed 28 individuals significant in themselves but they were less than half the age of the Dmanisi fossils and were more fragmentary as it was with the Rising Star cave finds in the area known as the Cradle of Manin South Africa. Second, each Dmanisi individual has an almost unique morphology, displaying a set of cranio-facial characters that separate one from another presenting a wide range of facial morphology among this early dispersal grouping. They all have a very archaic appearance that has drawn some arguments suggesting some may be related to *H. habilis*, and their small cranial capacities could support that idea. Also, their cranial capacity is generally smaller than might be expected for erectines, ranging between 550 and 750 cm^3. Third, they were small bodied hominins and displayed some post-cranial differences from later *erectus* people. Finally, their condition makes them some of the best-preserved fossils of their type and age yet found, thus offering the best evidence for what the first hominins to move out of Africa, perhaps before 2 million years ago, looked like (Fig. 2.4).

The group included a sub-adult (D2700) and an older edentulous individual (D3444) and each had skeletal scaring indicating they had led active but harsh lives in competition with carnivores. One interesting anatomical feature is that they were incapable of full rotation of the shoulder and proper striding or running, all capabilities attributed to later *H. erectus* and, important for migrating great distances. Such differences make them poor candidates to have reached East Asia and Indonesia, but it looks more and more likely they did! Their morphology and other capabilities do, however, place them close to a very early *erectus* or late habiline group. It is also interesting that the Dmanisi site shows

FIGURE 2.3   **Mode 1 stone artefacts from Dmanisi (1.75 Ma).** *(Steve Webb)*

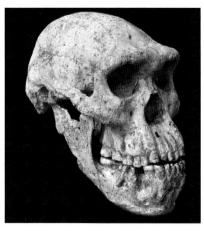

FIGURE 2.4   Dmanisi Skull 5 (D4500) has the smallest cranial capacity (546 cm³) of the group, a prominent supra-orbital torus and a long, prominent prognathic face and large teeth.

early hominins did accomplish a long journey to an area between the Caspian and Black Seas, 2000 km from Africa. That indicates a feisty hominin in a fully explorative mode going somewhere, but how far did they get? The Lower Pleistocene dates for early hominins from China certainly suggests that they could reach that far, so it seems likely it was either Dmanisi hominins or someone else and I cannot say who that could have been. Incidentally, the Indonesian *H. floresiensis* could also have had them as an ancestor.

Even though the skeletal characteristics and cranial capacity of the 'Dmanisi five' reflect hominins like *H. habilis or rudolfensis* and even late australopithecines, could it be possible that they made it all the way to China, as the earliest dates there seem to suggest, and Indonesia. As mentioned earlier, it has been suggested both an australopithecine and *H. habilis* could have been suitable ancestors of *H. floresiensis*. Already possessing a small brain capacity, just a little subsequent 'island' or 'insular dwarfing' might have maintained or slightly diminished a Dmanisi hominin forging a population that eventually gave rise to *H. floresiensis*, even though I do not necessarily favour insular dwarfing operating on hominins in the past. The possibilities for change among our early ancestors with a shove one way or another from *9EM* should never be underestimated. Indeed, underestimation of possibilities is a mistake made many times in the past, although that is easy to do when dealing with hominins that far back. After decades of being involved in palaeoanthropology I am no longer shocked or disappointed by what is discovered about our ancestors only pleasantly surprised, and sometimes delighted when it is clear we have underestimated them and their capabilities. However, one guess about their capabilities is as good as another although hopefully made with good evidence. Moreover, we are inclined to think in the straight conker lines with one hominin giving rise to another longitudinally. It is clear that is not necessarily the case when it comes to sorting out early hominin groups. It may be time to think laterally, that is, of different hominins groups migrating into various regions and either forming their own evolutionary pathway and/or mixing with previous indigenous hominin groups to form new ones with all that being driven along by the *9EM*.

As I have already hinted, the range of very archaic Dmanisi cranial features forms a 'mosaic' and that leads to some confusion about their place in the human story. It is for that reason they have been given the species name *Homo georgicus* even though their wide range of variation might suggest they do not represent a homogenous population or single lineage (Gabunia et al., 1999). If their combined group morphology indicates that they represent an earlier pre-*erectus* hominin, such as *habilis* or *rudolfensis*, that would mean one or both those species were the first to leave Africa. In fact, beautifully preserved Skull 5 contains a mixture of features reminiscent of three hominin species: the large teeth and small brain case (546 cm^3) of *H. rudolfensis* (KNM-ER 1470); a long face like *H. erectus*, and a small brain case like *H. habilis* (Fig. 2.5). Only one good fossil of *H. rudolfensis* exists dated to between 1.9 and 1.8 Ma, slightly earlier than the Dmanisi hominins, but it had a cranial capacity of 770 cm^3. Without its very large mandible it's lateral profile, I believe, it reminiscent of a late australopithecine.

The range of cranial capacity among the five Dmanisi hominins is 546–780 cm^3 with an average of 645 cm^3 which besides KNM-ER 1470 also reflects that of O.H. 24 from Tanzania (1.8 Ma) that had a brain size of 590 cm^3. Their morphological variety suggests that while they were not all contemporary they were close and might reflect an origin close to or just after the time of the *habilis/erectus* transition. Whichever is the case, when they left Africa, they took with them a range of transitional morphological traits that their isolation maintained and perpetuated perhaps for longer than it might have done back in Africa living in a larger gene pool. They may also represent a Georgian variety of early *erectus*, again through the machinations of the *9EM*, particularly elements like isolation, founder effect and being part of a very small population. Among a small group in isolation, characteristics can emerge through variable or biased gene frequencies from the parent group. So, such a population develops some characteristics and they can become more dominant than they were in the original gene pool (Africa) where a greater frequency of intermixing and genetic assortment would take place.

The concept that *H. erectus* travelled from Africa across the world setting up populations from which modern humans eventually emerged was always an inviting one, principally because it seemed logical, easy to understand and the progression from *erectus* to modern humans seemed possible using the only fossils available at the time, although some imagination was required and plenty was provided. The pathway from *erectus* to us still holds true but the intervening evolutionary pathway is really the opposite and is not as well understood as we might like. Even *erectus* leaving Africa is open for discussion with suggestions that it could have evolved *outside* Africa, perhaps in Europe, and then

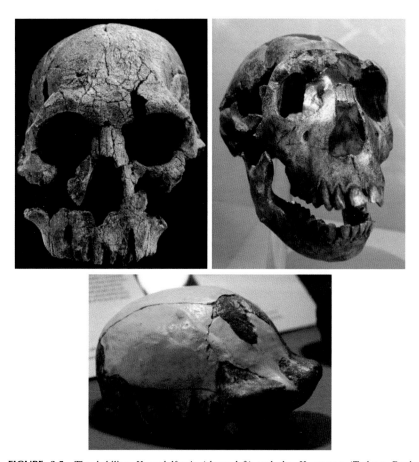

FIGURE 2.5 The habiline *H. rudolfensis* (above left) and the *H. erectus* 'Turkana Boy' KNMER-45000 above right. Below is the OH 9 calvarium with its distinctive heavily developed brow ridge that strongly contrasts with the much smaller brows on the above hominins. *(S. Webb casts)*

moved east, or it emerged in East Asia and moved west to Europe later entering Africa (Dennell, 2011). The possibility of *H. habilis* or a similar hominin leaving Africa prior to *erectus* muddies the waters of that theory somewhat as well as casts doubt on what might or might not be *erectus* outside the African continent. These ideas cover an each-way bet but the problem might finally be solved if DNA can be extracted from *erectus* fossils. I have mentioned that before and return to it below.

At present, we must assume the first European *H. antecessor* shared a common ancestor with an African *H. erectus* or, indeed, it was one. Its arrival in Europe could be an '*erectus* exit' in the AE2 period as opposed to being in the AE1 'pre-erectus exit' which included the 'Dmanisi five'. Several African *erectus* fossils dated to around 1.5 Ma could be candidates for the ancestors of the earliest AE2 arrivals in Europe. Examples include: KNM-ER 3883 (1.6 Ma),

ER 42700 (1.55 Ma), KNM-WT 15000 (1.5 Ma), and OH 9 (1.5 Ma) (Fig. 2.5). They represent a wide morphological variety and all were more evolved and capable of going to Europe at least 250–500,000 years after the Dmanisi hominins. *H. habilis* and *H. erectus* overlapped in time and although the former was the smaller of the two one would have thought it was just as capable of reaching Europe or anywhere as it was reaching Georgia. The earliest hominin yet found in Europe is *Homo antecessor*, originally *erectus*. It is a sub-adult, aged between 10 and 12 years, found in Grand Dolina cave, Atapuerca, Spain, and dated to 1–1.2 Ma (Carbonell et al., 2008) (Fig. 2.6). It has the mixed features of *erectus* together with those of slightly more modern people. Its modern features, however, can change with skeletal development and maturity. Therefore, until an adult *antecessor* is found we do not really know if the more modern looking features persisted into adulthood or just reflect the sub-adult skeletal form. Its *erectus* features are important and suggest an *erectus* ancestry prior to 1.2 Ma. The Georgian hominins have been a special find even among other great palaeo-anthropological discoveries. The possibility of a Late Pliocene or early Lower Pleistocene movement to Asia or Europe is no longer fanciful. After reaching Dmanisi it is intriguing to think that those first out of Africa hominins were faced with a choice of going east or west. They were at a T-junction of choice in the Caucasus, it looks likely that some went east but did others go west? Their existence in the Caucasus shows how resourceful, physically tough and resilient they were to reach that remote region. Therefore, it would not be surprising to find them further into Europe because they seemed to reach the Far East, even living at a moderately high latitude. They must have been capable creatures because they did not stop moving until they met oceanic and climatic barriers

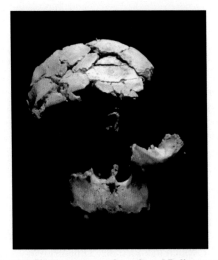

FIGURE 2.6 The sub-adult Homo antecessor from Grand Dolina cave, Atapuerca. *(S. Webb)*

that they could not overcome. Those barriers were big enough to bar them from entering the Americas, probably because of the extreme cold of north-eastern Siberia, and embarking on the wide ocean crossing required to reach Australia. They persisted with their journeys, however, because at least one group came quite close to Australia about the same time that the Dmanisi people lived in Georgia, but there is no suggestion it was one of them.

INDONESIA

It is one of the greatest ironies in the history of paleoanthropology [sic] that the most surprising and revolutionary discoveries which shed real light on the origin of man should come from the Old World. These are the very regions which, according to the theory of the old school, were the least fitted to provide evidence on this most discussed problem. (Franz Weidenreich, 1945, p13).

Franz Weidenreich was one of a very small handful of prominent palaeo-anthropologists in the 1930s and 1940s and like Davidson Black working in Peking, he was so enthusiastic he died at his desk and in mid-sentence writing his detailed monograph *The Morphology of Solo Man* (1951). Another of his manuscripts is *Giant Early Man from Java and South China* in which, he tentatively describes the enigmatic giant hominin *Meganthropus palaeojavanicus* (Sangiran 6) found by G.H.R. von Konigswald. He was tentative because he did not wish to upset the discoverer of the jaw fragment before Weidenreich had fully described it. That behaviour was an early reflection of the '…it's their fossil…' syndrome that still pervades the palaeoanthropological profession, both for professional and possessive reasons. The fossil consisted only of a massive section of mandible that contains a first molar (M_1) and two pre-molars (Pm_1 and Pm_2). According to Weidenreich's measurements it is basically equal in size to that of a male gorilla. Because of that, it continues to defy categorisation and placement in the fossil record although it is often compared with other hominins.

Indonesia was indeed an early *erectus* empire lying 5000 km from Dmanisi, but in many ways it has been just as important for hominin discoveries as Africa has. For example, the very first *H. erectus* was discovered in Java in 1891. After becoming a physician, Eugene Dubois joined the Dutch army and deliberately went to Indonesia to find early humans. He believed that it would be in the tropics where an ancestor between apes and humans would be found. His ideas arose from lectures given by Ernst Haeckel while at university. Haeckel was reflecting the thoughts of Charles Darwin with whom he corresponded. Dubois began working with a group of convict labourers loaned to him by the Dutch colonial authorities, and it was during this work that a *H. erectus* calvarium (Trinil 2) was found, near Trinil village on the Solo River, together with a molar tooth. The team also found a femur with a severe bony outgrowth (exostosis) along the medial side of the upper shaft. That probably resulted from a very painful, severely torn *vastus medius* tendon that had largely parted from its anchor point on the bone. Since then almost 200 hominin fossils have been discovered on Java particularly in the eastern half of the island near the Solo River (Indriati, 2004). I have always been surprised that none have been found on the larger island of Sumatra to the northwest that *erectus* must have reached before entering Java. Java's main sites include Sangiran, Sambungmanchan, Kedung Brubus, Mojokerto, Patiayam, Trinil and Ngandong. Combined, they represent 75% of all *H. erectus* fossils recovered in Indonesia. The region is dominated by active strato-volcanoes that form a line through central and eastern Java and are associated with the activities of the nearby subducting Indo-Australian plate boundary that runs just south of and parallel with the Javan coast.

In 1936, a child's cranium was found on Java near Mojokerto on the Solo River. Both its personal and temporal age have been equivocal, attracting much argument. Personal age estimates range from 6 months to 6 years but its temporal age has been even more contentious. Originally dated to 1.81 Ma, it became the oldest hominin in eastern Asia (Swisher et al., 1994), More recently, a date of 1.49 revised it downward (Morwood et al., 2004), but some still believe that it is older possibly lying somewhere between 1.8 and 1.6 Ma (Huffman, Shipman, Hertler, de Vos, & Aziz, 2005). The fossil has been difficult to date because of establishing its exact stratigraphic position and despite all the effort put into it, we may never know its age with any accuracy. However, if the date is anywhere near accurate the 'Mojokerto child' represents not only a very early hominin exit from Africa but a broad and successful dispersal of hominins far into Southeast Asia, pushing as far as their ingenuity could take them. We might also consider what was said earlier about Dmanisi hominins coming from the east rather than Africa and Java is closer to China than Africa.

Java today is a tropical landscape and because it lies only a few degrees south of the Equator, it is not much different from that through which *H. erectus* roamed. The fluvio-volcanic derived Sangiran Dome uplift in central Java consists of several major formations that have become famous because of the 80 or so *H. erectus* fossils that have emerged or been found in or on them. Outside the dome the main lower sediment layer is known as the Pucangan Formation. Inside the dome, the older layer is termed the Sangiran Formation and that is overlaid by the younger Bapang Formation, somewhat confusing, but it is from those that some of the names of different fossils have been taken (Larick et al., 2001). The Sangiran Formation contains several volcanic Tuff layers that usefully lend themselves to being dated using the $^{40}Ar/^{39}Ar$ method. A narrow Grenzbank layer separates the Sangiran and Bapang layers but some argument surrounds the age of the Sangiran/Grenzbank transition in which the oldest hominin fossils in several assemblages have been found dating to ~1.6 Ma. So far, no hominins have emerged from below the Grenzbank layer. Hopefully, I have tried to make this stratigraphic sequence sound simple. However, if there is one thing about Javan sedimentary sequences and accurately provenancing fossils that emerge from them is that the job is very complex, thanks to millions of years of volcanic activity, sedimentary upheaval and the tropical climate.

Further east along the Solo River are several fossil localities that have given their names to the fossils found in them. Three are Mojokerto, Trinil and Sangiran, now either famous as salient individual hominins or collections. Fossil discoveries emerged in Java during a geological survey of the region carried out in 1931–1933 by geologists W.F. Oppenoorth, Carel Ter Haar and G.H.R. von Koenigswald. Since then, over 100 *erectus* fossils have emerged, many consisting only of variable sized cranial and post-cranial fragments. The Solo River runs roughly southwest–northeast across eastern Java cutting a deep incision through Lower and Middle Pleistocene sediments from which hominin and faunal fossils emerge. During the Monsoon season, high rainfall flooding the river washes fossils from the banks and in some instances, redeposits and remixes them with fossil faunal bone. They then undergo re-deposition further downstream, either remaining exposed or in some cases reburied and exposed again, perhaps many times, during many rainy seasons. So, between the down cutting of the Solo River and uplifting by the Sangiran Dome sediments become exposed and the fossils they contain with them emerge in a confusing sedimentary washing machine. These taphonomic processes make placing hominins in secure temporal sequences extremely difficult if not impossible to interpret or date securely. The majority of Javan hominins come from this region but the poorly documented contextual records of earlier discoveries and the fragmentary condition of the bone leaves our understanding of their taxonomic and palaeobiological status less than complete. That is particularly frustrating because the Ngandong or Solo group of hominin crania discovered by Ter Haar have been particularly fascinating from the point of view of Australian palaeoanthropology.

The Ngandong collection consists of 12 individuals, all with missing faces, bases and mandibles. They are famous in Australia for being firmly implicated as contributors of the robust ingredient of early Australian fossil humans, something considered to be the case for many years, particularly within the context of the Regional Continuity or Multiregional Hypothesis. Oppenoorth, Ter Haar and von Koenigswald made their Ngandong discoveries at a similar time to those they made at Sangiran. The Ngandong fossils suggested that *erectus* was very successful in Indonesia making the transition from the Lower into the Middle Pleistocene. However, dating the Ngandong hominins has been problematic, as usual. Their poorly understood stratigraphic context has made it difficult to accurately date or place them within their original stratigraphic context. Dates produced so far for them have ranged from about 350 ka 30 years ago (a more subjective than empirically derived age) to as late as 50 ka and now they are thought to be between 550 ka and 150 ka with an average close to their original perceived age of 350 ka. The youngest date for them has always seemed too late because morphologically they should occupy the earlier half of the age range.

Some workers have suggested they are erectines, albeit late erectines, and their archaic appearance supports that, but that should put them well into the Middle Pleistocene. However, their archaic appearance also includes modern morphological features that tend to confound rather than enlighten our understanding of their age and who they really are. High cranial vaults (for an *erectus*) and their range of brain capacity (1013–1252 cm^3, $n = 5$) are two such features more in tune with a 'tropical' *heidelbergensis*. It can also be speculated that the Ngandong hominins may have moved into Indonesia from somewhere outside during the mid-late Middle Pleistocene, perhaps from India or Southeast Asia. These are important issues for consideration because the question of whether a Trinil 2 hominin later evolved into a Ngandong type is not even close to being settled and neither is the link between Sangiran erectines and Trinil 2. A direct lineage between these fossils seems far from proven at present but I return to that later.

Indonesia's hominin fossils are usually associated with large quantities of extant and extinct faunal remains and sometimes stone artefacts, such as the basically designed Patjitanian chopping tools that come from the significant geological sequence that extends from the Middle Pleistocene back to the Pliocene. The fossil hominins from there have yielded ages ranging from 1.6 to 0.4 Ma although the most famous individual, Sangiran 17, is ~1.25 Ma and the first *erectus* discovered with a face (see Fig. 7.5). Its rugged appearance of heavy facial buttressing provides a stark reality to what the erectines of the region looked like. The importance of the Sangiran collection is that it represents the only group of early hominins that we know of occupying dense tropical rainforest and tropical savannah (Fig. 2.7). As far as we know, no hominin leaving Africa could have ever dwelt in such a lush and densely forested environment because there is no comparable environment in Africa except for the Congo and we have no idea whether erectines ever lived there. However, movement through South East Asia and Sunda on their way to Indonesia may have acclimatised early hominins to rain forest conditions and conferred some adaptive traits on them prior to their arrival in Java but that could depend on how long they took to get there and how long they lived there. The Mojokerto date and those from Sangiran suggesting a 1.5 Ma minimum age for some hominins, suggests only *H. erectus* reached Indonesia. However, who knows what the future holds for Indonesian palaeoanthropology, will an australopithecine or *habilis* turn up some day, some say the 'Hobbit' may be a distant testament to such an event.

Turning to the internal machinations of human evolution on Java, it is obvious that so much could have happened in the 700,000-year gap between the Sangiran 17 and Trinil 2 fossils: extinction, isolation and new arrivals for a start. Change, morphological or otherwise was inevitable during that time. It almost surely occurred among hominin dispersals in and around the region then and we know dispersals took place. One of those occurred around 1 million years ago as erectines continued their explorations of the archipelago by moving east. They crossed several open-water barriers in an island-hopping venture that took them from Java through Bali, Lombok and Sumbawa, as well as a few smaller

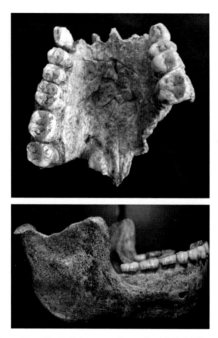

FIGURE 2.7   The Mauer Jaw, the holotype specimen for *H. heidelbergensis*. *(S. Webb)*

islands, to eventually reach Flores Island. We do not know precisely who made that journey, what they may have looked like or their relationship to the Sangiran fossils because no hominins of that age have been found on Flores. Their presence was indicated only by the discovery of stone flakes there a couple of decades ago below a volcanic tuff layer dated to 800 ka (Morwood, O'Sullivan, Aziz, & Raza, 1998). It has always been assumed they were made by *H. erectus*, possibly one similar to Sangiran 17. During glacial low stand Sumatra was joined to Singapore by a flat savannah providing *erectus* with easy access to Indonesia. But Flores was never joined by a land bridge to neighbouring islands whatever the sea level, leaving a permanent minimum 12 km water gap. The islands are steep sided volcanic submarine peaks lacking a continental fringing shelf that could reach out to join them as sea levels lowered. So, how did *erectus* make the sea crossings?

A presumed *erectus* presence on Flores, prominent enough to be detected archaeologically a million years later, suggests they were there in numbers, although what those numbers were, we don't know. The crossing distance and strong currents flowing between islands prevented *erectus* from swimming to Flores, and they would have had to do it several times between islands and we must allow several successful trips to form a presence. To cross the large water gaps (12 km minimum) is impossible for a human swimmer, particularly if you include sharks and estuarine crocodiles as natural hazards in those waters. The crossing needs a watercraft, but it is difficult to believe *erectus* could make the type of water craft required, but we are back to early hominin capabilities raised earlier. The craft had to be safe in open water and to make a direct crossing without being swept south into the Indian Ocean meant it had to be steerable. So, steering and a sail were required to overcome the strong north to south currents by tacking. The difficulty of accepting *erectus* had boatbuilding skills, prompted the eminent palaeoanthropologist, the late Philip Tobias, to once suggest *erectus* may have hitched a ride on an elephant. It just so happens that several species of elephant, such as a Stegodon (*Stegodon florensis insularis* or *sondaari* (pygmy stegodons) or *S. florensis* (normal size)). In the Pleistocene, stegodons inhabited Sulawesi, Sumba, Flores and Timor as well as the Philippines, Taiwan and Japan. Recent work in the Soa Basin of central Flores has produced a date of 1.02 Ma for *in situ* stone artefacts in association with large bodied *Stegodon floresiensis florensis* elephant fossils (Brumm et al., 2010). While erectines hitching rides on elephants sounds fanciful, it is no worse than invoking them building seagoing water craft, sail making, using steering and with the skills of tacking. Other suggested but unlikely methods assisting their journey include arriving via tsunami or on floating mats of vegetation that would be totally at the whim of currents and wind. In fact, in the face of these ideas, elephant riding is a great deal more logical.

The arrival of *erectus* on other islands via Inter-island crossings 1 million years ago is a fact. How they got there is for us to figure out. Such crossings are extremely difficult to explain if we cannot accept they could 'tame' elephants or tether themselves to them. Perhaps they could somehow gain the confidence of elephants, even small ones, enough to swim with or on them. Elephants are good swimmers and their original distribution along the island chain took place by inter-island swimming. A turnover of fauna on Flores due to a major volcanic eruption (~0.9 Ma) resulted in the loss of the small *S. sondaari* and arrival of larger *S. florensis florensis*. That repopulation must have occurred through the animals swimming from island to island both by the original and replacement species. It also indicates all these species could swim to Flores to repopulate and spread through the islands. Of course, we have no idea how tame small elephants were, but perhaps the tameability of Indian elephants, that existed at that time, as now throughout Southeast Asia, was a behavioural trait among extinct forms inhabiting the Indonesian archipelago. If the faunal turnover included the demise of smaller elephants around 0.9 ka, it must have been those that *erectus* used to make the crossing if, indeed, they used these animals in that way.

The discovery of *erectus* on Flores was sensational from the point of view that it was a quiet memorial to the ingenuity of these early humans (whichever way they crossed) as they continued to cross the world. It also meant the possibility of finding them on other islands in the region. And that is what has happened, *H. erectus* or another hominin is now known to have reached Sulawesi ~1.0 Ma at a similar time to reaching Flores (van den Bergh et al., 2016). That journey is particularly intriguing because of the distance to Sulawesi from the nearest exposed part of the Sunda shelf (~48 km) even at times of peak oceanic low stand. From the north coast of central Flores, the trip was close to 100 km and that required an island-hopping exercise to Salayar ending on the southern tip of Sulawesi. While that seems the route taken, another is entry from Kalimantan is the most likely route. Did they use elephants to make that crossing? Moreover, did they reach Timor using the same convenient 'water craft'? First, they had to reach the small island of Alor at the end of the Indonesian archipelago and that requires several more crossings. The last was to Timor, a narrower gap than some they would made to reach Alor, so to reach Timor would prove no more of a barrier than to reach Alor and elephants probably reached Timor. Australia was probably too far, even for an elephant, or else we might have had elephants in Australia by their own volition. However, an association between *erectus* and small or larger elephant species close enough or trusting enough to make such journeys is still not easy to accept even though I would suggest it is not impossible or as improbable as boat building. The fact is, *erectus* moved to Flores and Sulawesi and they did it by 1.0 Ma: end of story. How they did it is up to our imagination to work out, I doubt whether any archaeological evidence will ever emerge to show us exactly *how* they did it, but I do like the elephant story.

While we review the subject of Indonesian *erectus* capabilities, a recent discovery has provided startling evidence about those that lived on Java, and it is relevant to the previous discussion. It emerged during a study of freshwater mussel shells (*Pseudodon* sp. spp.) originally collected by Eugene Dubois during his excavations 125 years ago, that had been stored in Holland since (Joordens et al., 2014). The research shows rather startling insights into *erectus* culture and food gathering. Besides providing the first proof of *erectus* using shellfish in their diet, close examination of the mussel shell revealed some had been opened using a shark tooth. The tooth point was used to pierce the mussel shell close to the hinge, damaging or paralysing the hinge muscle that would then lose control of the lid, it was then easily opened and the flesh consumed. It also shows how *erectus* had worked out how the opening mechanism worked and could be neutralised. We might also ask how they got the shark teeth, was it by catching sharks or finding dead specimens? Moreover, if they caught them, how did they do it? It is worth mentioning that the absence of stone artefacts associated with early Indonesian hominins although some bone has cut marks. How they got there has been a bit of a mystery although thick marine shell flakes may have been used (Simanjuntak et al., 2010). Perhaps that evidence is now clearer with the above discoveries. The use of shark teeth also showed stone was not the only implement *erectus* used for a tool, strengthening a previously held belief that in the tropics they may not have relied totally on stone for tools. The use of shark teeth showed they probably carried these items around with them because a useful shark tooth is not something that you commonly find lying around. The evidence also implies *erectus* used multiple materials for tools using other natural materials including bamboo, wood and jungle vines for binding, making basic traps or for carrying objects. We might even go farther and suggest they could have used certain plants for poisoning animals and fish and they ate more vegetable, fruit and plant varieties than has been previously acknowledged. We will never know from the usual archaeological evidence; we have to be lucky.

There was luck with a further discovery in the research on the mussel shells, possibly the most astounding about *erectus* 'culture'. Close examination of the shell surface revealed a set of regular straight lines deliberately scratched into the surface in a definite geometric pattern (Joordens et al., 2014). These lines constitute a deliberate pattern interpreted as an artistic representation and with the shells dated to between 540 and 430 ka goes back half a million years. Similar enigmatic straight lines scratched on animal bone found at Bilzingsleben in Germany, dated to between 430 and 350 ka, are like those found in Java (Mania & Mania, 1998). The Bilzingsleben date suggests, however, that the hominins were not erectines but *H. heidelbergensis*. Whoever it was, it seems that the cognitive abilities of hominins at that time were not as limited as we may have previously imagined and we are back to what capabilities were available to what hominins when! These pieces of evidence from opposite sides of the globe show an interesting parallel in behavioural development between two widely separated hominin groups at about the same time. Could that mean human neurological development was reaching a new threshold throughout the species wherever hominins lived? Javan *erectus* now seem to have been the first to use 'boats' (or tame elephants), as well as being possibly the first among early hominins making abstract designs. They were also the first people not only to gather shellfish but have the anatomical knowledge used to paralyse it! Perhaps another first could be they were also the first shark catchers. It does not take much to speculate they had some sort of language also.

Finally, the date for the mussel shell dates the Trinil 2 calvarium associated with the shells, so its new date (540–430 ka) has halved its original age of 900 ka (Joordens et al., 2014). Importantly, that places a 700-ka gap between it and Sangiran 17 (1.25 ka) making a direct lineage connection between them tenuous at best. Sangiran 9 and 10 (1.2 Ma) had a combined average brain size of 850 cm^3 while their slightly older close contemporary Sangiran 17 had one of just over 1000 cm^3, 60 cm^3 more than Trinil 2 that lived over 700,000 years later. Trinil 2 cranial capacity is small for its time although not unique with two others widely separated, relatively contemporary and much younger individuals, Zhoukoutien VI in China (850 cm^3) at 450 ka and Salé (880 cm^3) in Morocco at 400 ka, both cerebrally challenged for the time (Rightmire, 2004). So, cranial capacity does not necessarily follow a linear scale through human evolution or adhere to spatial trends and it varies across continents. However, the idea of a continuous Javan *erectus* lineage, from the Lower to the Middle Pleistocene, represented by Sangiran hominins and Trinil 2 hominin, respectively, is in my view unlikely.

A similar issue is the place of the Javan Ngandong hominins. Their equivocal age range of 540–150 ka, using three dating techniques (^{40}Ar/^{39}Ar, ESR and Uranium Series) amounts to saying they are mid-late Middle Pleistocene and that could be safely said just by looking at their cranial morphology (Indriati et al., 2011). At their oldest, they could be contemporaries of Trinil 2, so were they related? On the other hand, if they lived around 300–400 ka or later, they may represent the arrival in Indonesia of another archaic hominin dispersal without any association with previous Javan erectines. They may, for example, represent a *'heidelbergensis'* type from somewhere in Asia, India or China, or perhaps an archaic hominin representing a mix of two other archaics or an unknown archaic. If Ngandong lie at the recent end of the proposed age range around 200–150 ka and they are erectines, which many believe they are, it would mean they were the last of that species but very late indeed overlapping with Neanderthals, *heidelbergensis* and early modern humans. So, were they being displaced and moved on by modern people who left Africa early, crossing Eurasia and reaching East Asia? At their youngest proposed age, they were contemporaneous with the first modern humans entering the Levant. Modern humans leaving Africa that early would raise the possibility of being able to mix, theoretically at this stage, with several hominin species (*H. erectus*, *H. neanderthalensis*, Denisovans), including Chinese archaics and others that I will introduce later. That would make a very complex set of combinations and possibilities for intermixing in East and South East Asia and for the composition of ongoing dispersals. That could possibly produce hybrid populations and morphological varieties among further migrations as well as produce migratory groups of people morphologically distinct in themselves. Lastly, assuming the Ngandong dating is correct; it is difficult to believe they descend from Lower Pleistocene Javan hominins or a Trinil 2 type. It seems much more likely that Ngandong may represent a mid-late Middle Pleistocene archaic dispersal from elsewhere in Asia. This is where palaeogenetics might help.

HOMO FLORESIENSIS, A PERSONAL VIEW

There has been much written about and many arguments over what *H. floresiensis* represents as a hominin and what its origins are. I don't intend to add very much to the many papers describing and hypothesising what this extraordinary fossil might be and what it might represent but I do have some observations to make for the record.

My only direct association with it was 13 years ago, when I invited its discoverer Michael Morwood, whom I had known since our student days at the Australian National University, to my home during one of his visits to Brisbane, Queensland. My invitation was prompted by his recent discovery of *erectus* on Flores dated then at around 800 ka and I eagerly wanted to know more. When I spoke to him on the phone he told me of further possible discoveries to come including the more than likely presence of *erectus* on Flores at least 1.0 Ma and on Sulawesi, he was cagey about that then but he was right (van den Bergh et al., 2016). Our meeting included my wife, the Indonesian palaeoanthropologist Fachroel Aziz and my dog that, to Fachroel's amazement, was called Solo. We talked about the future of the Flores work, and Mike invited me to join it and other preliminary work he had begun on Sulawesi. I had research commitments of my own in Australia and regular scheduled teaching and only a narrow research window for anything else, but envied the open opportunity. Towards the end of his visit, Mike opened his briefcase and extracted several sheets of 35 mm slides. He told me he wanted to show me something but was clearly being cautious and somewhat hesitant. Before showing me the slides he asked me not to reveal what I was about to see. I was slightly taken aback by that but intrigued and I assured him I would not divulge whatever it was that he was concerned about and needed to be kept secret.

The slides he showed were of a rather peculiar skull, including an X-ray of it. I immediately thought it looked pathological and with my palaeopathological background I felt confident from looking at the slides that it was pathologically misshapen. The skull was a very different and unusual shape and apart from being pathological I could not think of what else could produce such a misshapen head. Then Mike told me the cranial capacity was ~400 cm^3. It was equal to or less than that of an australopithecine but this skull was securely dated to the Upper Pleistocene. I had no answer. I was looking at the now famous 'Hobbit', *H. floresiensis* so was my wife. I maintained my silence in the weeks after Mike left and never mentioned the fossil to anyone. The rest is Hobbit history as they say. I never saw Mike again, and I miss his energetic and enthusiastic work and the fact he trusted me to remain quiet and show me his discovery. He gave a lot of himself to his work and investigating areas within and close to Australia where for many reasons others feared to tread. He believed as I do that neighbouring islands will reveal many surprises in regional human evolution particularly concerning *H. erectus* and other hominins. Fortunately, others have taken up Mike's baton and I wish them well in their work in this fascinating region.

The discovery of the Flores 'Hobbit', as it has become known, was first published in 2004 but years of work had led up to it (Brown et al., 2004; Morwood et al., 1998, 1999, 2004, 2005). Reasons for the diminutive Hobbit morphology have included pathology (cretinism, microcephaly etc.), and that it was descended from an *erectus* that experienced insular dwarfing, sometimes called 'Island Dwarfism': a reduction in body size observed in some mammals confined to small islands like elephants; reptiles often grow bigger. Dwarfing, it was suggested, had happened to *erectus* eventually producing *floresiensis*. I would argue Flores is not a particularly small island and I would like to know how long the dwarfing process takes to occur in hominins and how confined a mammal must be, particularly a hominin, as far as I can make out we do not know. Such body and brain shrinkage has no precedent elsewhere in the hominin fossil record as far as I am aware so, for me, dwarfing has been difficult to accept as a cause. Moreover, no pathology I know of satisfies the observed range of changes in *floresiensis* and that occurs in all individuals found so far.

Explaining the evolutionary origin of *floresiensis* has been difficult. We know *erectus* was on Flores by 1.0 Ma from the stone artefacts found there (Morwood, Aziz, van den Bergh, Sondaar, & De Vos, 1997; Brumm et al., 2010).

But was it a normal-size *erectus* or a diminutive one or something between? We have no *erectus* fossils from that time from Flores; so, until one turns up we cannot know, but we do know Sangiran fossil hominins were not dwarves of any kind at 1.2 Ma. Recent work at the Hobbit site of Mata Menge in the So'a Basin of central Flores discovered a partial jaw, three adult molar teeth, a cranial fragment and two deciduous teeth dated to 700 ka (van den Bergh et al., 2016). Comparison of these fossils with those of *H. floresiensis* has seemingly established these remains as ancestral to the recent Liang Bua *H. floresiensis* fossils although the link is tentative at best without more cranial and post-cranial skeletal parts for comparison. Two erectines living on Java, Sangiran 9 and 10, at around 1.2 Ma had a combined average brain size of 850 cm^3 and Sangiran 17 (1.25 Ma) had one of just over 1000 cm^3. That would mean the Hobbit brain dwarfed to less than half size (350–400 cm^3 in *floresiensis*), as well as its other body proportions, in less than 500,000 years, some might think that possible but I find it difficult to accept. An australopithecine or *H. habilis* is the best guess as an ancestral to *H. floresiensis but that is also difficult to believe*. Perhaps, if anything what is more remarkable about *floresiensis* than anything else is its regressive reduction of cranial capacity. A literal shrinkage of brain size from something around 800–900 cm^3 to between 350 and 400 cm^3 is in my mind extremely unlikely as a process in hominin evolution. However, it should be remembered that the Dmanisi individuals with brain sizes not much bigger had already left Africa far behind by 1.75 Ma. That could mean they were also capable of crossing the Old World and reaching places like Indonesia even if we allowed them millennia to achieve it as well as all the luck in the world. But, does it mean they could have moved so far and crossed water gaps to Flores (remember the elephant requirement)? I would venture it is no more incredible than the proposition of hominin dwarfing. But if insular dwarfing can occur in hominins, including cerebral shrinkage (which I doubt), *H. habilis* and/or Dmanisi hominins, already possessing small body and brain size, would have had a head start (sorry) to be an ancestral *floresiensis*. We should also not underestimate their capabilities to travel long distances, they had already ventured 2000 km from Africa to be in Georgia. I am no longer shocked by what is discovered about our ancestor's abilities only amazed. I accept that one guess is as good as another when it comes to addressing some of the difficult or unsolvable issues concerning hominins. If the recent find is an ancestral *floresiensis*, it suggests a diminutive body and brain size established itself very early, well before 700 ka. The new finds seem to have taken pathology, as a cause of *floresiensis* general size and cranial shape, off the table but, other than accepting a human species can undergo extraordinary size reduction over time, it is difficult to believe we are really any closer to a good explanation for the *floresiensis* morphology except pathology is out and an a *habilis* ancestor is the most likely explanation. To add more fuel to this discussion *H. naledi*, mentioned earlier, might also be in the mix. Its existence at a similar time to *floresiensis* with a similar cranial capacity might represent an African version (although much taller) of an early hominin left-over living a quiet life in a remote area of South Africa just in the same way as *floresiensis* was on Flores Island. If true the world of human evolution and the players in it is becoming stranger every day, not to mention making those interested in this subject very unsure of what they might find next.

With hominins on Sulawesi at 1.0 Ma, it begs the question who were they? Is it possible they were ancestral Hobbits, or non-dwarfed erectines, who came to Flores about the same time either from there or Java? Sulawesi is much larger than Flores and consequently not a place where island dwarfing might be expected to occur. With Sulawesi closer to Kalamantan than Flores, it could mean that the hominins who made the artefacts found there came via Borneo that at ocean low stands was connected to Asia. Hominins could also reach Borneo from Java, both being part of the Sunda shelf. We naturally assume *erectus* made the artefacts found on Sulawesi but another hominin could equally have done so, such as an ancestral *floresiensis*. A sea crossing from Sulawesi to Flores might be possible but that would mean crossing a 100 km gap and island hopping on a much bigger scale than that needed to reach Flores from Java which makes taking that route much less likely. Therefore, Flores and Sulawesi possibly represent two dead ends to earlier hominin movement if no movement east of Flores was possible. However, until securely dated hominin remains are found on Sulawesi or islands east of Flores we cannot know the answers to these questions. One final question: could small people ride small or even large elephants? Why not, anything seems possible in palaeoanthropology.

The Indonesian hominins present a special case in the story of human evolution if for no other reason than the many questions and issues yet to be solved. Indonesia has more fossil hominins than anywhere else and they present a puzzling array of types, with issues. They vie for being some of the earliest hominins out of Africa and their story does not anymore involve Java alone, it extends to neighbouring islands like Sulawesi, Kalimantan, Borneo, Flores and, presumably, the islands of Lombok and Sumbawa that lie between Flores and Java. And all that activity occurred at least 1 million years ago, probably longer. It was not just an active volcanic region it was a very active hominin region. Moreover, it appears erectines on Sulawesi, Kalimantan and Borneo by inference means they had spread across much of the exposed Sunda shelf possibly before hominins of any kind had reached Europe: we are a tropical animal indeed Charles Darwin! Sunda was a region where, seemingly, hominins were moving in all directions, travelling wherever they chose to but their relationships to one another and their arrival timing before 1.0 Ma are a mystery. The temporal difference between Sangiran 17 and Trinil 2 and the morphological differences between Trinil 2 and the Ngandong group seem uncertain. The latter may represent an arrival of archaic groups 200–300,000 years after Trinil people who may have gone extinct. These issues need resolving, but *floresiensis* is probably still the greatest enigma in the region whose origins are yet to be solved with certainty. Perhaps close behind is still the possibility of an archaic hominin extending its empire to the end of the Indonesian archipelago and even beyond? It would not be surprising if that had occurred we just need the evidence.

CHINA

China was another outpost of *H. erectus* but *erectus* was not the first to reach China. The Longgupo Cave site at Chongqing, Sichuan Province, was discovered in 1984. There, fossil mammal bone and primitive stone artefacts were recovered together with 16 *Gigantopithecus* teeth showing the broad range of this huge ape. Hominin teeth found there displayed pre-*erectus* characteristics and initial dating of the site suggested it was 1.98–1.78 Ma (Wanpo et al., 1995). Other evidence for the presence of early *H. erectus* in China has come from redating of loess palaeosoles associated with the Lantian or Gongwangling cranium found in 1964 (Ju-Kang, 1966). Like some other fossil hominins, Lantian has had several dates attributed to it over the years as different dating methods have been tried and improved. Those dates include ~700 ka, and 1.1 Ma. But the most recent work shows this early *erectus* calvarium to be much older at ~1.6 Ma (Zhu et al., 2015). That places it close to the dates for the Dmanisi hominins and seems to suggest an early group of hominins dispersed across Asia to China. These results also coincide with the date for stone artefacts found at Majuangou in the Nehewan Basin, north China (Zhu et al., 2004).

Manjuangou and several other sites in north and south China at Yuanmou and in the Nihewan Basin also date to around this time (Norton, Gao, Liu, Braun, & Wi, 2010Shen, Zhang, & Gao, 2016). However, the primitive-looking teeth had characteristics that were like OH 6, 16, 39 as well as the habiline KNM-ER 1813. Recent uranium series (U-series) dating has been able to improve on the original date producing an average age for three hominin teeth of 2.48 Ma. They are now the oldest hominin evidence outside Africa sitting at the Plio-Pleistocene boundary and strongly support a pre-*erectus* habiline African exit that reached East Asia (Han et al., 2017). It seems an extraordinary age to find hominins outside Africa but before this, the presence of primitive Pliocene bolas at Dongyaozitou in Yunxian had been claimed to be dated to between 3.0 and 2.48 Ma (Tang, Chen, & Chen, 2000). These implements may be even more extraordinary, because they were found near the base of the Dongyaozitou Formation and that could put it closer to 3.0 Ma than 2.48 Ma which leaves open the maker may have been a pre-*Homo* hominin. Could any of the early Chinese hominins have reached Indonesia and been the ancestor to *floresiensis*?

Twenty-eight years after Dubois' discoveries in Java, archaeological work began at Zhoukoudian (then known as Choukoutien) outside Beijing (old name Peking). In 1929, Pei Whenzhong began excavations at a site called Locality 1. There he found the first ever *H. erectus* skull in China, then called *S. pekinensis*. The work continued from the 1920s to the outbreak of World War II during which time 14 skulls, 16 mandibular pieces and 147 individual teeth were recovered together with faunal fossils and stone artefacts (Gao & Dennell, 2016; Shen et al., 2016). None of the skulls were complete with '…five represented by fragments of a single facial bone, four by more than one isolated piece and five are calvaria with or without basal parts…' (Weidenreich 1943:7). The Chinese discoveries showed several things about *H. erectus*. It had not just reached Java but spread to the farthest part of East Asia, so setting up two settlements. It also proved the success of the long journeys this hominin had undertaken and it also indicated they had thrived and adapted to both widely separated and environmentally contrasting outposts for probably hundreds of thousands of years. Quite understandably the presence of *erectus* in Java and China, 5000 km apart, naturally gave rise to the idea that modern populations had emerged independently as regional lineages from *erectus* founders. Franz Weidenreich, who took over the Zhoukoudian work after the sudden death of Davidson Black in March 1934, began to formulate the idea that modern humans may have originated in that way. His ideas were later taken up by Franz Weidenreich who published them in his book, *Apes, Giants, and Man* in 1946. They eventually formed the corner stone of the Regional Continuity or Multiregional Evolution hypothesis that explained, for him, and later for others, how modern humans had emerged across the planet. Weidenreich also wrote a large and very detailed monograph describing the Zhoukoudian hominins that is still regarded as a classic of its genre: *The Skull of Sinanthropus Pekinensis,* published by the Chinese Academy of Sciences, *Palaeontologia sinica* in 1943.

I have always been intrigued how erectines survived in the way they did and for so long. There is a panoply of morphological examples associated with *erectus* and that almost certainly reflects their broad distribution, their morphological variety and the many environments they occupied worldwide. They lived in small populations, some isolated, in which over time the *9EM* constructed the variable *erectus* morphologies we see in the fossil record. Isolation led to the extinction of some with replacement by others. They did not all become extinct at once but in a staggered pattern evolving into other hominin types, spatially and temporally throughout the Middle Pleistocene with some possibly hanging on till the early Upper Pleistocene and that may have been the case in both Indonesia and China.

Erectines led a tough and extremely dangerous existence in China for hundreds of thousands of years. They competed with much larger and more varieties of large carnivore than exist today and, in northern parts of China, there was a cool to very cold climate to cope with. An example of that life has emerged in recent re-examination of 700 ka *erectus* remains from the Zhoukoudian cave system. Scars and wounds on the bony remains seems to show that while erectines may have lived in and around the area for some time they probably did not permanently occupy the cave system they are famously associated with and in which their remains were found. Instead, they may not have occupied the caves as a Stone Age fortress and defended them against vicious predators. Seemingly, their remains found their way into the cave by being taken there by packs of predators, specifically the giant cave dwelling hyena (*Pachycrocuta brevirostris*). The research suggests it is likely that the hyena preyed on *erectus* people outside the caves and took their remains inside, probably in chunks where they devoured them. Instead of being occupiers of the cave, *erectus* was a meal for the real occupiers. It is worth considering how these hyaenas and similar creatures not only affected early hominins in their day-to-day survival but also during erectine and later modern human migrations.

Tragically, all the *erectus* fossils found before WWII were lost at the beginning of the war in the chaos of the Japanese invasion of China, ironically, it was as they were being transported to the United States for safe keeping. Fortunately, since 1949 similar fossils have been found in many parts of China as well as at Zhoukoudian. However, even with predatory setbacks, the discovery of more *erectus* hominins as well as other archaic hominins species in various parts of China, shows that the Chinese *erectus* groups successfully lived at various times throughout China for many thousands of years. It seems *H. erectus* explored the region, eventually setting up small groups in different parts of China and they lasted for some considerable length of time as further discoveries have shown. They show that *erectus* was also capable of living at high latitudes. Those living around 600 ka during the MIS 15 glacial had to cope with a very cold environment and had acquired cold adaptations including suitable behaviours.

Several later Chinese hominins still require explanations in terms of their ancestry and possible descendants as well as what part they played in East Asian human evolution. One of those is the Dali cranium, found in 1978 in Shaanxi Province, north central China. It too has been recently redated using several optically stimulated luminescence methods that now revised it upward to around ~260 ka (267.7 ± 13.9–258.3 ± 14.2 ka) from its previous age of 210 ka (Son et al., 2017). Dali's cranial capacity (1120 cm^3) is close to 1140 cm^3 of the Zhoukoudian 5 *erectus* cranium and the 1160-cm^3 average brain size of five Ngandong crania that could be a similar age. Its cranial capacity seems to show an evolving brain among indigenous Chinese hominins, particularly when compared to earlier Chinese *erectus* crania one of which was only 850 cm^3 (Zhoukoudian 6). However, the question can be asked as it was for the Javan 'lineages': can we be sure Dali is an evolved *erectus* in a lineage stemming from Zhoukoudian or any contemporary erectines? Dali also contrasts with the Beijing *erectus* group with its long, low cranial vault, broad, flat face and wide, almost Neanderthal looking, nasal aperture all of which might suggest local adaptive changes towards cooler conditions over the 350,000 years that separate it morphologically from Zhoukoudian hominins. Certainly, cold adaptation could have been a strong selecting force at that time especially in central northern China, and Dali was found only 800 km from the south-eastern edge of the Gobi Desert. Its morphology suggests cold adaptation but only a tenuous relationship to the Zhoukoudian or a Hexian *erectus* type and its cold adaptation could belong to the history of another lineage.

The Hexian cranium was found in 1973 on the north slope of Anjiashan Mountain, Hexian County, in eastern Anhui Provence (~270 ka). Although it is similar in age to Dali, Hexian has a smaller brain case (1025 cm³), close to the average cranial capacity of seven Zhoukoudian hominins, with whom it shares a close morphology (including the supra-orbital sulcus) contrasting it with Dali that lack a sulcus. Another individual, Jinniushan, was found in 1984 in northeastern China's Liaoning Province close to the North Korean border. It has a large, fully modern cranial capacity of 1400 cm³ and is dated to a similar time (260 ka) to Dali and almost contemporary with Hexian although its brain size is over 25% larger than the latter, a spectacular contrast (Rosenberg, Zuné, & Ruff, 2006). Jinniushan and Dali share a similar cranial shape and the difference in cranial capacity between the two may reflect sexual dimorphism with Dali possibly being female. Nice thought, but wrong: they are both female. Jinniushan is a young female with a comparatively modern face, around 168 cm tall, 20 cm taller than *erectus* females. She is only slightly shorter than *erectus* males and weighed around 78 kg, 50% and 40% heavier than Peking *erectus* females and males, respectively. She also had a larger and more rounded body shape with shorter arms that highlights the clear difference in body proportions between her type of hominin and earlier *erectus* types. Her height and weight suggests a female completely adapted to the colder conditions in which she lived and one that reflects a female robustness normally associated with Neanderthals who were also generally adapted as cold climate people. Jinniushan's northern location at 40° and Dali's at 35° 1300 km away, shows living at higher latitudes during the MIS 8 glacial, when world temperatures were −8°C below present, was eminently possible at least by them. But it was not the latitude they lived in that needs consideration, it is also the contrasting world in which they lived. Jinniushan's skeletal morphology reflects not only a fully modern brain size but body characteristics that follow Bergman's Rule which states that a larger and more rounded body shape with shorter limb proportions help retain body heat thus reducing heat loss. Such characteristics found among Canadian Arctic and Siberian tundra peoples, is a well-known adaptation to cold and one that would have enabled hominins to live in high latitudes and endure those harsh environments. But more than that, these two hominins may also be showing us something about their origins.

Are we seeing both morphological and neurological change in East Asian archaic populations, either as a development within a Chinese lineage or as the entry of a new lineage in which they arose? Therefore, was there an evolutionary development towards modernisation on the part of Dali and Jinniushan in a separate lineage, or was it just adaptive development among archaics like those that took place among modern hominins? It would seem both these are one-and-the-same thing: human evolutionary adaptation takes place no matter in which hominin is occurs. Rather than an increase in body size accompanied by changes to cranial morphology over time, the Dali/Jinniushan type shows some fundamental differences from Hexian and other older Chinese erectines.

Therefore, they may be unrelated to the latter, although Hexian may have been. The morphological dichotomy between Dali/Jinniushan and Hexian seems to indicate the latter was part of a largely unchanged relic lineage stemming from ancestral Zhoukoudian erectines or a similar population that continued spreading south perhaps driven by strengthening glaciations or extremely cold temperatures in the north during the Middle Pleistocene. It might also show a rapid 9% increase in brain capacity in the region by 260 ka, represented by the Jinniushan cranium. Could that increase reflect new genes entering the region brought by people fully adapted to cold conditions moving into northeast China from colder areas in the north and west? At present, those hominins could only originate from the west or northwest not directly north in eastern Siberia because as far we know nobody lived there.

The Altai in the northwest may be an origin for migrant entering northeast China following the Western and Eastern Sayan and Hetyin Mountains, passing north of the Mongolian Plateau and Gobi Desert to the south. That route could funnel hominins into northern China, and at present, only one group lived in that region, Denisovans. (I have mentioned Denisovans previously, they feature in Table 2.1, and I discuss them again in Chapters 3 and 4.) We don't know what Denisovans looked like but they must have been a larger bodied people, like their Neanderthal ancestors, and adapted to the cold climate of the Altai. Perhaps the deepening MIS 8 glacial that spanned 312–245 ka made it difficult for even them to live in the Altai at 50° north and some were forced south and east. Once they arrived in northern China, they could have maintained a discrete population or mixed with indigenous late *erectus* types like Hexian and perhaps that is where the mysterious archaic DNA found in their genome came from. One thing is certain, Dali and Jinniushan show a distinctive archaic cranial morphology unlike hominins emerging elsewhere at that time or those living in China before then. They have been referred to as East Asian *heidelbergensis* and that might not be a bad guess. If they are, indeed, Denisovans it means they were related to *heidelbergensis* who was their direct ancestor.

In 1958, a partial cranium was found in a cave at Maba, Guandong Provence, southern China. Originally, it was thought to be around 130–120 ka on associated faunal dating, but since then, it has been assessed as double that, between 230 ka and 278 ka (Shen et al., 2014; Xiao et al., 2014). Even though the face and left orbit are missing, it reflects a combination of modern and archaic features including a rather high, oval cranium and relatively thin cranial walls (a rather modern morphology), but it has a distinctively shaped, prominent brow ridge (an archaic trait). Its appearance is unlike the earliest African modern humans and in some respects, resembles Neanderthals and could be an Asian version of them (Howells, 1973). Originally, that seemed fanciful, if for no other reason than Maba's fragmented condition basically consisting of a calotte with a partial right eye socket. Above the eye is a section of prominently arched brow ridge that has a distinctive shape but unlike that of its, till now, assumed predecessors Dali and Jinniushan. Maba's high cranial vault is different from their long, low vaults and suggests that once again there is distinctive variation among late Middle Pleistocene East Asian archaic hominins. That suggests a local or regionalised evolution taking place on the other side of the world from Africa. Alternatively, it could indicate mixing of archaic groups that possibly roamed around in East Asia at that time.

WHO FOLLOWED *HOMO ERECTUS?*

H. erectus forms a major cornerstone in human evolution, it consisted of a morphologically varied species that travelled the world and lasted over a million years. It stands as a giant among hominins but even its reign had to finish. What came next has become a confusing mish-mash of fossil hominins that cause arguments and handwringing among some but we can try and make some sense out of them. That might not be surprising with the morphological variety among erectines, however. The earliest African exits might be presented as pre- or early *erectus* exits (Dmanisi, China, Indonesia?) and later *erectus* exits; some reaching Europe covering a period between 1.8 and 0.7 Ma (see also Chapter 3). At present the dates we have indicate a European *erectus* arrival around 1.3 Ma, another around 1.1 Ma and possibly another ~800 ka with perhaps local and continental extinctions occurring throughout that time. Although there is no proof they are separate exits, or that there were many of them, the few available dates make the known hominin fossils look like that. There could have been a steady trickle of *erectus* hominins moving out of Africa or, indeed, coming from somewhere else, but it seems unlikely.

One route from Africa to Europe was through Turkey and an *erectus* skull cap found at Kocabas, dated to 1.11 Ma, suggests that route may have been taken at least once (Vialet, Guipert, & Alçiçek, 2012; Lebatard et al., 2014). A partial cranium found at Ceprano, Italy, in 1994 was originally believed to have been the oldest European dated to the end of the Lower Pleistocene (900–800 ka). Its age made it an *erectus* or because it was in Europe, *H. antecessor*. Again, recent redating, however, has halved its age to 430–385 ka (Mounier, Condemi, & Manzi, 2011), placing it within the time of *H. heidelbergensis*. There is much argument about Ceprano's morphology and how it contrasts with other European fossils. Consequently, that possibly reflects its origins. Whatever its affiliations are, some feel it deserves its own species name: *Homo cepranensis* (the splitters strike again) while others feel it could represent a bridge between *erectus* and *heidelbergensis* although closer to the latter (Mounier et al., 2011). Ceprano possesses traits found in the robust African Bodo (600 ka) and Kabwe (300 ka) crania, both of which are designated *H. rhodesiensis* (the African equivalent of *H. heidelbergensis*) although Bodo seems too old to be in that group, and in my view too archaic. However, Ceprano might share a common ancestor with these fossils that might not have lived too long before, perhaps around 700–600 ka which could point to a Bodo type as that ancestor. The redating of Ceprano aligns it with Atapuerca 5 and together they may represent a late *erectus* arrival from Africa in Europe around 600–500 ka and that could mean an AE2 exit.

H. heidelbergensis is interpreted as a pre-Neanderthal but *heidelbergensis* has become a grab-bag of specimens of great variety from all parts of Europe that lie within a broad chronology spanning roughly 600 to ~300 ka. Prominent hominins in the group include Petralona (Greece, 300–200 ka), Arago (France, 400–200 ka), Swanscombe (England, 500–400 ka?), Bilzingsleben (Germany ~400 ka), Steinheim (Germany, 300–200 ka), Atapuerca 5 (Spain, 500–350 ka) and the recently discovered Aroeira cranium from Portugal dated to between 436 ka and 390 ka (Daura et al., 2017). A large, robustly built mandible, known as the 'Mauer jaw', found near Heidelberg Germany in 1907, has become the type specimen of *H. heidelbergensis*. The robust jaw with well buttressed mandibular body and wide rami has recently been dated to around 609 ka (Wagner et al., 2010) (Fig. 2.8). It is hard to think why single jaw was made a holotype specimen without any idea of what the cranium may have looked like but *H. georgicus* and *H. antecessor* also have a mandible for their holotypes. European *heidelbergensis* fossils display a range of morphologies which could demonstrate an evolutionary expression of isolation, different ancestral sources and/or the mixing of different ancestral types in Europe. The uncertainty surrounding *heidelbergensis* and the imprecise parameters that seem to define it, perhaps require its revision particularly in the light of what future DNA work might turn up and the collection of fossils that have been put under the *heidelbergensis* umbrella. *Heidelbergensis* is a controversial species and often termed a pre-Neanderthal or the species that gave rise to Neanderthals (*H. neanderthalensis*). However, the fossils included in *heidelbergensis* display cranial and post cranial characteristics that are somewhat more modern than *H. erectus* with a larger brain case like those found among Neanderthal skeletons. They include: a wide nasal opening; an occipital bulge or bun; a retro-molar space between the third molar and the mandibular ramus; shovel-shaped incisors; rugged muscle attachment areas on their long bones; thick long-bone cortices and a large, heavily built pelvis. Those morphological

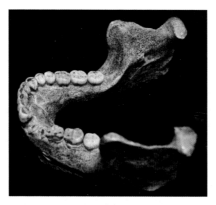

FIGURE 2.8 Original Sangiran 4 *H. erectus* maxilla (1.5 Ma) discovered by GHR von Koenigswald in 1936 showing little dental wear. *(Senckenberg Museum, Frankfurt)*

characteristics indicate European *heidelbergensis* gave rise to our esteemed Neanderthal cousins and, indeed, most features became highly developed in later classic Neanderthals.

The enigmatic Denisovans who inhabited the southern Siberian Altai and are only represented by a distal section of finger bone and a tooth, were the cousins of Neanderthals. When they were first discovered through the extraction of DNA from the finger bone, the close genetic signature between them and Neanderthals suggested they had split sometime from European Neanderthals probably ~500–400 ka. However, DNA extraction using other older bone changed things (Meyer et al., 2014). Twenty-eight *H. heidelbergensis* individuals have been discovered at the Sima de los Huesos (Pit of Bones) site at Sierra de Atapuerca in northern Spain (Arsuaga, Martinez, Gracia, & Lorenzo, 1997a; Arsuaga et al., 1997b; Carretero et al., 2012; De Castro et al., 2013; Huguet et al., 2013). All display Neanderthal derived cranial and dental features but mitochondrial DNA (mtDNA) extracted from the right femur of Atapuerca 13 indicates that Denisovans branched off before or evolved earlier than Neanderthals, probably from an early *heidelbergensis* ancestor, suggesting that *heidelbergensis* gave rise to two species. The Denisovan ancestors then made their way out of Europe and crossed to southern Siberia very early in the *heidelbergensis* story, in effect they became Siberian Neanderthals. Originally, their close genetic affinity with Neanderthals suggested they might look very much like them. The problem is, they might very well look like them and that will mean that to separate the two, we will need to verify that by examining their DNA and that is not always present in fossil bone. However, with the latest genetic information pointing to an early split from *heidelbergensis* as well as the workings of the *9EM* that would have operated so effectively on small populations like theirs, and their long temporal and spatial separation from their ancestors, it might be expected that they look like but at the same time are different from Neanderthals. Indeed, we might expect them to have developed a local (regional) morphology with adaptively evolved characteristics that were imposed by their journey, their area of settlement and time.

Several European archaeological sites in northwest Europe predate the Mauer jaw (600 ka) indicating they may be those of *erectus* (*antecessor*). Two occur in Britain, one at Pakefield (700 ka) in Suffolk (Parfitt, Barendregt, Breda, Candy, & Collins, 2005), and another earlier site at Happisburgh (1.0–0.78 Ma) (Ashton et al., 2014). Happisburgh has revealed the oldest and largest set of early hominin footprints outside Africa, probably made by people walking along the edge of a tidally influenced estuary that disappeared when sea levels rose. Boxgrove in West Sussex (500 ka) is a third site from the Middle Pleistocene although somewhat younger. Besides producing beautifully made spears, it has also revealed a human shin bone and two teeth that because of its age best qualifies these bones as *heidelbergensis* rather than *erectus*. These sites are special in their way because of their location in what was an extreme part of Europe's northwest when Britain was joined to Europe during a time of low sea level. It indicates, once again, the persistence of erectines as well as later hominins towards living in a cool to cold environment (~50–55°N) and during an Ice Age, which mirrors a similar adaptation to that among hominins living far away in China.

While trying to make sense of the background to the *erectus/heidelbergensis* transition in Europe and their relationship, it is an example of the many difficult areas that require unravelling in palaeoanthropology. I am not so concerned with either hominin in Europe because the Australian story does not begin there. But the evolutionary processes that occurred there reflect a similar situation that took place in Africa and Eastern Asia. There, several individuals do not fit easily into the taxonomic niches provided for them probably because of the requirements each of those regions imposed on the individual hominin populations living there. It is widely accepted that an African *heidelbergensis* equivalent, *H. rhodesiensis*, gave rise to modern humans (see below) and that story began with the discovery of a fossil cranium at Broken Hill in Zimbabwe in 1921, although its connection to modern humans was not known then. Originally called Broken Hill, its name was later changed to Kabwe and it has since become very important to our story (Fig. 2.9). It is now known under the species name of *H. rhodesiensis* reflecting the former British colonial name of Rhodesia that later became Zimbabwe.

Broken Hill was at first thought to be that of an *erectus*, but was then moved to the next rung of the evolutionary ladder and now most workers believe it to be an African equivalent of *heidelbergensis*. The name change reflects the same process that took place in Europe with the elimination of *H. erectus* when previously held 'erectus' specimens, such as the Arago and Petralona crania as well as others were placed in the *heidelbergensis* basket. It is problematic that we do not have an established connection between the European *heidelbergensis* and the African *rhodesiensis* except that in some respects they look similar and at some stage shared an ancestor. Therefore, the question emerges: is one or other a migrant to or from Europe, if so it would make them the same species. The

FIGURE 2.9 Broken Hill or Kabwe (*Homo rhodesiensis*) showing the pathology on the lower left temporal possibly caused by a spear point. *(S Webb)*

special status that Kabwe now has is that it is widely held to be the ancestor of early modern humans the emergence of which was a unique African event, whereas the European branch produced Neanderthals in whom we see so many parallels with us. Kabwe is not accurately dated but is believed to be between 300 and 160 ka, although generally regarded as closer to 300 ka if we take into consideration its general morphology. But who came before Kabwe?

The Bodo cranium was found in Ethiopia near Bodo village in 1976, together with associated Middle to Late Acheulean hand axe assemblages (Figs. 2.10 and 2.11). Bodo has been placed ancestral to the more recent Kabwe. Although Kabwe is classified by some as an Archaic *H. sapiens*, Bodo is definitely not. The sedimentary deposit in which Bodo was found has been dated to around 600 ka. The posterior part of the cranium is missing but it has a face like a bulldozer together with cut marks on the cranium that could indicate it was de-fleshed (White, 1986) or cannibalised. Bodo's face is broad and heavily buttressed with prominent brows, a large, almost circular, nasal aperture and

FIGURE 2.10 The robust Bodo cranium showing a deep scar embedded in the supra-orbital torus above the left orbit. *(S. Webb, cast)*

FIGURE 2.11 (Below) Bodo's massive brow ridges in left lateral view are reminiscent of OH9. *(S. Webb, cast)*

has several other *erectus*-like characteristics. It also has features seen on European and African *H. heidelbergensis* fossils, so it fits well as a transitional form between *erectus* and *heidelbergensis* or more correctly *rhodesiensis*. But what is Bodo's ancestry? One idea is that it might represent a migrant dispersal from Europe back to Africa! But a more logical view might be to see Bodo solely as an African derived individual descended from an earlier robust *erectus* perhaps something like OH 9 (1.4 Ma) which may also have given rise to the Kocabas *erectus in Turkey*. While OH 9's face is missing, the extraordinary robustness of its brows and those of Bodo it might have had a similar robust face as that of Bodo and these individuals represent a very robust form of *erectus* that developed alongside more gracile individuals such as KNM-ER 3883 and KNM-ER 3733. What we can say is that the continued *erectus* presence in Africa after 1.0 Ma is sadly lacking in representative fossils that might inform us of what African erectines looked like and how they might have continued to evolve during to the end of the Lower Pleistocene and into the Middle Pleistocene. Is it possible, for example, that there was a gracile and robust form of *erectus* in Africa and modern humans emerged from the more gracile form? Bodo is the closest hominin we have that could lead to *rhodesiensis*. But the continuum of human evolution in Africa from 1.0 to 0.6 Ma is not just unclear it is very foggy indeed. Moreover, the part *H. erectus* played in our evolution is also not as clear as we might have regarded it a couple of decades ago. It seems to be gradually suffering a diminishing importance as time goes on, certainly in other parts of the world, but I don't think that will continue. It is now also suspicious that *H. rhodesiensis* is our direct ancestor when more gracile forms of early modern people found recently in Morocco are contemporary with it or living even earlier.

The process of human evolution during the Middle Pleistocene is not clear either, but hominins may have left Africa many times during then and there was nothing preventing them from returning. It is always worth remembering if any hominin could leave Africa it could also find its way back again. Keeping our minds open, they also may have come from the east as has been suggested before (Hou & Zhao, 2010), moving from Asia to Europe and Africa. There is nothing to prevent that from occurring only our imagination that has somewhat been fixed on only considering an outward expansion from Africa and only a west to east movement therefrom. This whole area is full of questions and fraught with arguments that support different points of view, but generally these directions of travel dominate palaeoanthropological thought.

ARCHAICS AND THEIR DISPERSALS

The term archaics conjures up a vision of inferiority and incompetence among those so labelled; we cannot help it, it seems ingrained in us and we have Marcellin Boule from the Museum of Natural History in Paris to thank for that. His inaccurate assessment of Neanderthals as a stooped, round-shouldered pathetic looking caveman was based on his interpretation of the skeleton of the old man of La Chappelle aux Saints who as an old and very battered man, suffering osteoarthritis in various joints including the vertebral column had a stoop. But other characteristics such as very short legs, an imperfect gait and upright stance, and a large brain that was, in Boule's view, inferior, relegated Neanderthals to that below modern humans. But that was a time when anything that did not look Arian was considered inferior. He considered that the traits he noted showed Neanderthals had characteristics that marked them as savage, stupid and possessing animal strength (Stringer & Gamble, 1993). Neanderthals were definitely archaic so ipso facto archaic meant inferior. Boule had more to say, all derogatory and inaccurate, accept their great strength, but I leave the reader to pick up the drift of his heavily biased and very harsh assessment of an archaic species. It was, however, an activity that in 1913 the de rigueur in physical anthropology not only regarding archaic species but indigenous modern humans around the world, and we are now paying the price of that practice.

Racism is an emotive term and should not be bandied about lightly although the media does it every day, but the view of modern humans towards their extinct cousins today often borders on a covert racism that smacks of 'well you didn't make it and we did so perhaps you were inferior'. That is patently not the case because not only did archaic people live over a large area of the Old World, they were anything but failures. Indeed, they represented successful regional lineages that lasted a long time. The persistence of these hominins over hundreds of thousands of years signals their success and, moreover, morphological change occurred among them in a very real sense as well as in terms of evolutionary processes, adaptation and increasing brain size. This fact is not new and indeed it drove the Multiregional model of modern human origins. While today most people would agree that Africa is the only place where modern humans could have emerged, evolutionary change was taking place elsewhere among archaic populations living outside Africa, particularly in China. Those changes included bodily change and variation, genetic exchange, an increase in cranial capacity and changes in cranial morphology more generally. Perhaps that could be termed Regional Continuity.

South Asia comprises Nepal, Bhutan, Pakistan, Bangladesh, India and Sri Lanka, the latter two were joined during lowered sea levels. It is an area that is a vital part of the world in terms of early as well as later hominin dispersals. The reason is because it marks an area where at least two dispersal routes leading east began – it was a main corridor for these as well as for dispersals going the other way. Dispersals among early hominins and modern humans could split at several places along this corridor. The first is in Iran with the possibility to go northeast towards the Pamirs and others choosing a southern route along the Makran Coast (Fig. 2.12). Another is in India, south of the Himalayas, where some could trek south while others entered directly into the heart of the subcontinent. Another split could take place further north sending hominins either southeast and south of the Himalayas, while others might push even further

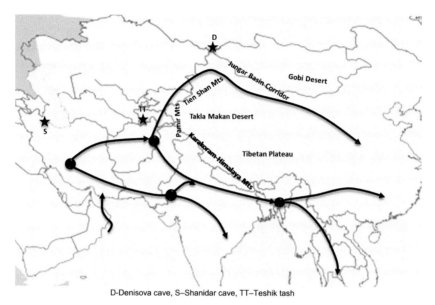

D-Denisova cave, S–Shanidar cave, TT–Teshik tash

FIGURE 2.12 Divide and conquer. AE1, 2 and 4 dispersals out of the Middle East through South Asia and into the Far East. D – Denisova Cave, S – Shanidar Cave, TT – Teshik Tash

north towards but west of the Pamirs and Tien Shen Mountains. Presumably, this is the route Neanderthals took when pushing into southern Siberia, leaving behind sites like Teshik Tash and occupying Okladinokov and Denisova Caves. There seems to be a distinct possibility that finding the Jungar Basin Corridor at the north end of the Tien Shen Mountains may have enticed hominins through it, eventually taking them into northern China as mentioned previously. Those moving south of the Himalayas and continuing east into Burma could have reached southern China or made another diversion south into South East Asia along the Malaysian Peninsula. But the evidence is non-existent for any of the actual dispersal patterns although we know that they did take place in a pattern similar those suggested.

The only hominin found so far in South Asia is an archaic cranium named Narmada (de Lumley & Sonakia, 1985a; Sonakia & de Lumley, 2006). It was found in the Narmada Valley of Madhya Pradesh in 1984 and is aged between 300 and 200 ka which could make it either a late *heidelbergensis* or early Neanderthal. Some suggest it may be as young as 160 ka making it very late for a *heidelbergensis*. As a fossil hominin, it is an anomaly because it tells us nothing about what was going on in India at that time except somebody was there during an uncertain time during the later Middle Pleistocene. It was originally identified as an evolved or late *erectus* but its 1290 cm^3 brain capacity suggests that if it was an *erectus*, it was a very late and much evolved one. A better allocation might be as an Indian *heidelbergensis* or even an Archaic *H. sapiens*, and that is what it is now favoured to be. However, if that is the case then early modern humans would have had to leave Africa ~300 ka. Whatever it is Narmada is yet another sign of a hominin population living between Africa and East Asia and that raises many more questions about its origins and what it represents than it can answer. Another question is whether it contributed anything to later Asian populations. We may also ask what the Narmada cranium is doing in India: to which dispersal did it belong; is it travelling west or east, or is it an indigenous hominin with a mysterious regional ancestry or does it represent a dead end. It must come from somewhere or represent someone but from where and who. Yet again, Indian archaeology shows hominins were spreading widely outside Africa long before modern humans left and they had made their way across vast territories successfully. Therefore, we can easily appreciate hominin adaptation, and see evolutionary and demographic change happening in Central and East Asia, something that in terms of human evolution is often ignored in the glory of light that shines out of Africa.

Another example of the broad spread of archaic humans and their adaptation is the accidental discovery by a mining company in 2006, of a calotte or skullcap at Salkhit in north-eastern Mongolia (Coppens, Tseveendorj, Demeter, Turbat, & Giscard, 2008; Tseveendorj, Gunchinsuren, Gelegdorj, Yi, & Lee, 2016). It is described as an Archaic *H. sapiens*, but apart from a distinct and well-developed brow ridge and certain features some claim it shares with Neanderthals, it consists only of the right and half the left parietal and a frontal bone. Like Narmada, it lacks a date, so not much else can be said. It does, however, flag the presence of an archaic hominin living in a tough, high latitude environment (48°N) between 300 and 200 ka. The Neanderthal site at Okladnikov and Denisova Cave lie in high latitudes which reinforce once again the idea that archaics were well adapted to cool to cold environments and seemed to occupy them during glacial phases as some Chinese hominins were.

The variety of fossil crania mentioned here shows a regional continuity operating that is doing its best to select for evolutionary modernity among the inhabitants. But is that the case? Is it possible that one or several lineages are represented in this process? Are Maba and Dali related and are they related in any way to Denisovans? Are they the result of Chinese archaics mixing together and/or with Denisovans and what could their relationship be to the Narmada and Salkhit hominins? Moreover, are Archaic modern humans entering the region?

The recent discovery of two cranial fossils (Xuchang 1 and 2) from Linjing in Henan province have added to the mystery of Chinese archaics (Zhan-Yang et al., 2017). Dated to between 125 and 105 ka, they display a range of archaic characteristics, such as brow ridges and nuchal crests, but both these features are less developed than might normally be expected among archaic crania. Their cranial vault shape is long, sagittally low and wide, the widest cranial width yet measured on hominins but that measurement occurs low down on the temporals, typical of an archaic hominin. But they have extraordinary cranial capacities of 1800 cm^3 (X1) and 1700 cm^3 (X2) which is the high end of Neanderthal cranial capacity, the holders of the cranial capacity record among hominins. It has been suggested they might be Denisovans, but Denisovans, as close relatives of Neanderthals, could be expected to have an occipital bun but the Xuchang crania don't have a distinctive occipital bulge or bun. However, Denisovans may not necessarily have had that feature. These new crania may also indicate mixing; either between different archaic groups or perhaps between archaics and the earliest modern humans to reach China. Their age overlaps comfortably with that suggested for modern human teeth found in Fuyan Cave (80–120 ka) (see also Chapter 4).

Another example of a mysterious Chinese archaic is the Penghu mandible recently recovered by fishermen in the Taiwan Strait. It is estimated to be between 200 and 10 ka which means nothing. The jaw has a shallow body that extends around the anterior arcade to the left lateral incisor socket. Two right molars and two premolars are present with the canine broken at the alveolar rim and all teeth with crowns show advanced attrition. But the important characteristic is that the mandible lacks a chin as well as a third molar. There seems little space for a third molar to have erupted and no sign of resorption of the socket. On this slim evidence Penghu seems to represent another archaic hominin that must be fitted into the picture somewhere but it is more mysterious than Denisovans. Its place of discovery in 60–90 m of water suggests it had ventured a long way out on the emergent continental shelf to at least that depth, placing it in either the MIS 6 or MIS 8 glacial given its estimated time frame. The latter may be too early but the time of lowest sea levels during MIS 6 (190–140 ka) may be the most appropriate time for this hominin to be venturing far out on the continental shelf.

The dates associated with Chinese archaics make them likely to have been around when the first modern humans arrived in China. The lack of any cranial remains makes further conclusions difficult but even this single jaw reminds us again of a seemingly independent regional evolution taking place in the Far East and beating time with its own drum. The result is that it is producing distinctive types of hominin some possibly regionally distinctive and probably some which we have yet to find. Certainly, the distribution of Chinese archaics shows they were spread across an extensive region, from close to the Mongolian border almost to Taiwan, reminding us to think about who else might be out there to find as well as the opportunities for variation among them. Were they then moving around within China or the greater region making contact that resulted in gene exchange, or did some become and became doomed to extinction?

With that in mind, one question that arises is: were they all related in some way or do they represent hominin outposts, distinct relict groups that have their own lineages that had little contact between them and who went extinct contributing nothing to other lineages? It is reasonable to think that contact between groups was inevitable at some stage and some hominins represent the mixing of archaic genes within China at least. That connection may have been sporadic, offering an opportunity for archaics to continue gathering local adaptations that fostered evolutionary change and allowing some to persist into the Late Pleistocene.

There seems to be enough morphological differences among the broader Chinese archaics to suggest at least some encountered one another and by doing so they did contribute to each other's gene pools but how extensive that was is anybody's guess at the moment. They may even have replaced other archaic hominins that had moved from elsewhere. The mysterious archaic genetic marker found in the Denisovan genome picked up around 500–400 ka must be a marker of such a connection between them and Chinese archaics because there was no one else around. They could have met as they entered northern China (Fig. 2.13). Another possibility is that with or without interbreeding, extinction may have occurred spatially and temporally across the region because of inbreeding, isolation and small, non-supportable populations. Genetic evidence showing modern humans and Neanderthals could interbreed indicates also people that were separated for hundreds or tens of thousands of years had not speciated in the true sense. Some male sterility may have occurred but even so viable offspring were produced as the carrying forward of Neanderthal genes into our genome testifies. Long-term separation does not seem to be a factor as it is when many other mammal species interbreed producing sterile offspring. Successful modern

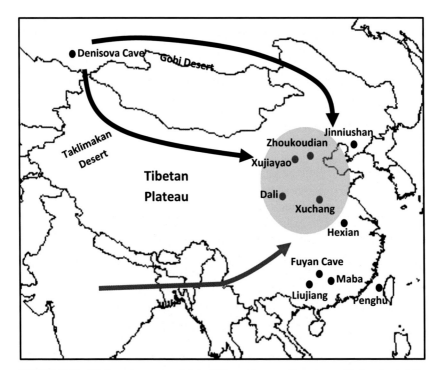

FIGURE 2.13 Denisovan movement into China using corridors to avoid deserts and the Tibetan Plateau (*black arrows*) initiating meetings with Middle Pleistocene Chinese archaics. Later modern human dispersals from the south (*blue arrow*) before 100 ka entered a general mixing region (*green area*) as a possible the site of the first mixing between Denisovans and earlier Chinese archaics.

human–Neanderthal mixing does, therefore, reveal the strong possibility of inter-archaic breeding success. Also with different archaic groups living in comparative proximity, it seems more than likely there was interbreeding between them as well as later between non-Neanderthal archaic humans in China and modern humans. Finally, a series of questions arise from the discussion. What happened to those archaic Chinese hominins? Did they continue in a different form? Were their gene pools swamped by later contact with others like them or modern humans coming in from the west? Or did they just go extinct one by one? What did they contribute genetically, if anything, to modern humans when the two groups met later or had they gone extinct by that time? Did they undertake further dispersals in their original or a mixed archaic–modern form, when opportunities arose or when glacial–interglacial cycles made it necessary to do so? One last question: could some have made it to Australia?

It seems human evolution was not static outside Africa even if modern humans were to eventually steal the show. A human pageant of life went on elsewhere during the Middle Pleistocene and early Late Pleistocene across a very large section of Asia. In fact, there does seem to be a kind of 'regional continuity' going on. Lineages were evolving in their own way and trends show they were moving towards producing a less archaic morphology although none that would conform to the standard model of an anatomically modern human. Brain size was increasing among some individuals and adaptations were also taking place in step with changing environments and their slow movement into marginal and harsher environments. East Asian archaics had continuity, at least throughout the Middle Pleistocene and possibly early Upper Pleistocene. Some were adapting to climatic and environmental changes as, presumably, they began to seasonally explore those marginal areas in higher latitudes. If that was the case, we might expect a likelihood of commensurate cultural adaptation moving with biological adaptation. East Asian archaics were demonstrably successful and had a population probably consisting of multiple bands with some form of social organisation enabling them to successfully survive for tens of thousands of years, and all that indicates the use of language. The same, of course, was happening in Europe with the dominant Neanderthal culture that had spread from Portugal to the Altai region of Siberia and possibly further into northern China as part of the East Asian archaic stage.

A distinctive trait both outside and inside Africa, however, was persistence of a certain amount of skeletal robusticity, particularly in East Asia. That robusticity may have been due to climatic factors, particularly living and adapting to colder environments as well as a combination of archaic genes. We can only guess at what part these populations eventually played in modern human migrations through Eurasia and Eastern Asia. It may be that they transformed/evolved to a stage where mating between indigenous groups and newly arrived modern humans from the west was not too difficult in terms of appearance as happened between modern humans and Neanderthals. There are so many questions and so few answers, but we can do a bit of guessing in the next chapter.

So, we now look at the spatial and temporal journeys of hominins out of Africa and trace the path they took. We look at exits taken to assess who went where, when and, if possible, why. We also look at who might have met who along the way, as suggested above, and the fossil evidence for those journeys along migration routes that eventually took humans across the world to Australia.

REFERENCES

Arsuaga, J.L., Martinez, I., Gracia, A., Lorenzo, C., 1997a. The Sima de los Huesos crania (Sierra de Atapuerca, Spain). A comparative study. Journal of Human Evolution 33, 219–281. doi: 10.1006/jhev.1997.0133.

Arsuaga, J.L., Martínez, I., Gracia, A., Carretero, J.M., Lorenzo, C., García, N., et al., 1997b. Sima de los Huesos (Sierra de Atapuerca, Spain). The site. Journal of Human Evolution 33, 109–127. doi: 10.1006/jhev.1997.0132.

Ashton, N., Lewis, S.G., De Groote, I., Duffy, S.M., Bates, M., Bates, R., et al., 2014. Hominin footprints from Lower Pleistocene deposits at Happisburgh, UK. PLoS ONE 9 (2). doi: 10.1371/journal.pone.0088329.

Brown, P., Sutikna, T., Morwood, M.J., Soejono, R.P., Jatmiko, Saptomo, E.W., et al., 2004. A new small-bodied hominin from the Late Pleistocene of Flores, Indonesia. Nature 431, 1055–1061.

Brumm, A., Jensen, G.M., van den Bergh, G.D., Morwood, M.J., Kurniawan, I., Aziz, F., et al., 2010. Hominins on Flores, Indonesia, by one million years ago. Nature 464, 748–752.

Carbonell, E., de Castro, J.M.B., Parés, J.M., Pérez-González, A., Cuence-Bescós, Ollé, A., et al., 2008. The first hominin of Europe. Nature 452, 465–469.

Carretero, J.-M., Rodríguez, L., García-González, R., Arsuaga, J.-L., Gómez-Olivencia, A., Lorenzo, S., & Quam, R., 2012. Stature estimation from complete long bones in the Middle Pleistocene humans from the Sima de los Huesos, Sierra de Atapuerca (Spain). Journal of Human Evolution 64, 242–255.

Coppens, Y., Tseveendorj, D., Demeter, F., Turbat, T., Giscard, P.-H., 2008. Discovery of an archaic *Homo sapiens* skullcap in Northeast Mongolia. Comptes Rendus Palevol 7 (1), 51–60.

Daura, J., Montserrat, S., Arsuaga, J.L., Hoffman, D.L., Quam, R.M., Zilhao, J., 2017. New Middle Pleistocene hominin cranium from Gruta da Aroeira (Portugal). PNAS 114 (13), 3397–3402.

de Castro, J.M.B., Arsuaga, J.L., Carbonell, E., Rosas, A., Martinez, I., Mosquera, M., 1997. A Hominin from the Lower Pleistocene of Atapuerca, Spain, possible ancestor to Neandertals and modern humans. Science 276, 1392–1395.

de Castro, J.M.B., Martinon-Torres, M., Balsco, R., Rosell, J., Carbonell, E., 2013. Continuity or discontinuity in the European Lower Pleistocene human settlement, the Atapuerca evidence. Quaternary Science Reviews 76, 53–65.

de Lumley, H., Sonakia, A., 1985a. Context stratigraphique et archéologique de l'Homme de la Narmada, Hathnora, Madhya Pradesh Inde. L'Anthropologie 89 (1), 3–12.

de Lumley, M.-A., Sonakia, A., 1985b. Première découverte d'un *Homo erectus* sur le Continent Indien à Hathnora, dans la moyenne vallée de la Narmada. L'Anthropologie 89 (1), 13–61.

Dennell, R., 2011. An earlier Acheulean arrival. Science 331, 1532–1533.

Gabunia, L., Joris, V., Justus, O., Lordkipanidze, A., Muschelisvili, D., Nioradze, M., et al., 1999. Neue Homininenfunde Des Altpalaeolithischen Fundplatzes Dmanisi (Georgien, Kaukasus) Im Kontext Aktueller Grabungsergebnisse. Archaeologisches Korrespondenzblatt 29, 451–488.

Gao, X., Dennell, R., 2016. Peking man and related studies. Quaternary International 400, 1–3.

Groves, C., Mazak, V., 1975. An approach to the taxonomy of the Homininae, gracile *Villafrachian hominins* of Africa. Casopis pro Mineralogii a Geologii 20, 225–272.

Han, F., Bahain, J.-J., Deng, C., Boëda, É., Hou, Y., Wei, G., et al., 2017. The earliest evidence of hominin settlement in China: Combined electron spin resonance and uranium series (ESR/U-series) dating of mammalian fossil teeth from Longgupo cave. Quaternary International 434, 75–83.

Hou, Y.M., Zhao, L.X., 2010. An archaeological view for the presence of early humans in China. Quartenary International, 223–224.

Howells, W.W., 1973. Evolution of the genus Homo. Addison-Wesley, Reading, MA.

Hublin, J.-J., Ben-Ncer, A., Bailey, S.E., Freidline, S.E., Neubauer, S., Skinner, M.M., Gunz, P., 2017. New fossils from Jebel Irhoud, Morocco and the pan-African origin of *Homo sapiens*. Nature 546, 289–292.

Huffman, O.F., Shipman, P., Hertler, C., de Vos, J., Aziz, F., 2005. Historical Evidence of the 1936 Mojokerto Skull Discovery, East Java. Journal of Human Evolution 48, 321–363.

Huguet, R., Saladié, P., Cáceres, I., Díez, C., Rosell, J., Bennàsar, M., et al., 2013. Successful subsistence strategies of the first humans in south-western Europe. Quaternary International 295, 168–182.

Indriati, E., 2004. Indonesian Fossil Hominin Dscoveries from 1889 to 2003 Catalogue and Problems. National Science Museum Monographs 24, 163–177.

Indriati, E., Swisher, III, C.C., Lepre, C., Quinn, R.L., Suriyanto, R.A., Hascaryo, A.T., et al., 2011. The Age of the 20 Meter Solo River Terrace, Java Indonesia and the Survival of *Homo erectus* in Asia. PLoS ONE 6 (6), e21562. doi: 10.1371/journal.pone.0021562.

Joordens, J.C.A., d'Errico, F., Wesselingh, F.P., Munro, S., de Vos, J., Wallinga, J., et al., 2014. *Homo erectus* at Trinil on Java used shells for tool production and engraving. Nature 518, 228–231.

Ju-Kang, W., 1966. The skull of Lantian man. Current Anthropology 7 (1), 83–86.

Kramer, A., 1994. A critical analysis of laims for the existence of Southeast Asian australopithecines. Journal of Human Evolution 26, 3–21.

Larick, R., Ciochon, R.L., Zaim, Y., Sudijono, Suminto, Rizal, Y., et al., 2001. Early Pleistocene ^{40}Ar/^{39}Ar ages for Bapang Formation hominins, Central Jawa, Indonesia. PNAS 98 (9), 4866–4871.

Lebatard, A.-E., Alçiçek, M.C., Rochette, P., Khatib, S., Vialet, A., Boulbes, N., et al., 2014. Dating the *Homo erectus* bearing travertine from Kocabas (Denizli Turkey) at least 1.1 Ma. Earth and Planetary Science Letters 390, 8–18.

Lee, S.-H., 2015. *Homo erectus* in Salkhit Mongolia? HOMO – Journal of Comparative Human Biology 66, 287–298.

Lisiecki, L.E., Raymo, M.E., 2005. A Pliocene–Pleistocene stack of 57 globally distributed benthic δ^{18}O records. Palaeoceanography 20, 1–17.

Lordkipanidze, D., Ponce de León, M.S., Margvelashvili, A., Rak, Y., Rightmire, G.P., Vekua, A., et al., 2013. A complete skull from Dmanisi, Georgia, and the evolutionary biology of early Homo. Science 342, 326–331.

Mania, D., Mania, U., 1998. Deliberate engravings on bone artefacts of *Homo erectus*. Rock Art Research 5 (2), 91–95.

Meyer, M., Fu, Q., Aximu-Petri, A., Glocke, I., Nickel, B., Arsuaga, J.-L., et al., 2014. A mitochondrial genome sequence of a hominin from Sima de los Huesos. Nature 505, 403–406.

Morwood, M.J., Aziz, F., O'Sullivan, P.B., Nasruddin, Hobbs, D.R., Raza, A., 1999. Archaeological and palaeontological research in central Flores, east Indonesia results of fieldwork. Antiquity 73, 273–286, (1997–98).

Morwood, M.J., Aziz, F., van den Bergh, G.D., Sondaar, P.Y., De Vos, J., 1997. Stone artefacts from the 1994 excavation at Mata Menge, West Central Flores Indonesia. Australian Archaeology 44, 26–34.

Morwood, M.J., O'Sullivan, P.B., Aziz, F., Raza, A., 1998. Fission-track ages of stone tools and fossils on the east Indonesian island of Flores. Nature 392, 173–176.

Morwood, M.J., Soejono, R.P., Roberts, R.G., Sutikna, T., Turney, C.S.M., Westaway, K.E., et al., 2004. Archaeology and age of a new hominin from Flores in eastern Indonesia. Nature 431, 1087–1091.

Morwood, M.J., Brown, P., Jatmiko, Sutikna, T., Saptomo, E.W., Westaway, K.E., et al., 2005. Further evidence for small-bodied hominins from the Late Pleistocene of Flores, Indonesia. Nature 437, 1012–1017.

Morwood, M.J., Soejono, R.P., Roberts, R.G., Sutikna, T., Turney, C.S.M., Westaway, K.E., et al., 2014. Archaeology and age of a new hominin from Flores in eastern Indonesia. Nature 431, 1087–1091.

Mounier, A., Condemi, S., Manzi, G., 2011. The stem species of our species, a place for the Archaic Human Cranium from Ceprano Italy. PLoS ONE. doi: 10.1371/journal.pone.0018821.

Norton, C.J., Gao, X., Liu, W., Braun, D.R., Wi, X., 2010. Central-East China—A Plio-Pleistocene dispersal corridor, the current state of evidence for Hominin occupations. In: Norton, C.J., Braun, D.R. (Eds.), Asian Paleoanthropology, from Africa to China and beyond. Springer, New York, NY, pp. 159–168.

Ogg, J.G., Ogg, G.M., Gradstein, F.M., 2016. A concise geologic time scale. Elsevier, Oxford, UK.

Parfitt, S.A., Barendregt, R.W., Breda, M., Candy, I., Collins, M.J., 2005. The earliest record of human activity in northern Europe. Nature 438, 1008–1012.

Raghavan, P., Groves, C., Pathmanathan, G., 2003. Homo erectus - Was Our Stone Age Engineer An Indigenous Species of Asia? Yes! He was A Product of Evolution in "Isolation". Journal of the Anatomical Society of India 52 (1), 16–19.

Richter, D., Grun, R., Joannes-Boyau, R., Steele, T.E., Amani, F., Rue, M., McPherron, S.P., 2017. The age of the hominin fossils from Jebel Irhoud, Morocco, and the origins of the Middle Stone Age. Nature 546, 293–296.

Rightmire, G.P., 2004. Brain Size and Ecephalization in Early to Mid-Pleistocene *Homo*. American Journal of Physical Anthropology 124, 109–123.

Rosenberg, K.R., Zuné, L., Ruff, C.B., 2006. Body size, body proportions, and encephalization in a Middle Pleistocene archaic human from northern China. PNAS 103, 3552–3556.

Shen, G., Tu, H., Xiao, D., Qiu, L., Feng, Y.-X., Zhao, J.-X., 2014. Age of Maba hominin site in southern China: Evidence from U-series dating of the Southern Branch Cave. Quaternary Geochronology 23, 56–62.

Shen, C., Zhang, X., Gao, X., 2016. Zhoukoudian in transition, research history, lithic technologies, and transformation of Chinese Paleolithic archaeology. Quartenary International 400, 4–13.

Simanjuntak, T., Sémah, F., Gaillard, C., 2010. The paleolithic in Indonesia: Nature and chronology. Quaternary International 223–224, 418–421.

Sonakia, A., de Lumley, H., 2006. Narmada *Homo erectus* – A possible ancestor of the modern Indian. Comptes Rendus Palevol 5, 353–357.

Son, V., Hopfe, C., Weis, C.L., Maffessoni, F., de la Rasilla, M., Laleuza-Fox, C., Rosas, A., Meyer, M., 2017. Neanderthal and Denisovan DNA from Pleistocene sediments. Science 27, http://dx.doi.org/10.1126/science.aam9695.

Stringer, C., Gamble, C., 1993. In search of the Neanderthals. Thames & Hudson, London.

Swisher, III, C.C., Curtis, G.H., Jacob, T., Getty, A.G., Suprijo, A., Widiasmoro, G., 1994. Age of the Earliest Known Hominins in Java, Indonesia. Science 263, 1118–1121.

Tang, Y., Chen, W., Chen, C., 2000. Discovery of a Pliocene stone tool at Yunxian, Hebei Province. Chinese Science Bulletin 45 (4), 380–383.

Tchernov, E., 1988. The age of the Ubeidiya Formation (Jordan Valley Israel) and the earliest Hominins in the Levant. Paléorient 14 (2), 63–64.

Tseveendorj, D., Gunchinsuren, B., Gelegdorj, E., Yi, S., Lee, S-H., 2016. Patterns of human evolution in northeast Asia with a particular focus on Salkhit. Quaternary International 400, 175–179.

van den Bergh, G., Li, B., Brumm, A., Grün, R., Yurnaldi, D., Moore, M.W., et al., 2016. Earliest hominin occupation of Sulawesi Indonesia. Nature 529, 208–211.

Vialet, A., Guipert, G., Alçiçek, M.C., 2012. *Homo erectus* found still further west, Reconstruction of the Kocabas, cranium (Denizli Turkey). Comps Rendes Palevol 11, 89–95.

Villamoare, B., Kimbel, W.H., Seyoum, C., Campisano, C.J., Di Maggio, E.N., Rowan, J., et al., 2015. Palaeoanthropology. Early Homo at 2.8 Ma from Ledi-Geruru, Afar, Ethiopia. Science 347 (6228), 1352–1355.

Wagner, G.A., Krbetscheck, M., Degering, J.-J., Shao, Q., Falgueres, C., Voinchet, P., et al., 2010. Radiometric dating of the type-site for *Homo heidelbergensis* at Mauer, Germany. Proceedings of the National Academy of Science 106, 19726–19730.

Wanpo, H., Clochon, R., Yumin, G., Larick, R., Qiren, F., Schwarcz, H., et al., 1995. Early Homo and associated artefacts from Asia. Nature 378, 275–278.

Weidenreich, F., 1943. The Skull of *Sinanthropus pekinensis*: A comparative study on a primitive hominin skull. Palaeontologia Sinica, New Series D, No. 10. Lancaster Press, Brooklyn.

Weidenreich, F., 1945. Giant Early Man from Java and South China. Anthropological Papers of the American Museum of Natural History 40 (1), 1–134.

Weidenreich, F., 1951. Morphology of Solo man. Anthropological Papers of the American Museum of Natural History 43 (3), 205–290.

White, T., 1986. Cut Marks on the Bodo cranium, A Case of Prehistoric Defleshing. American Journal of Physical Anthropology 69, 503–509.

Xiao, D., Bae, C.J., Shen, G., Delson, E., Jin, J.J.H., Webb, N.M., et al., 2014. Metric and geometric morphometric analysis of new hominin fossils from Maba (Guangdong, China). Journal of Human Evolution 74, 1–20. doi: 10.1016/j.jhevol.2014.04.033.

Zhan-Yang, L., Xiu-Jie, W., Li-Ping, Z., Wu, L., Xing, G., Trinkaus, E., 2017. Late Pleistocene archaic human crania from Xuchang, China. Science 355 (6328), 969–972.

Zhu, R., Potts, R., Xie, F., Hoffman, K., Deng, C.L., Shi, C.D., et al., 2004. New evidence on the earliest human presence at high northern latitudes in northeast Asia. Nature 43, 559–562.

Zhu, Z.-Y., Dennell, R., Huang, W.-W., Wu, Y., Rao, Z.-G., Qui, S.-F., et al., 2015. New dating of the *Homo erectus* cranium from Lantian (Gongwangling), China. Journal of Human Evolution 78, 144–157.

Chapter 3

Leaving Africa

After a 10-year journey the *New Horizons* spacecraft has visited Pluto. It has sent back detailed pictures of the little known ex-planet in the same way the *Mars Exploration Rover* robot vehicle is doing as it trundles across the Red Planet and the *Rosetta* spacecraft and *Philae Lander* have done, visiting a comet for the first time, although now silent. We are seeing more star clusters and galaxies than ever, and discovering exoplanets with the possibility of life on them, although I would venture we don't seem to be able to take care of this planet as we race to look at others! Nevertheless, we are at it once again, this time using robotics, advanced technology, astrophysics and all manner of scientific knowledge and techniques. Clearly, we are exploring again, beyond our planet this time, although we have not quite explored this one yet. These endeavours and our spacecraft mark the human talent and inquisitiveness to explore. Perhaps we have an '*exploration*' gene somewhere in our DNA, handed down through tens of thousands of generations. Our earliest explorations took place inside the African continent a long time before we ever left it, but it was always only a matter of time until we did.

Why did hominins leave Africa over 2 million years ago, that question is almost on the level of a 'why are we here' question; it's impossible to answer definitively. Was it a subconscious act with no real planning or ambition to achieve something, more of an accident really, or was it curiosity, subliminally or unconsciously undertaken. Was it just curiosity, the need to know what was ahead or over in the next valley. Did we even have curiosity of that level or that sort of consciousness back then? What did hominins think at that time? Could they plan, or did they appreciate, at any level, the world around them? Did they

Made in Africa. http://dx.doi.org/10.1016/B978-0-12-814798-6.00003-0

think in terms of beauty when looking at a sunset or another hominin? Did they possess the notion of love or a special closeness between individuals? One look at a chimp or gorilla mother holding her baby might give us a clue about these things, but what is that mother really thinking?

What bonded hominins together beyond the need to be safe from predators, nothing? Was success felt as an achievement or was it a priority in anything more than shear survival, and did they feel a sense of pride, hatred or vendetta, chimps seem to have vendettas? Were their lives and behaviour like other anthropoids or was it always something different and was it that something that helped us depart from our anthropoid ancestors? Did they recognise climatic signals, seasonal changes, different animal tracks and if not when did these skills and reasoning emerge? Besides size, how different was the hominin brain in its thought patterns and logical thinking then compared to ours? Did some hominins stand out as leaders, thinkers, planners or fighters? Was it when we made the radical change from Mode 1 to Mode 2 tools that we began to develop language, building on sign language and complex vocal signalling, that made trans-continental dispersal easier and successful, and did the act of moving outside Africa switch on or enhance skills they already had supporting the old saying that travel is good for the mind?

We might be able to make good guesses for any or all the above questions but we will never answer them precisely. But, I would suggest knowing them would make our predictions of where and how fast they moved lot easier to know and understand. Who knows exactly where we went and when we did it during the 4 million years before we first left Africa. Perhaps we can regard those very early journeys as the ones on which our earliest ancestors cut their 'exploration teeth' before moving outside the continent. Early hominins, particularly australopithecines, who we claim as our direct non-*Homo* ancestors, are found in northeast and southern Africa with the former having the oldest dates (4.2–2.3 Ma) that overlap with the younger southern australopithecines (3.6–1.8 Ma). Do those dates suggest they originated from a stem hominin population that possibly lived in neither place. Perhaps it was a population that divided and the two groups then went their own ways moving to places where they found a home and where their fossils are now found.

The trouble with talking about human evolution and the journeys of various groups is that we seek answers to questions that in themselves produce another set of questions that need answering first. It is said *the Devil is in the detail* and it certainly is in human evolution but that detail is very elusive. Human evolution is like nuclear physics: you know it takes place but it consists of fundamental particles many of which are theoretical, some we have not seen yet or are so rare we don't know where they fit in but there is so much still to find out. In other words, the more we understand the less we seem to know. We are still trying to understand all the mechanisms involved in human evolution and the details required to make it work even though we understand that we do not know about all the constituent parts: like a Meccano set with only a few nuts and bolts and half the parts missing including the picture on the front of the box. Human evolution takes place at the macro and cellular level together with the actions and reactions of Earth systems like the climate, environment, biological processes, volcanics and tectonics. In the case of hominins, it might even be said that the more we think about the detail required to fully understand Lower and Middle Pleistocene hominin dispersals from beginning to end the more difficult it is to contemplate them at all.

WHY LEAVE, THERE ARE NO MAPS?

The usual *reasons* for expansion among mammalian groups are population pressure, over exploitation of resources and enviro-climatic change. So why did the first hominins leave? The above reasons are valid but mammals usually move to where there is space and resources and leave it at that. Why did hominins move on and on, for the same reasons? Perhaps the answer is in the exit itself. Contemplating the AE1 journeys brings many questions that cannot be answered with any certainty. Moreover, what made early and later hominins leave Africa in the first place and what played a part in the direction they chose to go are primary questions. Perhaps curiosity drew them in one direction or another or they were responding to climatic or environmental factors pulling or pushing them on, or perhaps it was just random behaviour, mere chance of direction taken! It is likely we will never know for sure.

When hominins first left Africa I like to think they were *exploring* but that requires curiosity and in some instances a goal. Exits have been described as 'migrations' and 'dispersals', I'll use these terms interchangeably but I like to believe they were really *exploring.* Perhaps it's the old romantic in me, but I choose to see their journeys as explorations into a human-empty world driven by curiosity. If true, that is one question explained and something that makes them like us. Possessing a curiosity or inquisitive mind would make them closer to us across an almost incomprehensibly long-time gap. Animals are curious to a degree, so why not these first people to a greater degree? People have always been curious, feeling the need to go and see something for themselves whatever it might be that takes their fancy. I would expect AE1 travellers to be curious also, needing to know what lay over the horizon. There is no need for a formal language to 'feel' that the grass might be greener on the other side of the mountain range or further down the valley and go and see if it is, instincts can do that. Such curiosity could have swum around in the minds of our earliest ancestors as it does in us today. Nevertheless, whatever drove them on they were prepared to move over the horizon to see what was there.

The process of *gradual range expansion* is a natural phenomenon. It is a good mechanism to explain early human explorations and which removes them from any deliberate or systematic decision to march in a particular direction, but it is a more complex idea than one might think. Goals, as such, were almost certainly not part of their lives but survival was and if that required moving this way or that, they did it. Range expansion is a subliminal means of exploring using a natural, randomly directed, spreading behaviour. It takes place through day-to-day movement derived from living close to the environment, following and hunting animals, fleeing from predators, foraging, gathering and scavenging and looking for water. That lifestyle required constant probing and sampling of the environment in different directions to find and follow food on the hoof or the remains of those killed by other carnivores. It also required following carnivores, very discretely, using them as proxy hunting weapons.

A reason for beginning range expansion may not be the lack of food but that too many people are sharing what there is. This mechanism was described 20 years ago, suggesting hominins increased their home territories in small increments as a gradually growing population required more territory (Walker and Shipman, 1996). I doubt whether anyone 2 million plus years ago, thought about population reduction so people moved out forging territorial expansion and gradually filling a hominin-empty world and there was plenty of room out there. Expansion takes place when a range or territory becomes too small for survival meaning there is not enough carrying capacity. Therefore, territorial expansion takes place by spreading out or a group budding-off. That requires a single or several bands to gradually move away from their parent home range to find resources. Thus, more room is provided for those left behind and new territory opens. The process also avoids conflict but it causes multiple episodes of founder effect to occur. An inevitable result was that hominin bands expanded away from their original territory, out of sight of that territory and eventually out of the African continent.

It is very likely that early hominins had no real purpose of mind or plan beyond being occasionally inquisitive about distant features and their immediate surroundings. Attractions that drew them forward were not necessarily geological features such as mountain ranges, they sought game to follow in whichever direction they moved and their movements would have probably been seasonal. Entering new regions new species could be encountered some perhaps easier to hunt than others and their tracks were then followed and so on. Seeking out water sources was a constant for those not living close to rivers, streams, lakes, soaks and other waterholes. Hominins stopped when they found water and moved on when they did not and following game may not have always been to seek food rather than water. Crocodiles were a danger along travel routes from Africa to Indonesia. Ambushes could occur in swamps, marshes and along river banks, and were not uncommon in tropical regions at a time when these animals grew to large proportions and were more common.

Another danger was contracting disease. Moving into western Asia and beyond meant endemic diseases like malaria, sleeping sickness, elephantiasis, leishmaniasis, encephalitis and others must have been encountered. Many diseases were present (endemic) among various animal and primate populations that acted as vectors for them (T-W-Fiennes, 1978; Bar-Yosef and Belfer-Cohen, 2001). Most of the above diseases also impose high mortality rates on primate populations with whom wandering hominins shared their environment across the Middle East and Asia. Encountering disease was not such a problem in higher latitudes for ecological and environmental reasons and the lack of transmission vectors. But in those areas through which hominin dispersals took place, there was a good chance of an encounter with them.

Niche spotting may have been another driver of forward momentum when fruitful habitats were encountered supporting game. These could be seen at a distance and were another directional sidestep, like game trailing, preventing movement in a straight line; something that was never taken. Hominins would move like the animals they followed without attachment to a particular place except merely being aware that one environment contained what they wanted and another did not. Occasionally they had a reason to move on, perhaps as local circumstances changed for the worse or they encountered too many carnivores or harsh terrain and that would either shift them on quickly or push them to one side. This type of stochastic, erratic travelling pattern involved equally erratic behaviour, as we might see it, making travel slow, unpredictable and difficult for paleoanthropologists to assess and, especially, predict. Our earliest hominin voyagers certainly had no attachment to one place. They had no affinity or feeling that a certain place was their 'home'. Home was where they slept that night, either on the ground or, much safer, in trees. Possessing fire was an added benefit as a deterrent against night marauders that cast a flicker of daylight towards the unknown darkness. For the travelling hominin, there were no plans or route maps to follow other than the basic driving forces fuelled by the need to thrive and survive and by finding the necessities of life. Curiosity or good hunting could have driven them over the horizon or along a valley to seek opportunities that benefitted them just as we might do.

These are some likely elements of exploration but beyond those it is difficult to know what drove them forward if not something like range expansion with the complexities that involved. Explorers of the last 500 years often had stories or even crude maps of places that were supposed to exist beyond the horizon, but all the African exit (AE) groups had nothing. Their journeys were random with outcomes that can be misconstrued as purposeful. However, there was nothing purposeful about the evolutionary driver of exploration during AE1, just the sheer luck of survival that allowed enough of them to pass on their genes. It is always difficult to put us today in the minds of those travellers and make sense of the direction they might have taken and speed of travel that took them across the world. There may be reasons for their journeys we have no real idea about and it is likely that we may never be able to make a guess at what those journeys were really like.

Hominin range expansion as described above would not have had a formal *beginning*. It was a continuation of the natural passage of early hominin life and something that they had always done inside Africa and continued as they unknowingly moved out. It resulted in a slow creeping movement across the landscape that generally took them forward as an integral part of living without planned ambition, even if on some occasions, it resulted in people circling back on themselves. No doubt the basic forward driver was a need to seek better circumstances, move away from dangerous places or from enemies or predators and seek food and water, always food and water.

STAGES OF AES

This chapter and the next describe those Pleistocene hominin exits and their explorative dispersals as far as it can. It tries to make a cohesive story of those explorations and some sense out of who, when, where and why they made those journeys. But the tale is complex, patchy and sometimes confusing and perplexing. There are no definitive answers, however, it is based only on what we know with some 'intelligent guessing' and theoretical tying of loose ends where appropriate, where it seems logical to do so and where the fossil evidence has been found. Model building of that kind helps to focus the mind on possibilities and probabilities. This chapter provides a background to the protagonists; the hominins that played on the migration stage for hundreds of thousands of years. It describes what their migration patterns may have been like and where and when they moved along them. To relate that we need to begin with the very first hominins we know of who left Africa.

Exiting Africa was not just the privilege of modern humans, many left before us. The first exits by early hominins I have termed AE1. The exits made by late *erectus* and/or *heidelbergensis* type of hominin during the early and mid-Middle Pleistocene I term AE2 (Table 3.1). There are differences of opinion between genetic studies on modern human genomes regarding exit timing (Endicott et al., 2009; Malaspinas et al., 2016; Mallick et al., 2016; Pagani et al., 2016). But the AE1 and AE2 stages are useful for demarcating two broad dispersals covering large blocks of time as well as separating pre- and early *erectus* dispersals from those of later *erectus* and *heidelbergensis* types. The AE3 stage is the time when modern humans and archaic *sapien* people moved within Africa, although there is no good reason why exits also did take place then. The AE4 stage covers that of modern human exits and I have divided that into *Early, Middle* and *Late* phases. Those divisions are not fixed, there has always been one hominin or another moving somewhere outside those nice boundaries, but by dividing modern human dispersals it covers the controversy between genetic and archaeological data, that at one time were out of phase by as much as 50–60,000 years, and the confused timing of those exits. I explain more about that is later chapters.

WHO LEFT FIRST?

Whatever species made the first exit (AE1) it is easy to feel they were always *Homo*: or us to some degree. While australopithecines may have been the first, there is no evidence for them doing so and they are ignored for now. However,

TABLE 3.1 Hominin exits out of Africa showing four main phases (AE1-4) with north or south exit points for each and the three sub-phases of modern human exits with suggested dates.

AE	Exit point	Hominin species	Date
AE1	North	H. habilis/H.rudolfensis/ H. erectus	2.4–1.7 Ma
	South	H. habilis/H.rudolfensis/ H. erectus	?
AE2	North	H. erectus/H. heidelbergensis	1.7Ma–300 Ka
	South	H. erectus/ H. heidelbergensis	1.7Ma–300 Ka
AE3 (AMH Transition)	Intra-African Dispersals	Archaic Homo sapiens/ Homo sapiens	300–135 Ka
AE4 (Early)	North	Homo sapiens (MIS5e)	135–114 Ka
	South	Homo sapiens (MIS5e)	135–114 Ka
AE4 (Middle)	North	Homo sapiens (MIS5d, c, b, a)	114–80 Ka
	South	Homo sapiens (MIS5d, c, b, a)	114–80 Ka
AE4 (Late)	North	Homo sapiens (MIS4)	80–60 Ka
	South	Homo sapiens (MIS4)	80–60 Ka

AE, African exit.
North – Along Nile Corridor-Levant.
South – Southern Red Sea exit (Bab Al Mandab Strait)-along Southern Arabian Peninsula.
() - Marine Isotope Stage and sub-stage.

a spectacular picture is conjured up by imagining an *Australopithecus* building and sailing a watercraft to reach Flores Island, to begin the 'Hobbit' story, unless they were also good with elephants. With that in mind, we can begin to ask who, when, where made the first exit(s). Although we will probably never know exactly who left first, we can make some reasonable suggestions.

AE1 travellers were probably early *Homo erectus, Homo habilis* or *Homo rudolfensis*, the last two (habilines) being basically the same hominin. Whoever they were, they were equipped with highly developed senses even though habilines had some cerebral limitations and I will come back to that later. To be successful they needed to be tough, resilient and were likely to be more cunning than other animals around them. They also had to be equipped with adaptive capabilities and acute and highly responsive senses. Their abilities were supplied by the biggest brain of any terrestrial mammal of the time, almost twice the capacity of any australopithecine brain depending on which species was involved. Australopithecines lacked the anatomical framework of body shape and size and lower limb morphology and proportions that certainly allowed *erectus* and probably habilines to *stride* properly. *Homo erectus* postcranial remains show they were effective runners and trotters, able to cover great distances and run-down game. They had lost the barrel chest and small stature of australopithecines and had a trimmed-up body replete with a larger brain, all requirements for a successful migration. However, refinement of those anatomical qualities was probably not achieved until the Lower Pleistocene, although we may yet be seeing the first exit occurring at the very end of the Pliocene. Another disadvantage for australopithecines, they were vegetarian. Regular meat eating empowered development of a larger brain in early *Homo* providing cognitive abilities that travelling required. However, while they needed to work out what plants were edible as they travelled through different environments, hunting and eating the animals that fed on the plants growing in those environments, some of which may have been poisonous or have unpleasant side effects, was a better idea and a distinct advantage for travelling but and something achieved only later by *H. erectus*. Therefore, limited australopithecine dietary flexibility could hamper their ability to move through various types of environment. Its worth remembering, however, that australopithecines did live in northeast and southern Africa from at least 3.6 Ma and that suggests they did spread on that continent successfully, although they could have probably achieved that with moderately limited adaptive capabilities.

If a more developed tool kit was required to allow hominins to leave Africa, that alone might have left out the possibility of an australopithecine exit. However, we might be putting too much emphasis on brain size in this discussion and might have to give australopithecines more credit for being able to make a trans-world journey than we do. After all, a hominin with 546cc cranium reached Georgia! Like Chimpanzees, perhaps australopithecines were not always vegetarian, just most of the time. Journeying could be done with a comparatively small brain, but a larger one would certainly have enhanced the possibility of success just a little more. They did, however, lack other anatomical necessities and skills possessed by bigger built, larger brained hominins all of which would give those having these characteristics an advantage in the migration stakes. That brings us to arguments regarding how well those early hominins were physically and mentally equipped for a passage across the world. AE1 suggests that while being only a little more evolved than australopithecines, the early hominins who left were capable enough to do it. In other words, to some extent they may have developed a level of 'socio-cultural' capabilities to make long, successful journeys in cohesive bands beyond what we might expect of Chimpanzee groups, for example. That may have been comparatively far more advanced for early erectines with their basic but utilitarian tool kit (hand axes and choppers), the skills in making of which they must have passed on to each other beyond mere copying in silence, so that big brain may have provided them with a rudimentary language and/or complex verbal/visual signalling. It is likely they also had the ability to recognise seasons, co-operate communally when tracking animals and to organise rudimentary band cohesion and organisation. In combination, these skills would have given them an edge for travelling that bit farther, be more successful and have the capability to overcome obstacles and dangers that might have stopped earlier hominins.

A STRING OF EXITS

There was more than one exit during the AE1 stage but we can only guess how many. It is often the case that when a new fossil hominin is found somewhere it is either lumped in with others in that region or taken to represent the signal of the arrival of a new dispersal. We generally know much less about AE1 than the later AE2 dispersal stage that, it seems, consisted of several more traceable exit pulses over hundreds of thousands of years. As for AE1, it seems likely that more than one exit was needed to place hominins in Georgia, Java, and China between 2.0 and 1.8 Ma. Moreover, fossils that fall within that window and making it successfully to far off regions not only confirms multiple exits, it might also indicate exits before 2.0 Ma and the evidence from China certainly suggests that might be the case.

Ridged demarcation between AE phases is probably not a wise thing to try and enforce but I have ignored my own advice and constructed four because we need limits to organise large time slices although I recognise the opacity of the boundaries (Table 3.1). Most workers in the area would suggest that the first exit was by habilines around 2.4 Ma, thus marking the earliest date for AE1 and the latest around 1.7 Ma, corresponding to a minimum date for the combined China–Java–Georgia settlements. Another, later exit ended with hominins entering Europe between 1.5 and 1.2 Ma shown at Atapuerca in Spain, meaning by then they were spreading themselves northwest. It also shows other exits followed in the AE2 phase. That pattern of settlement is interesting because it shows early hominins reached Europe long after reaching the farthest corners of the world in the same way modern humans did much later. AE2 covers almost 1.5 million years from 1.7 Ma to the most recent dates for *H. heidelbergensis* around 300 ka. It is a very long period and covers the dispersal time of those who contributed to our emergence, including later *H. erectus* and *H. heidelbergensis*. It also covers the first hominin entry into Europe; the emergence and the dispersal of Neanderthals into Asia and the time when hominins may have returned to the west from the east. Therefore, it covers a very eventful time.

AE3 and AE4 phases extend from 350 ka to 60 ka, covering the emergence of modern humans and their spread across Africa and the world to Australia. AE3 (350–135 ka) covers the intra-African spread of modern humans to the northwest Atlantic coast and to the tip of southern Africa. It was an exploration of the continent, although the recovery of a *H. erectus* at Tighenif (Ternifine), Algeria (?700 ka), that is also termed *Homo mauritanicus*, and a *heidelbergensis* at Salé in Morocco (400 ka) shows modern humans were not the first to enter Africa's northeast. This period stops at a slightly extended time for a proposed earliest modern human AE around 135 ka (see below). AE4 has three substages marked *Early*, *Middle* and *Late*. The *Early* one marks the first modern human entry into the Levant (180–114 ka) and the Arabian Peninsula (135–114 ka), although it may have been somewhat earlier if we have learned the lesson from the Misilya Cave (Israel) discovery Rolf and coworkers (2018). Recent discoveries in China suggest they reached there by 120 ka (Bae et al., 2014; Liu et al., 2010, 2015). That evidence has now been joined by other discoveries at Zhiren Cave in southern China pointing to the possibility of modern humans being there by at least 100 ka (Cai et al., 2017). These mutually supportive discoveries indicate modern humans entered China before 100 ka raising the possibility it might have been as early as terminal MIS 6 (Marine Isotope Stage).

The *Middle* substage takes in MIS 5d, c, b comprising two stadials and two interstadials, although the interstadials were not much warmer than the stadials. They were all cool to cold with average world temperatures −4°C to −5°C lower than present and seas −20 to −60 m lower. After 80 ka, the world entered a short Ice Age and the *Late* substage of AE4 that lasted 20,000 years to 60 ka. In MIS 4 temperatures dropped to below −8°C and sea levels reached −80 m below and modern humans moved east across the Arabian Peninsula to India and the Levant (Armitage et al., 2010; Buffler et al., 2010; Groucutt et al., 2015; Rose et al., 2011). Other exits reached East Asia and Australasia. In later chapters I attach dates to the exits times in the substages, describe the northern exits through Sinai, the Levant, south across the Red Sea, through the centre of the Arabian Peninsula and along its southern coast (see Chapter 4).

AE1

While there was a *very first* AE1 it is extremely unlikely the evidence for it will ever be found. Exits probably began once hominins were poised on Africa's doorstep, environmental conditions were right and they possessed the necessary anatomical, neurological and behavioural development to go, then nothing could stop them. Knowing the number of exits is not particularly important but it would be nice to know when they took place because exit timing would provide evidence for who was involved. It is likely that the earlier it was the fewer hominins were involved, but we will probably never know. Those exits achieved different and very distant end-points across the world. There were no barriers to long haul journeying then. It may have taken many attempts, a very long time and more than a handful of hominins to accomplish such broad world dispersal. That raises the question of whether hominin evolution had reached a stage that allowed such long-distance travelling, or was it the case that those early settlers originated from regions outside Africa? I won't go there, but it is a question that has emerged before in palaeoanthropology. While we assume several pulses of hominins set out at different times during AE1, not all could be expected to have successfully crossed the world. Perhaps there was an advantage or naturally easy way by taking a certain direction at different times with advantages afforded by enviro-climatic influences at the time.

AE points are geographically limited. They only consist of two places: a Northern and a Southern route. The former would be through Nile River corridor, across the northern Sinai Desert or along the Mediterranean coast, and north into the Levant. The Sinai crossing presented a choice. In glacial or drier times, the coastal route was the best choice. During moister interglacials they might have turned southeast, pushing diagonally across the Arabian Peninsula and choices might change due to conditions for each exit. Wetter conditions than those at present would allow movement into what is now desert country when the region supported more game, water was available in streams and lakes and the environment was more benign. Crossing southeast across the Sinai meant entering the Arabian Peninsula. Present conditions would prevent a south-eastern crossing leaving no choice but to take the northern route through the Levant. That may have been how hominins reached Georgia. However, artefact evidence from the Saudi Arabian interior and Nefud Desert, shows some did move into what is today a very hostile environment. While we have no firm dates for the many artefact scatters across the peninsula, Mode 1 (Olduwan artefacts) show that AE1 hominins were active at some time in the interior (Petraglia et al., 2012). For a few small and vulnerable bands, the odds of success in reaching East Asia through this country were slim but this route may have been taken by some bands who eventually reached there. If so, might that tell us there were more of these bands making the trip than often assumed, because the more there were the more likely it was they would eventually cross Asia.

A southern route out of Africa meant a water crossing at Bab al Mandab Strait on the southern end of the Red Sea. Complete closure of this narrow neck during glacial low stands was not as likely in the Lower Pleistocene as it was during the Upper Pleistocene AE4 exits. The reason for the difference is that glacial cycling was not as deep or protracted during the Lower Pleistocene and sea level low stand was not as great as it was in the Upper Pleistocene. (The possibility of a dry foot crossing at Bab Al Mandab is discussed in Chapter 4). Even short sea crossings to places that could be seen from the African side of the strait were probably out of the question for habilines and early *erectus* and remained so until the later Lower Pleistocene at least. However, it is possible that there could have been very short-term dry foot crossings from time to time during the Lower Pleistocene, perhaps lasting for a few years or a couple of decades, just long enough to allow a crossing assuming there were hominins there ready to take advantage of it. Our macro view of glacial fluctuations could mask such micro-events, otherwise it seems that an AE1 southern route was out of the question.

The length of time those first journeys took to cross the world is the biggest puzzle. Did they take hundreds, thousands or tens of thousands of years to complete or did hominins move faster than we might reasonably expect them to? It might be safe to assume the time it took to reach East Asia was thousands of years but that would require very long-term survival of early hominin groups. On the other hand, to expect a very fast time might imply they knew where they were going and their route was almost a straight line, and we can safely say that neither of those was the case. There was a real possibility, however, that environmental constraints and geographical features may, in effect, have directed hominins along narrow paths that confined their movement to a corridor-like space. A mountain range could do that by presenting an obstacle thus pushing them in the easiest direction to follow along valleys and foothills. A valley could funnel both animals and hominins between ranges reducing the choice of alternative routes and effectively quickening their pace.

A journey that took thousands or even hundreds of years suggests several possibilities. Firstly, that hominin bands were very successful, they did not totally succumb to any dangers they encountered and were very adaptable. Secondly, they adapted to environmental, climatic and geographic change all of which were no doubt very different from those they were used to in their original homeland. They were in effect undergoing adaptive radiation in the same way as any other small to medium-sized mammals might have done. Many animal species have dispersed between continents over millions of years as natural behaviour often under pressure of environmental change or an opportunity to cross land bridges. Perhaps these early hominins might be looked at in the same way. Finally, if the journey was slow, as might be expected, then subsequent generations would not have known their original homeland and its conditions and this picture is also true for later modern human migrations. Early hominins would be aware only of their day-to-day existence and the region where they were living. One thing we can be sure of is that these adventurers did not take a straight line to a targeted destination which would be required if their explorations were short term. As mentioned earlier, a journey that has no goal, no notion of a target and lacks a precise or planned direction is most likely to have been slow and rambling.

If the journey lasted thousands of years, environmental conditions could change considerably in some regions they passed through. Alterations in sea level, rainfall patterns and environmental and faunal composition, all affected food resources. The world's geography altered during the Lower Pleistocene although not to the extent it did in the Late Pleistocene. But even small climatic changes would have challenged early hominins altering their chances of moving and surviving. During glacials, crossings to places normally separated by higher sea levels opened, providing onward movement. An example includes closing of the Sunda Strait which allowed *H. erectus* to move to Sumatra and Java possibly as early as 1.8 Ma.

During glacials, deserts expanded requiring long circuitous negotiation in some regions that took time and stopped further movement possibly for generations. While refuges can provide a home for a while, they can also be places of entrapment eliminating hominins and animals alike. Hominins could become tethered to such places where shrinking resources slowly brought about their demise because, by definition, refuges are surrounded by hostile regions. If they begin to succumb to enviro-climatic change there is nowhere to move. Hominin journeys must have slowed in some places and sped up in others when they were forced out of a region from the lack of food and water, assuming they did not succumb. Random trails took up time as they meandered in various directions avoiding poor resource areas and making journeys convoluted. Convoluted patterns of movement also resulted in broad exploration of a defined region almost subliminally and without a primary purpose beyond survival. Such journeys would certainly have brought evolutionary pressures directed by a principle of *learn fast or perish*, which, in turn, had the potential to hone hominin capabilities and eliminate those that could not adapt and change their behaviour. Something that these groups did not encounter, which modern humans did, was other hominins.

Water and food were kerbstones forcing movement, even if it was a convoluted movement. Fine views were only useful for finding food. Standing on a high cliff or accessing a high hill that presented, on an extended plain or valley floor, an opportunity to spot a meal on the hoof or already dead and suitable for scavenging. The same view could also show where the nearest water was, in the form of lakes, rivers and streams, and predators. These early explorers may have calculated the risks of crossing rivers but when it came to an ocean that was different. Just looking at the ocean's fury, its currents and crashing surf would show it was beyond their capabilities. Moreover, the lack of an opposite bank or sight of land on the near horizon told them there was nowhere to aim for. We can assume that while their level of intelligence may not have been high, *erectus* must have had a good instinct for survival when weighing risk, whether earlier hominins could do the same is questionable but there was always instinct. While the open ocean presented barriers far too challenging to attempt, *erectus* did make the first ocean crossing that we know of as discussed in Chapter 2.

The AE1 was as successful in its own way as later ones were, with people reaching the other side of the world. But the discovery of the Dmanisi hominins has become a salient marker in early hominin exits and indicates at least one group chose a northern route through the Levant. They must have crossed the Sinai, then north to the Caucasus between the Caspian and Black Seas. They either became extinct somewhere in the Caucasus region or are out there somewhere waiting to be found, perhaps their stone implements have been found as far away as China. Whatever happened, we do not see signs of very early hominin settlement again till it appears in Java and China. Making it as far as Georgia was testament to their explorative prowess, the ability to go on can only be supposed by archaeological evidence from China's Boas Basin (1.75 Ma) and the Mojokerto and Sangiran fossil hominins (1.8–1.6 Ma) in Java. But an actual link or continuity between these places of settlement by the same hominins with the same origin lacks evidence and is still not settled.

It is worth considering that it was not an African *erectus* that reached East Asia but a descendent. They were 5000 km and many generations from their African forebears. That suggests the ancestors of the earliest East Asian hominins, that may have been early *erectus* or *habilis/rudolfensis*, ended up as two quite separate populations living outside Africa for perhaps hundreds of thousands of years and consisting of thousands of generations. We can only wonder at how much evolutionary selection and morphological change might have taken place during that time and in different places. In all likelihood, there were other AEs that over time met hominins descended from earlier exits and they mixed together, what a confusing morphological picture that could make. They could then have dispersed further across Asia in various directions. Replacement and/or mixing together with the evolutionary mechanisms of isolation and separation brought about changes that do not fit into any recognisable category for the three basic *Homo* species mentioned above and their descendants. It is possible that a series of dispersals took place throughout the Pleistocene originating from such population mixes both inside and outside Africa. Although yet to be found, their skeletal remains must be scattered across western and southwest Asia and indeed evidence for them and their journeys in the form of their artefacts has emerged.

MISSING LINKS AND AE1 STOPOVERS, HERE AND THERE

Middle East

Earlier I mentioned a possible route from Sinai going southeast to the interior of northern Saudi Arabia. I also suggested that while normally seen as inhospitable it underwent enormous environmental change during moister times. Artefacts linked to early hominins have been found in the Nefud Desert pointing to their entry perhaps from the northern end of the eastern Red Sea coast and moving in a south-easterly direction. Various types of artefacts have been found across the 2,500,000 km^2 Arabian Peninsula setting in place an archaeological record that reveals its part in the out of Africa story (Petraglia, 2003; Petraglia et al., 2012). Unlike the spectacular hominin fossil evidence from Georgia, the Arabian evidence is wholly in the form of large artefact scatters. They offer stark evidence not only for the expansion of hominins but also the direction they took when environmental conditions were very different from today.

Any discussion of hominin movements needs to consider the climatic and environmental changes that were taking place at the time because they dictated possibilities for movement as well as onward direction. The environmental contrast that existed when the earliest hominins moved through the region, included an abundance of water and fodder for a variety of animals that in turn attracted hominins to move with them. It was a time when humans could feed themselves and those environmental changes occurred more than once in the Lower Pleistocene. Therefore, the archaeological evidence likely shows there were many dispersals throughout the Arabian Peninsula. People did not always follow coastlines; they moved opportunistically anywhere inland if the climate was favourable. The distribution pattern of archaeological sites shows that internal travel across the peninsula was largely dictated by the lakes and rivers that formed during wetter times. They provided freshwater highways, opportunities to move and explore, refuges and islands of resource that eased the passage of people and provided natural avenues for travel as well as short-cuts. Movement across the peninsula, then, was probably in pulses connected to periodic climatic amelioration, higher rainfall and an associated upturn in environmental conditions indicated by water flow and pooling in those ancient river and lake systems.

Evidence for multiple forays into and across the Arabian Peninsula is un-equivocal, shown by the myriad of artefact scatters of different typologies and densities that pepper the north-western, western and southwestern Arabian provinces. In fact, most areas of the vast region display a rich collection of stone artefacts of one type or another. The earliest typological forms include crude Olduwan hand axes, choppers and pebble tools typical of the Mode 1 form normally associated with early *erectus* (Antón and Swisher, 2004) (Figs. 2.2 and 2.3). They occur in more than one site and could also sug-gest a hominin presence possibly at the Plio/Pleistocene boundary suggesting a *habilis/rudolfensis* dispersal predating the Dmanisi dispersal by as much 700,000 years: a very early AE1. Similar artefacts appear in widely separated sites from the southwestern tip of Yemen to the northern edge of the Nafud Desert close to Iraq's southern border, a distribution suggesting the tool mak-ers wandered widely throughout the region. Later hand axes typical of both early and later Acheulean forms of the late Lower and early Middle Pleisto-cene (Mode 2) were probably associated with *erectus* in the region. Artefacts include axes, discoids, cleavers, knives and blades of various shapes, quality and size. Many sites also occur in Yemen's Hadramaut Mountains, Oman and Wadi Fatimah. The trouble with almost all sites is that while tools of an Oldu-wan and Acheulean typology indicate the presence of early hominins, frustrat-ingly, they cannot be accurately dated. The reason is they are surface scatters and lag deposits where the arid desert environment causes surface erosion by strong wind action and ground desiccation, which causes deflation and slumping with compaction of different layers one on top of another, mixing together typologically earlier and later artefacts. These taphonomic processes are common in arid environments, it occurs in the Willandra region. Frequent windstorms remove and shift sediments that originally separated different sedimentary layers and both expose previously buried artefacts and bury those that were previously exposed. The result mixes artefacts from different time sequences so that early and later types lay side by side. Therefore, typological dating is the only method that can be used on these sites. However, it is not satisfactory for separating sequences and identifying the appearance of dif-ferent hominin movements into and across the region. Encouragingly, some success in site dating has recently been made on modern human sites in the southern Arabian Peninsula and even in the 'Empty Quarter' of the peninsula (Armitage et al., 2010; Groucutt et al., 2015).

Evidence for an AE2 northern route out of Africa comes in the form of the Ubeidiya site close to the southern end of the Sea of Galilee in Israel, dated to around 1.5 Ma (Tchernov, 1988). That seems to offer some proof of *erectus* passing through the Levantine corridor after the Dmanisi group. The presence of a late *erectus* also occurs at a younger site in the region. Well-crafted Acheulean hand axes, elephant butchery and fire have been found at Gesher Benot Ya'aqov (QBY) on the Jordan River dated to 780 ka, indicating another example of an AE2 movement (Goren-Inbar et al., 2000). The hominins that occupied these sites seem to have been following the same northern route to Eurasia as the Dmanisi people. The archaeological evidence from QBY also provides an indication of the abilities of the group not only in stone tool production but also in the hunting of elephant and use of controlled fire also indicating *erectus* people. Dispersals during the late Lower Pleistocene and the boundary between the Lower and Middle Pleistocene may also indicate a natural exit strategy of hominins choosing a northeast route from Africa, but that may have been forced on them by unfavourable environmental conditions preventing them taking a south-easterly route across the Arabian Peninsula. The evidence from Israel, Georgia and the Arabian Peninsula shows those hominins were travelling at a time when environmental conditions on the Arabian Peninsula were unfavourable thus keeping them to the north.

Some of the undatable artefact scatters across the Arabian Peninsula may eventually prove to have originated from the Dmanisi dispersal or perhaps even earlier. The evidence is probably not far away showing there was not a constant trickle of people but pulses, initially few and far between but becoming more regular later, initiated by enviro-climatic conditions. What we can say for now is that there is an excellent possibility that all types of early hominin may have moved easily into and across the Arabian Peninsula leaving at least some of the chopper and hand axe sites found there. It is reasonable to suggest that once hominins reached the *Homo* stage they were equipped with the skills and anatomical and neurological ability to explore, and they did so whenever and wherever they could if conditions were right. The artefacts might show multiple dispersals from the beginning of the Pleistocene and possibly earlier, albeit occasionally, separated by large slices of time. Environmental contrasts throughout the Pleistocene highlights vast climatic swings that grew in intensity as glacial cycling intensified incorporating greater temperature swings and longer glacial cycles producing times when it was not a matter of hominins not wanting to travel it was just that they could not do so.

The evidence early hominins were moving across and exploring the Middle East is not at question, the question is who were they. Who made the Mode 1 tools is an intriguing question, although there is little doubt that an early hominin made them, probably *erectus* or even a predecessor like *habilis/rudolfensis*.

But the sequence of these early hominin populations entering the region and how they spread across it eludes us. The question of how those early hominins might have changed during their explorations, particularly their skeletal morphology, is another interesting question. That might depend on the length of their journey as well as any isolation they experienced or, conversely, who they mixed with on the way. Once again, we must invoke those evolutionary stalwarts: founder effect, isolation, genetic drift and random mutation coupled with morphological change and adaptation. An example of these processes at work may be responsible for the different cranial morphology among the Dmanisi individuals. It can also point up the possible variability that might be found in similar hominins that might be found in China and Java in the future. On a similar theme, although I mentioned earlier that there were no other hominins the earliest hominins could meet outside Africa, unlike modern humans, well that is not quite true. The plethora of sites and large stone scatters in Arabia could signal an area where successive generations of hominins found a home. That would have led to erectus or other hominins being able to meet with anyone who still occupied the region. More intriguing is the possibility that they then mixed and if they did, what the outcome of such mixing would produce.

Before moving on I want to mention that while I have not talked about stone artefacts in any depth I have my reasons for not doing so. Probably the main reason is that I am no expert in the area, but also, this book is essentially about humans and not the tools they made. However, while there is a great range of stone artefact types that appeared over the last 2 million years, there have been some distinct forms. Some are not without their controversies, that not only give them a unique label they are also associated with certain stages in human evolution. The Mousterian and Neanderthals are one such linkage, the Solutrean and later modern Europeans is another. Arguably, the most distinctive and longest serving of all stone tool type in human evolution is the *Acheulean* or Mode 2 assemblage marked by its distinctive leaf-shaped hand axe (Figs. 3.1 and 3.2). The Mousterian and Solutrean stages of cultural development are named after their places of discovery, Le Moustier and Solutré, in France. Likewise, the first distinctly designed and manufactured hand axes and cleavers, discovered at Saint Acheul, a suburb of Amiens in northern France, are also connected with that culture (Fig. 3.1). Acheulean hand axes are morphologically distinct and while they are synonymous with *erectus* people, appearing in Africa around 1.6 Ma, others, such as *H. heidelbergensis* also made and used them.

The human story relies on finding hominin remains, particularly cranial remains. Because of environmental conditions and their fragile nature, they do not always last, even when fossilised. The lack of hominin remains along likely dispersal routes, therefore, has hampered our ability to confirm the routes taken and the people who took them and confounds our best efforts to knit the story of hominin dispersals together. We can be thankful for the comparatively few fossil items we have but those large gaps on the map force us to make leaps of faith in putting the story together and leave the bad taste of an unsubstantiated

FIGURE 3.1   An Acheulean biface cleaver from the type site of St Acheul, France. *(Source: Photo S. Webb.)*

theory in the mouth. At least when we do find a hominin fossil we can say that it did get that far and others likely passed by. However, as we have seen earlier in this chapter there are the tell-tale scatters of refuse and artefacts that, thankfully, dispersing hominins left behind to help us. As in Saudi Arabia, tools indicate hominins and their basic design (Mode 1 or 2) suggests who the manufacturers were. To go further, the presence of an Acheulean hand axe not only tells us something about its maker, its design can also provide a limited idea of its age. A certain style of hand axe and its production technique, core preparation and material chosen can all assign the artefact to either Early or Late Acheulean. So, Mode 2 tools show us who the maker might have been and from that a time frame for its production, albeit a very crude one. That is typological dating, but it is not very accurate. Generally, Acheulean artefacts are found in a Lower to early Middle Pleistocene tool kit. However, they were being made in some areas until around 300,000 years ago, although they gradually disappeared between 600–300 ka. By following the trail of Mode 1 and 2 artefacts it is possible, to trace AE1 and AE2 dispersals to some degree as they 'cross' an area that has no fossil hominin evidence for them ever being there. Comparatively speaking we see little skeletal evidence of early hominin dispersals from the Middle East to South and East Asia. Therefore, many fundamental questions arise from the discovery of an isolated fossil like Narmada, sitting on its own in western India. What does it represent? Where did it come from? Was it passing through or

FIGURE 3.2   A very large Acheulean hand axe 32 cm long (*left*) and (*right*) the elegant design of a typical Acheulean hand axe. *(Source: Photo S. Webb.)*

does it represent an indigenous hominin group? We do know, however, it was a latecomer to the region and was not the first hominin type to pass that way.

South Asia

South Asia forms the midway point between Australia and Africa and it played a vital role in early hominin as well as later modern human dispersals. As pointed out in the previous chapter, crossing India south of the Himalayas could allow hominins to move in two directions, either east to China or southeast into the Malaysian Peninsula, a prime route for early hominins to reach Java and later modern humans to reach Australia. Unfortunately, there is little fossil skeletal evidence of hominin travellers in South Asia to chart those who passed through over the last 2 million years or when they might have been there. It is strange the Narmada cranium has been the only specimen found in such a vast area besides a jaw fragment found close to it. Narmada represents a comparatively late arrival (300–200 ka) along what must have been a highway of sorts between Western and Eastern Asia for possibly 1.5 Ma. Another peculiarity is that it is one more hominin than has been found on the Saudi Peninsula suggesting that their respective environments may be playing taphonomic tricks eliminating bony remains. However, the two regions do have plenty of artefacts showing hominins certainly were there and that they had spread themselves out across arid, temperate and tropical environments. Over the last 30 years or so those artefacts have been found on several sites around India consisting of Mode 1 and 2 cores, choppers, flakes, hand axes and similar tools showing hominins were there during Lower and Middle Pleistocene, although it may not have been on a continuous basis. Some have doubted suggestions that they were there before 1 million years ago, because dating problems or stratigraphic interpretation made age assessment of sites uncertain or controversial. That has frustrated attempts to prove conclusively pre-*erectus* or even early *erectus* reached South Asia even though regions such as Java and China have older dates than India for hominins and their fossil remains.

One claim for a Lower Pleistocene Indian site is near Isampur on the Deccan Plateau. That came from an electron spin resonance date on bovid teeth showing an average date of 1.27 Ma (Paddaya et al., 2002). The bovid teeth were associated with 81 hand axes, cleavers, knives and other artefacts. The date has been questioned and many do not accept it, but I return to it below. Interestingly, the only sites with similar artefact assemblages are in Africa (Gaillard et al., 2010). Previously, 2.0 Ma had been claimed for artefacts from Riwat and Pabbi Hills (2.2–0.9 Ma) both in Pakistan close to the western Himalayas. Farther east along the Himalayan chain in the Siwalik Hills other Mode 1 where artefacts have been found that were naturally assumed to be Lower Pleistocene in age. The geographical position of these sites naturally elicits the route taken by their Lower Pleistocene owners after leaving the Saudi Arabian Peninsula, perhaps indicating they had moved southeast along the Makran coast that took them out of Western and into South Asia. That route took them to the Indus River delta where modern Karachi is today presenting them with the choice of either crossing the river or avoid it by leaving the coast and trekking north following the Indus River valley. With a string of mountain ranges on the Iranian Plateau and the Hindu Kush mountains to the west and the Thar Desert on the eastern side, they were probably contented to remain in the valley which provided food and water. Choices here also depended on the Indian Monsoon and the status of glacial–interglacial cycling both of which controlled conditions for better or worse. Thus, the Indus made a natural, safe way travel similar to the Nile Valley. The time gap between the two events remains unknown but the journey between the two regions must have taken many decades perhaps centuries or even longer.

Indus River headwaters flow from the Himalayas. It was probably the Himalayan foot hills that marked the hominin limit of travel because of environmental conditions, particularly the cold. But their artefacts are found there at Potwar, Riwar and the Pabbi Hills. The environment in headwater country would have been cool at best, very cold as standard and with regular or possibly permanent snow in a tundra-like setting during glacial times. With the harsh environment of the Hindu Kush to the west, they could continue east following the curved base of the Himalayan foothills. Some may have tried going west but they would not have lasted long I suspect, giving up or going extinct because of the degree of difficulty and cold, arid conditions. At times, palaeochannels, like those on the Saudi Peninsula, traversed the Thar Desert, mainly flowing south (Gupta et al., 2011). The largest (the Saraswati) flowed just east of the Indus and ran parallel to it, being fed by the Gaggar palaeochannel together with two others all receiving Himalayan melt waters. Hominin movement east could fall upon the headwaters of these rivers and may have tempted them to go south. If the choice was to cross them and continue east following the sweep of the foothills they would eventually reach the Siwalik Hills over a 1000 km to the east and at that point they were just over half way across the subcontinent.

Routes south into present day Bihar State were available following any number of rivers that flowed from the Himalayas into northern India. Continuing east meant crossing the Koshi River before reaching the mighty Brahmaputra. As mentioned previously, river conditions varied with the presence or absence of the Indian Monsoon and the stage of the glacial–interglacial cycle, a situation that either stopped hominins or allowed them to cross the rivers they met along this route but I won't go into how they did it. Those not making the crossing could turn south, although they faced other Rivers like the Sone and Ganges. Going south would eventually take them to India's east coast and it is here at Attirampakkam (ATM) northwest of Chennai shows the rejected Isampur date was probably not aberrant. After many years of careful work at ATM the artefacts found there have been confirmed as Lower Pleistocene with an average date of 1.51 Ma, the earliest in the region (Pappu et al., 2011).

Apart from mid-Lower Pleistocene evidence for hominins in South Asia, the next time we see them is in sites that cluster close to or within the early Middle Pleistocene. Examples include Amarpura (800 ka), Bori (670 ka), Morgaon (>780 ka) and Chirki (>780 ka). The latter two are palaeomagnetic dates showing artefacts in sediments with a reversed magnetic polarity corresponding to the Matuyama Chron (781 ka) (Gaillard et al., 2010). Another site at Potwar could be a similar age or younger (700–400 ka). Acheulean artefacts found in the Chennai region dated to 1.0 Ma are similar to those found in Africa as are some dated to 1.27 Ma at Isampur. Is this a coincidence, or can more be read into it? Do they reflect, for example, a group moving quickly enough to retain its Acheulean style of hand axe making, carrying it from Africa to India or perhaps it was a slower migration but were unable or unwilling to change the style that had had served them well for so long? I have pointed out several times that isolated stone artefacts at least offer some evidence of the presence of hominins at various times and India has several. However, without fossil evidence of the owners the tools can only say so much about the history of the region and the people involved, it says merely: 'a hominin was here'. Naturally we would like to know: what hominin?

The pattern and age of South Asian artefact sites suggest two basic visitations occurred during the Lower Pleistocene (>1.5 Ma and ~1.2 Ma) with the possibility of minor incursions between. However, it seems to show that the earliest Lower Pleistocene visit might coincide with evidence from the Levant. The two sites mentioned earlier, at Ubeidiya (1.4 Ma) and GBY (780 ka.), could represent an AE1 hominin pulse moving through Israel and reaching South Asia represented by the ATM site (Fig. 3.3). That conclusion might look like drawing a long bow but dating Mode 1 and 2 Saudi Arabian artefact scatters will help with this possibility by placing a date between Africa and India. Till then, it might be one of those times when drawing ends together needs a working hypothesis, however tenuous. So, using the same logic, it might be the case that the GBY site, not far from Ubeidiya, represents an AE2 pulse in the early Middle Pleistocene that reached South Asia represented by the cluster of sites there aged to just after that time. Could this be when the Narmada hominin's ancestors arrived? There could have been several Middle Pleistocene dispersals moving into and through South Asia some we are yet to identify. An increase in dispersals might be expected as hominin populations slowly grew and naturally needed to expand.

India is no longer the missing link in the story of early hominin dispersals, that dubious tag now belongs to the west such as Iran and Iraq. Whatever the South Asian story is, the pattern of early hominin sites there may tell us something about the movement of modern humans through the region during the Upper Pleistocene AE4 and that story continues in the next chapter.

FIGURE 3.3 An Acheulean chopper (1.0 Ma) from Chennai (*left*) compared to two similar choppers from Olduvai. *(Source: Photo S. Webb.)*

GLACIAL CYCLING AND AE OPPORTUNITIES

The Late Pliocene and Lower Pleistocene climate favoured early hominin excursions into northern desert regions because glacial cycles were short and not extreme but also the climate changed enough to make such places habitable. That was not the case for the AE4 explorations where long glacial cycles with deep, long cooling stages radically changed environmental conditions for up to 60,000 years at a time. They included extreme aridity punctuated by narrow windows of environmental amelioration during an interglacial bringing warmer and wetter conditions. They were excellent for dispersals but they were comparatively short although they could still be measured in thousands of years or hundreds of generations. Glacial–interglacial cycling brought both opportunities and setbacks for hominin movement especially at Africa's gateways. In many places between latitudes 20 and 30 degrees north glacial cycling caused enormous environmental change that increasingly grew more extreme during the late-Lower and Middle Pleistocene glacials. With the inherent strengthening of glacial events biased to deeper and longer periods of cooler weather, aridity and reduced rainfall (sometimes drastically reduced), enviro-climatic forces ushered along reduced opportunities for long hominin dispersals. The Nile River corridor, an exit route often quoted, runs between the Eastern and Western Deserts in its north and the Eastern Sahara and Nubian and Danakil Deserts in the south. Its place as a source of food and water would have made it a likely exit route and a funnel to the Mediterranean Sea and the outside world.

Glacial cycling was not confined just to the major dichotomy of glacial and interglacial episodes. Both these basic phases contained smaller reversals, eg. stadials (warmer) and interstadials (cooler) that brought their own tipping points superimposed on the major phases. Moreover, there were even smaller climatic fluctuations superimposed on stadials and interstadials. All these climatic fluctuations came with reciprocal environmental changes that affected different regions to varying degrees. Each provided hominins with opportunities to remain or stay where they lived. Better times brought the freedom to move but poorer climatic conditions meant confinement to refuges where life could be maintained until conditions improved. Even the shortest of climatic variations occurred over many decades or hundreds of years during which time refuge and sustenance were required to maintain life. Both early hominins and modern humans would have taken advantage of the best times to move on when opportunities presented themselves but this was not a planned thing, more of an unconscious appreciation of the very slow changing world around them. In this way climate in conjunction with accompanying environmental change acted like a pump. When times were favourable, during interglacials, some interstadials and even smaller episodes of climatic change, hominins moved into a region to stay or move on. Environmental deterioration during glacials and stadial events could strand them in a refuge. Occasionally, bands would become tethered and perished when they became isolated in an area under gradual deterioration. Others found refuges where they could remain sheltered from deteriorating conditions and then an opportunity to move on would arise by an ameliorating climate. This climatically controlled pump pulsated behind human explorations, moving people around and forward and it worked constantly and relentlessly throughout the Pleistocene, becoming a more powerful force in the Middle and Late Pleistocene. It is, therefore, more than likely that some or most hominin and later modern human decisions to move in a certain direction were determined by the need to survive. However, a peculiar but entirely natural phenomenon occurred during these times and one that could act as a lifeboat for migrations and dispersals. During glacials when many regions become unliveable, lowered sea levels brought an opportunity to escape and take a leap onward, as at the southern end of the Red Sea when Bab al Mandab almost became joined and the Gulf of Hormuz dried opening a wide plain for movement to Asia.

AE2

In some respects, we know less about AE2 dispersals than those of AE1 but it was AE2 that brought the first hominins to Europe. As usual, we have limited knowledge about that event, even less about movements farther east and none for any reaching East Asia at that time. Originally, the Georgian hominins used a route north through the Levant. That same one could have been used during AE2 dispersals when *H. erectus* or *Homo antecessor* moved into Europe for the first time. The earliest firmly dated evidence of that arrival comes from Sierra de Atapuerca a multi-level Karst system, near Burgos in northern Spain. Atapuerca's most important sites are those of Gran Dolina Galeria, Sima de los Huesos and Sima del Elefante in which the earliest, as well as later, hominins lived during the mid-late Lower to the Middle Pleistocene. Other major sites are identified in the main cave section of the Atapuercas, including Galeria del Silex, but Sima de los Huesos is an archaeological challenge because its fossils lie 500 difficult, tortuous metres from the cave entrance. It has yielded over 6500 hominid bones mostly subadults and children (hence its name) more than any other archaeological site in the world. Indeed, the whole Atapuerca complex is easily the most productive and longest chronologically well-sequenced Pleistocene site in the world ranging from 1.5 Ma (Mode 1 artefacts) to the Bronze and Romano period. It is from there that a *heidelbergensis*/pre-Neanderthal yielded clues to the separation timing of the mysterious Denisovans. Excavations have been able to trace developments in the evolution of *erectus*, *heidelbergensis*, pre-Neanderthals and Neanderthals made possible through the discovery of the most complete set of fossils covering that period. It was in Sima del Elefante that the oldest hominin found so far, tentatively identified as *H. antecessor*, has been dated to between 1.2–1.0 Ma, (see Chapters 2 and 3). Twenty-eight

H. heidelbergensis or pre-Neanderthals, among 6500 fossil hominin bones, have been found in the Sima de los Huesos cave dated to between 800 ka and 400 ka (Arsuaga et al., 1997; Carbonell et al., 1999, 2008; de Castro et al., 2013). One hominin, Atapuerca 5, is the most complete skull of its age found so far (Fig. 3.4). Hyoid bones and other possible cultural discoveries suggest basic language may have been used by these pre-Neanderthals during the Middle Pleistocene among the inhabitants of Atapeurca and that is given some support by clues of certain burial practises and other possible symbolic behaviour that requires language to fulfill.

The presence of *antecessor* in northern Spain indicates it and even earlier hominins entered Europe by either crossing the Mediterranean at Gibraltar, or coming in from the east through the Balkans. A third route may have been island hopping during low glacial sea levels when Sardinia joined Corsica and Sicily joined the toe of Italy. The ocean gap between Bizerte in Tunisia and the western tip of Sicily (Sicilian Channel) was reduced as was the distance across the Straits of Gibraltar but both required a watercraft to make the journey. The northern and southern Mediterranean continental shelves moderately extended at maximum low sea levels but not enough not to require a boat or elephant (small or large) to reach the mid-Mediterranean islands. Could erectines moving to Europe do the same as those crossing to Flores and Sulawesi about the same time? If elephants were the 'boat of choice' that choice was possible in the Mediterranean where, similarly, various species of mini and medium-sized elephants of the genus *Palaeoloxodon* and *Mammuthus* lived.

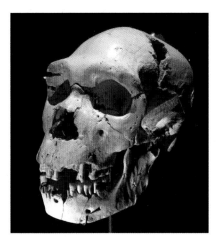

FIGURE 3.4   Skull 5 from Sima del Elefante site, Atapuerca is the most complete hominid skull found dated to the middle Pleistocene (550 ka). *(Source: S. Webb)*

They included *Elephas antiquus* spp, *falconeri, melitensis and Palaeoloxodon mnaidriensis* (Sicily and Malta), *Mammathus lamamorai* (Sardinia), *Mammuthus creticus* (Crete) and *Elephas cypriotes* (Cyprus) (Herridge and Lister, 2012). Some species became smaller through insular dwarfing on some Mediterranean islands, such as *M. creticus* which stood 1.3 m. These animals reflected those living on the Indonesian archipelago and they lived there for over a million years.

The above is one example of glacial cycling presenting movement opportunities mentioned before. While severe glacial conditions prevailed in the Sinai preventing hominin movement there, the same conditions opened possibilities for travel into Europe. Spreading aridity in North Africa then provided an incentive to move north, away from increasingly poor conditions encroaching from the south pushing hominins onto the exposed North African shelf, squashing them onto the exposed coastal strip perhaps with other animals including elephants. However, proving a Mediterranean crossing took place at that time is impossible unless fossil hominins are eventually found on the islands. It is far more parsimonious to imagine an easier and far safer, albeit much longer, passage through the Levant and Turkey to Western Europe. Those hominins would keep close to the ice-free lands of southern Europe. But where is the evidence for that route?

An *erectus* skull cap found at Kocabaş in Turkey (1.11 Ma) is contemporary with the Atapuerca *antecessor* (1.11 Ma) and may indicate an entry from the east that continued across southern Europe (Vialet et al., 2012; Lebatard et al., 2014). But, there is a prime example of one hominin becoming a dispersal. The fossil resembles classic African erectines such as OH9, Daka (Bouri) and ER 3733 with very prominent and similar shaped browridges. There are also morphological similarities between the Kocabaş cranium and Dmanisi specimens D2280 and D2282 as well as certain Zhoukoudian erectines. The presence of the Turkish *erectus* could indicate at least it had got that far travelling through the Levant and then moved on into Eastern Europe via the Balkans. However, a direct association between Kocabaş and the earliest Atapuerca ATD 6-69 is yet to be firmly established because Kocabaş is an adult and the Atapuercan ATD 6-69 is a subadult of 10–12 years. One last point to make is that an argument exists for a possible arrival in Europe and even Africa of *erectus* from Asia and that has come from comparisons between Chinese and Indonesian erectines with individuals like Kocabaş and Daka. If true that would lay open the very fabric of understanding AE2 dispersals as solely an African Diaspora and the AE would then stand for Asian Exit. An additional piece of evidence of a movement through Turkey besides the Kocabaş cranium has emerged on the Anatolian Peninsula of western Turkey in the form of a hard-hammer flake. It was associated with fluvial sediments from a buried palaeochannel, the Gediz River, and dated between 1.24 and 1.17 Ma using the 40Ar/39Ar method (Maddy et al., 2015). The date is almost contemporary with the Kocabaş cranium and a pathway from Asia through this area is just as likely as one from Africa via the Levant, but for now AE still means African Exit, although it might be wise to keep Asian Exit on standby.

The AE1 dispersals were probably aided by better environmental conditions than the AE2's that would have endured harsher times during glacials as they continued to slowly intensify and lengthen. The early Lower Pleistocene did not feature the increasing climatic extremes of the late Lower Pleistocene or those of even greater severity during the Upper Pleistocene. They brought deep, cold glacial cycling to high latitudes that pushed glaciers south across half of Europe and imposed extreme aridity across the Sahara, sub-Sahara and Middle Eastern landscapes. Even with limited skills, erectines could make their journey under somewhat better circumstances that included a wider exit window provided by shorter and less severe glacial cycles. But they missed out on the lowest sea levels that occurred later. So, with the enviro-climatic differences between Lower and Upper Pleistocene exits it is natural to suppose that dispersal patterns in terms of timing and direction were influenced by increasingly lowered sea levels as well as widening envelopes of aridity.

The large body of evidence for modern humans discovered over the last few decades shows their dispersals were more complex than those of AE1, although that might be because we know much less about the earliest exits. But the questions surrounding the two are similar. Viewing human evolution and hominin journeys requires seeking answers to questions that produce further questions. While trying to understand all the mechanisms involved and thinking we know what is required to make them work, the reality is that we don't. To add to this our empirical findings often throw us; pushing us back to rethink again. The three issues that occur constantly in our contemplation of the various dispersals are: how many exits were there; when did they take place and what routes did modern humans take. Discussions in previous chapters raise many questions but offer few answers. It would seem genetics would say one thing and intuition says another and fossils condemn both, leading to some overlap between the two but, in effect, leading nowhere.

It is easy to talk about the earliest hominin movements out of Africa and the direction they may have taken but we also should think about the possibilities of a continuum of movement into the Middle Pleistocene and when and where that took place. It is reasonable to expect that hominin exploring capabilities increased over time as they evolved and through experience. If that was so, survivability would also increase. But in what direction might that take them? Did it mean they moved in all directions west to east and east to west? The likelihood is that the freedom hominins had allowed them to move anywhere and everywhere they wanted to, they were not confined to one direction but moved multi-directionally as conditions, their skills and survival requirements dictated.

REFERENCES

Antón, S.C., Swisher, C.C., 2004. Early dispersals of Homo from Africa. Annual Review of Anthropology 33, 271–296.

Armitage, S.J., Jasim, S.A., Marks, A.E., Parker, A.G., Usik, V.I., Uerpmann, H.-P., 2010. The Southern route "out of Africa", evidence for an early expansion of modern humans into Arabia. Science 331, 453–456. doi: 10.1126/science.1199113.

Arsuaga, J.L., Martínez, I., Gracia, A., Carretero, J.M., Lorenzo, C., García, N., Ortega, A.I., 1997. Sima de los Huesos (Sierra de Atapuerca, Spain). The site. Journal of Human Evolution 33, 109–127.

Bae, C.J., Wang, W., Zhao, J., Huang, S., Tian, F., Shen, G., 2014. Modern human teeth from Late Pleistocene Luna Cave (Guangxi, China). Quaternary International 354, 169–183.

Bar-Yosef, O., Belfer-Cohen, A., 2001. From Africa to Eurasia: early dispersals. Quaternary International 75, 19–28.

Buffler, R.T., Brugeemann, J.H., Ghebretensae, B.N., Walter, R.C., Guillaume, M.M.M., Berhe, S.M., et al., 2010. Geoglogic setting of the Abdur Archaeological Site on the Red Sea coast of Eritrea, Africa. Global and Planetary Change 72, 429–450.

Cai, Y., Qiang, X., Wang, X., Jin, C, Wang, Y., Zhang, Y., et al., 2017. The age of fhuman remains and associated fauna from Zhiren Cave in Guangxi, southern China. Quaternary International 434, 84–91.

Carbonell, E., Rosas, A., Diez, J.C., 1999. Atapuerca, Occupaciones Humanans y Paleoecologia del Yacimiento de Galeria 1–390. Junta de Castilla y Leon. Zamora.

Carbonell, E., de Castro, J.M.B., Parés, J.M., Pérez-González, A., Cuence-Bescós, Ollé, A., Arsuaga, J.L., 2008. The first hominin of Europe. Nature 452, 465–469.

de Castro, J.M.B., Martinón-Torres, M., Blasco, R., Rosell, J., Carbonell, E., 2013. Continuity or discontinuity in the European Lower Pleistocene human settlement, the Atapuerca evidence. Quaternary Science Reviews. 76, 53–65.

Endicott, P., Ho, S.Y.W., Metspalu, M., Stringer, C., 2009. Evaluating the mitochondrial timescale of human evolution. Trends in Ecology and Evolution 24 (9), 515–521.

Gaillard, C., Mishra, S., Singh, M., Deo, Su., Abbas, R., 2010. Lower and Early Middle Pleistocene Acheulean in the Indian sub-continent. Quaternary International 223, 223–224.

Goren-Inbar, N., Felbel, C.S., Verosub, K.L., Melamed, Y., Kislev, M.E., Tchernov, E., et al., 2000. Pleistocene milestones on the out-of-Africa corridor at Gesher Benot Ya'aqov, Israel. Science 289 (5481), 944–947.

Groucutt, H.S., Scerri, E.M., Lewis, L., Clark-Balzan, L., Blinkhorn, J., Jennings, R.P., et al., 2015. Stone tool assemblages and models for the dispersal of Homo sapiens out of Africa. Quaternary International 382, 8–30.

Gupta, A.K., Sharma, J.R., Sreenivasan, G., 2011. Using satellite imagery to reveal the course of an extinct river below the Thar Desert in the Indo-Pak region. International Journal of Remote Sensing 32, 5197–5216.

Herridge, V.L., Lister, A.M., 2012. Extreme insular dwarfism evolved in a mammoth. Proceedings of the Royal Society B 279 (1741), 3193–3200.

Israel, H., Weber, G.W., Rolf, Q., Duval, M., Grün Rainer, Kinsley, L., Mina, Weinstein-Evron, 2018. The earliest modern humans outside africa. Science 359 (6374), 456–459. http://dx.doi.org/10.1126/science.aap8369.

Lebatard, A.-E., Alçiçek, M.C., Rochette, P., Khatib, S., Vialet, A., Boulbes, N., et al., 2014. Dating the *Homo erectus* bearing travertine from Kocabaş (Denizli, Turkey) at least 1.1 Ma. Earth and Planetary Science Letters 390, 8–18.

Liu, W., Jin, C.-Z., Zhang, Y.-Q., Cai, Y.-J., Xing, S., Wu, X.-J., Wu, X.-Z., 2010. Human remains from Zhirendong, South China, and modern human emergence in East Asia. PNAS 107 (45), 19201–19206.

Liu, W., Martinón-Torres, M., Cai, Y-j., Xing, S., Tong, H-w., Pei, S-w., Wu, X-j., 2015. The earliest unequivocally modern humans in southern China. Nature 526, 349–352.

Maddy, D., Schreve, D., Demir, T., Veldkamp, A., Wijbrans, J.R., van Gorp, W., et al., 2015. The earliest securely-dated hominin artefact in Anatolia? Quaternary Science Reviews 109, 68–75.

Malaspinas, A.-S., Westaway, M.C., Muller, C., Sousa, V.C., Lao, O., Alves, I., et al., 2016. A genomic history of Aboriginal Australia. Nature 544, 207–214.

Mallick, S., Li, H., Lipson, M., Mathieson, I., Gymrek, M., Racimo, F., et al., 2016. The Simons Genome Diversity Project, 300 genomes from 142 diverse populations. Nature 538, 201–206.

Paddaya, K., Blackwell, B.A.B., Jhaldiyal, R., Petraglia, M.D., Fevrier, S., Chaderton, I.I., et al., 2002. Recent findings on the Acheulean of the Hunsgi and Baichbal valleys, Karnataka, with special reference to the Isampur excavation and its dating. Current Science 83, 641–647.

Pagani, L., Lawson, D.J., Jagoda, E., Mörseburg, A., Eriksson, A., Mitt, M., et al., 2016. Genomic analyses inform on migration events during the peopling of Eurasia. Nature. doi: 10.1038/nature19792, Advanced online publication.

Pappu, S., Gunnell, Y., Akhilesh, K., Braucher, R., Taieb, M., Demory, F., et al., 2011. Lower Pleistocene presence of Acheulean hominins in South India. Science 331, 1596–1599.

Petraglia, M.D., 2003. The Lower Paleolithic of the Arabian Peninsula, occupations, adaptations, and dispersals. Journal of World Prehistory 17, 141–179.

Petraglia, M.D., Ditchfield, P., Jones, S., Korisettar, R., Pal, J.N., 2012. The Toba volcanic super-eruption, environmental change, and hominin occupation history in India over the last 140,000 years. Quaternary International 258, 119–134. doi: 10.1016/j.quaint.2011.07.042.

Rose, J.I., Usik, V.I., Marks, A.E., Hilbert, Y.H., Galletti, C.S., Parton, A., et al., 2011. The Nubian Complex of Dhofar, Oman: An African Middle Stone Age industry in Southern Arabia. PLoS One 6 (11), e28239. doi: 10.1371/journal.pone.0028239.

Tchernov, E., 1988. The Age of the Ubeidiya formation (Jordan Valley, Israel) and the earliest hominins in the Levant. Paléorient 14 (2), 63–64.

T-W-Feinnes, R.N., 1978. Zoonoses and the Origins and Ecology of Human Disease. Academic Press, London.

Vialet, A., Guipert, G., Alçiçek, M.C., 2012. *Homo erectus* found still further west: Reconstruction of the Kocabaş cranium (Denizli, Turkey). Comptes Rendes Palevol 11, 89–95.

Walker, A., Shipman, P., 1996. The wisdom of the bones: In search of human origins. Alfred A. Knopf, New York.

Chapter 4

AE3 and AE4: On the Road Again

'Goin' places that I've never been. Seein' things that I may never see again...'
(Willy Nelson, 1980)

The story of our evolution is no longer a slender tree with a few branches through which we descended. It is clear that human evolution is a burgeoning bush of intertwined stems and branches on which all the branches are dead except ours. Phenomenal improvements in analytical techniques and methods over the last 40 years have helped improve our understanding of the shape and spread of that tree, its growth, history and the life forms that grew on its branches. The methods now commonly used in palaeoanthropology have emerged from technological breakthroughs in physics, chemistry, biology, genomics and computer science that were not developed with anthropology in mind. They allow us to date archaeological sites and hominin age much more precisely and over a much longer time slice, and extract knowledge about human evolution in a way and from archaeological materials that we could never imagine when I was studying at university. But we still cannot directly date painted art, open sites, stone tools and fossil bone when found out of context. But, ironically, the technical breakthrough listed have made tracing our journey more difficult

Made in Africa. http://dx.doi.org/10.1016/B978-0-12-814798-6.00004-2

because continuous discoveries and more and more details constantly emerge and change the ground beneath our feet, producing ever more questions. That is not a bad thing, but it does show us that previous evidence on which we built models was wrong and much of what we thought took place 30 years ago, was also wrong. But, it is now clear, what we thought we knew was built on very flimsy evidence. However, not to worry because that is the nature of the work and flimsy evidence (although we did not know it was flimsy) it was all we had. At least some kind of story could be constructed and even if it was wrong it provided the impetus to push on to find out what was more than likely. Imagine a million-piece jigsaw puzzle. We have perhaps 20–50,000 pieces and a piece is made available increasingly regularly but randomly. Occasionally two or three pieces come together, what do we make of the picture on our jigsaw: at best, it is always changing but it is vastly incomplete, that is our problem. But at least we know it!

We left the last chapter contemplating the influences on modern humans as they traversed the Old World. This book is essentially about their journey to Australia, but that is a big story and reaching here is only the end piece. It also describes the salient fossils we know about both here and abroad. To understand the story, I began by looking at the original early hominin world journeys and the main hominin fossils involved in it to try to understand who they were, how they dispersed around the globe and what that might tell us about later dispersals. But it could be said that the more we learn about those ancient travellers and our origins, the less we know about them mainly because of the unanswered questions that arise from ongoing studies. As we look deeper at our evolution and learn more and more about ourselves so the subject looks more and more complex, uncertain, tangled and more questions than answers emerge. New discoveries take us in multiple directions that at one time we had no idea we would be taking. I liken the study of human evolution to a fractal with ever more facets blossoming before us the deeper we look. Modern humans are deeply configured by their own past, their origins and their geographic distribution. Trying to understand the many facets of us, we become deeply entangled in the data, some contradictory, that have emerged from the archaeological record and different methods of assessment over the last few decades. There can be little doubt that there were many journeys taken out of Africa. Each had its own story and set of experiences for those involved and they formed the basis of our story, but we don't know very much about any of them. Some no doubt vanished and we will never know about them. For one later group, their experiences led to a group of people that eventually took rafts across the ocean and landed here on the smallest continent perhaps enticed by fire (Dortch and Muir, 1980).

I am often asked by friends and students: 'If evolution takes place constantly what will we look like in thousands of years' time'. I don't know the answer to that. Except for the nutritional and environmentally controlled secular growth trend that has affected human height at maturity, the modern human skeleton has not changed significantly for at least 200,000 years or more. There has been very little change in the cranium, for example, other than a reduction in jaw size during the Holocene that has become the bane of young people's lives causing impacted third molars. Little else in our cranio-facial skeleton has changed certainly in the last 30,000 years or so. Most people that lived through the last glacial maximum were like we are today in terms of their skeletal build and cerebral capabilities but a considerable honing of our capabilities must have been due to our journeys and explorations and we must have learnt fast.

MY COUNTRY

Humans like to move and the evidence shows that goes back a long way. Moving is one of the things that characterises us. In deference to the popular song titles by Willie Nelson and Canned Heat, the title of this chapter sums up the situation of modern human explorations and dispersals. To exemplify what I mean, 244 million people were on the move around the world in 2015 (ref *United Nations Population Fund*) and many thousands died attempting their respective journeys. The situation for most of those people is different from the Pleistocene however. Today's immigrants are economic refugees and those fleeing life threatening situations that include war, poverty, disease, environmental changes, persistent drought, starvation, persecution, the breakdown or complete loss of their home governance and law and just the hopelessness of their situation. People make those journeys today to protect their personal safety, others with children do it in hope that those children will have a better life. Imminent death is a common denominator putting them to flight and they often put up with considerable deprivation to achieve their goals. However, while modern forms of greed, intolerance, fanaticism and overpopulation have generated many of these factors, some also initiated the movement of people in the distant past. The main difference is that today's immigrants (or refugees), had a home and many never wanted to leave it. Can we say that was the case in the past? Almost certainly not, hominins and modern humans just explored, almost subliminally, in the same way they just breathed, with little or no attachment to a perceived 'home'. In the Pleistocene, there were no barriers, borders, checkpoints and fences. The world was an oyster for anyone who wanted to face it, and people just helped themselves. It was open for complete unfettered movement unless an unfriendly group of people or animals stood in the way, but there were probably not many of the former.

The AE4 journeys of modern humans marked another chapter in hominin movement but this occurred almost 2 million years after the AE1 journeys. This time modern humans would not stop until they stared out across the implacable waves of the Southern and Pacific Oceans, Tasman Sea, Beagle Channel, East China Sea and North and South Atlantic. Much later, great sea voyages in giant canoes took them to the scattered islands and atolls of the Pacific Ocean. Our initial explorative journeys across the world effectively ended when people reached the remotest Pacific islands and New Zealand a little over 1000 years ago. Since then, world exploration has been essentially a *re-exploration*. Except for Antarctica, humans had already explored or at least visited all the world's continents and reached the most remote islands long before any galley, dhow, junk, longship, caravel, galleon or bark had left its respective home mooring. Moreover, the 'discovery' of native people in various parts of the world by European explorers and anthropologists from the 15th century onwards was patently no discovery at all. All those indigenous peoples knew where they were without others telling them or 'finding' them. They were in 'their country', their home with their culture and language, lore and beliefs systems. They knew that home intimately: the land, environment and everything in it as well as the natural, spiritual and cosmological entities that oversaw it and joined them to their natural surroundings and spiritual world. Modern humans, even the widest ranging hunter-gatherers, eventually 'settled' in a home somewhere and that was 'their country'.

It seems certain that modern humans came-of-age during the AE3 and AE4 dispersals. In those days, they had no real *attachment* or connection to *country* in the same way that Australian Aboriginal people and other hunter-gatherers do today. Such 'owning', 'custodianship' and 'belonging' to a special piece or country was something that arose much later, probably during the formative beginnings of territorial settlement when people decided to live on a special patch of territory and eventually claim it. That eventually produced tribes and tribal affiliations and the concept of tribal territory as others settled 'next door' and pronounced *their* 'ownership' of *that* land becoming tribal neighbours. It was at that point that it became necessary to defend or make well-known what you believed was *your place* and that required complex social systems and ways of defending your land and people. Possession of such feelings as 'my place', or 'my country' or even that a region or area was somehow yours through natural right, longevity of tenure or a belief system that dealt with a spirit world that had given you that country and now connected you to it, did not exist during AE4 times. Putting a stamp on country prevented people from just moving away and continuing to explore the world and patently that was not the case. If a notion of permanence of tenure was a concept among dispersing modern humans, they would never have made their transcontinental journey to Australia. Such ideas of possession or being part of a particular region were, seemingly, a late development in human society. Perhaps it took place as ritual, ceremony and belief systems became more complex linking people to a special place, through local and regional meaning and attachment through the spirit world. Perhaps the cave painters of Europe or anywhere else were, in fact, staking a claim to that area, saying *this is my place*. Whatever it was, a bourgeoning spiritual and ceremonial culture underpinned links with the land and the animals that inhabited it. A feeling of belonging to a place was aided by the developing complexity of language growing hand in hand with an increasing colourful and complex tapestry of ceremony, ritual and belief systems and, more mundanely, by slow population growth. It was during the process of later settlement that art emerged that told others and their own people that place was considered somebody's 'place'.

It was a demonstration of their spiritual world as a place of permanence as far as 'permanence' went 30–40 ka in Europe and elsewhere as Palaeolithic art in caves, rock shelters and other places emerged. The emergence of true artistic image reflecting the world around was the true beginning of the 'written' record and it was at that time around 40 ka that art also emerged in Australia.

The origins and development of modern humans were argued about for many years. The multiregional model was one of those arguments. Its followers were not a fringe group of crackpot scientists but people who interpreted the data differently from others and that is common among researchers. They cited and interpreted the data from the regions they were concerned with and based their arguments on the fossil evidence as it then was. Nobody could have foreseen what genetic evidence would reveal in the future. The multiregional model was the most parsimonious idea for the origin of modern humans and the evidence pointed to parallel hominin development into modern humans that seemed to occur in China, Africa and Europe. The point that was missed in Australia, and could not have been known at that time, was that some of the modern human fossils believed to be descended from archaic hominins in China, were, in fact, modern humans that had not evolved there but had come from Africa or descended from people who had. Before looking at the archaic evidence in other parts of the world it is worth revisiting the genetic work introduced above.

When Africa was proposed as the only place modern humans could have emerged it left, for some, many unresolved questions and some of those may never be resolved. For example, the original idea based on mitochondrial DNA (mtDNA) required complete replacement of all previous populations (archaic hominins) presumably by bands of 'murderous' modern humans. The out of Africa theory then required the elimination of all females. If that was not the case, archaic mtDNA would show up in modern humans and it does not. Well, we know now it does. Also, there was no evidence of mass murder of archaic groups and to eliminate all the archaics would require a much larger modern human population emerging out of Africa than could be seriously contemplated to overwhelm them on such a scale. Therefore, the *replacement theory* did not make sense then and it still doesn't in the form it was proposed. That puzzle was answered, to some degree, however, when it was discovered that at some time in our early journeys we did not eliminate people like the Neanderthals, we probably bread them out against a background of their much smaller population that was sparsely distributed. Genetic evidence now shows that except for African people we all carry 2%–4% Neanderthal nuclear DNA, but none of their mtDNA. That might suggest only Neanderthal males and modern human females interacted sexually. But we were told that no archaic mtDNA occurred among modern humans, so what happened to the mtDNA that must have been passed on from Neanderthals? Perhaps the scepticism among some with the early DNA work was justified to some extent after all. It seems that the archaics may not have gone extinct by spear and club rather they went extinct by sexual intercourse.

HOW DID WE BEGIN?

It is often said that we (modern humans) emerged 200,000 years ago, but that has always been misleading. It is true that, at present, the earliest individual with a lightly built, modern cranio-facial skeleton like ours called Omo 1 (Omo Kibish) is dated to around 195 ka. But it is natural that older modern humans will be discovered because Omo 1 is neither unique nor was an instant creation. I say that not for those familiar with human evolution but to qualify the statement *'we have only been around 200,000 years'* that is often quoted and might be taken literally. Naturally, others came before Omo 1 and they either looked the same or similar, although they would increasingly be more rugged-looking the farther back they emerged. It is a matter of how far back modern features go and where we put the boundary between archaic and modern humans that defines where we in our present form began. What makes Omo 1 a modern human is a group of features that collectively only *Homo sapiens* displays. They include: a chin or projected mental region of the mandible; little or no browridges; a fully oval cranium; thin cranial walls; few or no prominent areas of muscle attachment on bone and little buttressing around the skull, especially on the face and suboccipital area. These features are synonymous with modern humans and Omo Kibish complies, although it is rather fragmentary and difficult to confirm it has them all. A later more ruggedly built individual, who some believe is a modern human subspecies named Herto (*Homo sapiens idaltu*, 160 ka), is more ruggedly built with browridges and rugged facial features. Nevertheless, it is defined as modern human and together with Omo bookend the morphological range of variation of the earliest modern humans found in Ethiopia. Our emergence took a long time, tens of millennia, as we slowly morphed into us from our archaic forebears. We assume it was a slow process that probably took place differentially among various emerging populations, something to be expected in the evolutionary development of any complex creature. Therefore, when 200 ka is given for our emergence, really it means the date of Omo 1. However, we must have been a culturally and biologically emerging species long before that, but it has been difficult to pinpoint where we separate from archaic predecessors because there isn't one really. But the main reason for not knowing the process of our emergence, or the appearance of Omo 1, has been we have not had the fossils that predate it. (I assure the reader this paragraph was written before the discovery of the Moroccan archaic moderns late in 2017 (see below)).

Although the earliest direct ancestor of modern humans is assumed to be Bodo, its robust build, browridges and low cranial vault makes it difficult to accept as our direct ancestor even though it is around 600 ka. But robusticity would have been natural feature of any of our hominin ancestors whoever it was. Later individuals like Saldanha from South Africa (~400 ka) and Kabwe are somewhat less robust than Bodo and do make the transitory link a little easier to accept. It is worth noting also that the Saldana browridge reflects a similar morphological structure to that of Kabwe. Bodo and Saldana have

1250 cc crania, slightly smaller than the modern human average (1350 cc). A contemporary of Saldana is Ndutu an early or transitionary cranium from Tanzania (~400–300 ka), it has an even smaller cranial capacity (1100 cc) suggesting variable cranial capacities among later Middle Pleistocene hominins that preceded modern humans. The geographical position of Saldana and Ndutu spans nearly two-thirds the length of Africa and may also indicate the widespread emergence of their kind. We may never know the exact timing of our emergence as modern humans or our exact birthplace because of the gradual incremental process of emergence, as we slowly slid from one morphology to another, sometimes subtly, in different places and at different times to achieve our modern appearance. However, we would be right to say we must have begun to emerge before 300 ka because we needed that time to make the transition from earlier archaic stock to a modern form around 200 ka.

Until now it has been assumed that we emerged in northeast Africa (Ethiopia) where Omo I, the more rugged Omo II, Singa (>130 ka), Herto, and the Ngaloba-Laetoli Hominin 18 (150 ka) from Tanzania were found. The northeast African origin for modern humans, however, has now changed after recent discoveries in the Jebel Irhoud region of Morocco 60 km west of Marakesh. Jebel Irhoud first came to prominence as a hominin site when a cranium, Jebel Irhoud 1 (JH1), a calvarium (JH 2) and a child's mandible, all thought at the time to be 200 ka, were accidentally discovered by mining operations in the early 1960s. The same site has now yielded further hominin remains as well as stone artefacts dated to between 281 and 349 ka that includes by inference the previous hominin fossils (Hublin et al., 2017; Richter et al., 2017). The remains have been determined as modern human or a transitionary stage just prior to our emergence. The fossils are somewhat rugged in appearance and reflect the morphology of the earlier JH1 cranium but collectively they all seems to show that a population of early modern people were living far from northeast Africa on the Atlantic coast on the other side of the Sahara Desert around 300 ka and predating the northeast African evidence by at least 100,000 years. Florisbad (259 ka) from central South Africa (1400 cc) has also been termed a modern human so perhaps its geographical position might, together with the Jebel Irhoud evidence, suggest early modern humans were spreading themselves widely inside the African continent confounding our idea of a specific area or origin (i.e. northeast Africa) for them. The African evidence for early modern people now suggests that if they were travelling internally on the continent there was no reason they should not leave also.

Early modern people present as a cranially varied group. Some look very modern (high oval crania) while others carry archaic features, such as long and low cranial vaults, which the Jebel fossils retain as well as a prominent supraorbital region in some cases. On the robust end of the modern human spectrum fossils can have a generally rugged cranium with a thickened cranial vault, pronounced browridges, broad robust facial bones, raised areas of muscular insertion on the face and cranial base and in most cases a pronounced nuchal torus with long, low vaults. That pattern, however, generally contrasts with that of fully modern humans over the last 140,000 years. While cranial features vary, some fossils show emerging modernity to different degrees at different times and in different geographic regions that confounds a nice neat pattern of development that most researchers would love to be case. Till now, skeletal evidence for early modern humans was limited, particularly postcranial remains, but now we are beginning to have an idea about variation among them that can swing from a rugged to not-so-rugged appearance.

Our visual modernity has no precise separation point differentiating us from earlier *sapiens* except when viewed as a collective suite of features like those listed above. Viewed separately each feature can be variable and difficult to cite as representing or not representing a modern or archaic *sapien* person. Ideally, we would require a large set of fossils spread between 200 and 300 ka to properly trace and record the exact timing of morphological changes during our transformation. With such a sample, it would be possible to see the way in which temporal change took place, although it would not account for spatial change in different regions. The slide from an archaic to a modern human form was not uniform either across time or space. But it was even more complicated than that and the Herto and Omo 1 and 2, specimens alone demonstrate this. For example, Omo 1 and 2 lived in south-western Ethiopia at approximately the same time, both had fully modern crania of 1350 cc and 1400 cc, respectively, and Herto had one of 1450 cc. While the Omo 1 face is fragmentary and lacks many features, it nevertheless has a high, oval cranium with thin cranial vault bones and a chin. Those features are missing in earlier archaic *sapiens* and Neanderthals. Omo 2 is only represented by a calotte that is more robust than Omo 1 but not as robust as Herto. The extraordinary thing about these three is their contrasting morphologies: While Omo 1 is gracile, Herto is robust with prominent brows, a broad face and a wide maxilla and Herto and Omo 1 lived 35,000 years earlier than Herto. The Omo 2 face is missing but its calvarial morphology certainly suggests a robust individual. So, there is contrasting morphological variation within the Omo region as well as between widely separated places. Herto's robusticity suggests robusticity persisted among early modern humans living in widely separated places; among those living close together, and in those separated over some considerable time, all of which seems to show a mosaic of the two morphologies across Ethiopia, a situation similar among European *Homo heidelbergensis*. Wherever a fossil might be found does not necessarily reflect where they lived all their lives. They could have travelled extensively in all directions and their birth place might be somewhere else entirely. The very essence of this book is hominin movement and that is what they did.

Cranial variation is often used to define species and subspecies among hominins, but variation, such as that between Omo 1 and 2 or between Omo 1 and Herto, may have nothing to do with them being separate species or subspecies and everything to do with the mixing of different ancestors, geographical isolation, selective adaptation and their life's travels. Therefore, separating species in this context should be considered the last resort till ancestry is well known or at least better understood. Jebel Irhoud 1 was once mistakenly believed to be a late Neanderthal. The rugged appearance of Florisbad has found it in the *H. heidelbergensis* basket, although it shares some features with Herto living 100 ka later (Fig 4.1). If anything, Florisbad might qualify as a late *heidelbergensis*-transitionary modern just the sort of fossil we need to enlighten us about transitionary anatomy and help translate the changes that took place in the process of the emergence of modern

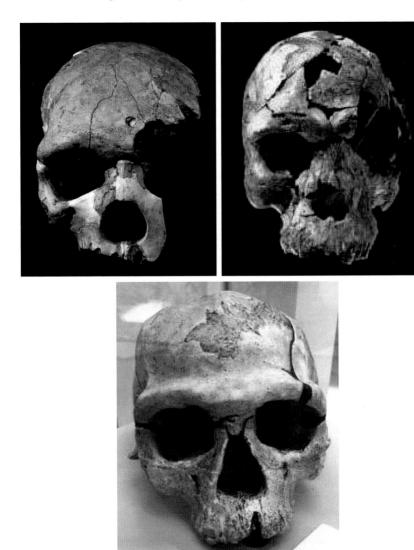

FIGURE 4.1   Florisbad (*left*), Herto (*right*), Jebel Irhoud 1 (*below*).

humans. To exemplify this, Florisbad may represent the stage before Omo 1 but found a long way from Ethiopia. However, instead of accepting variation occurring among transitionary forms, Florisbad's appearance required taxonomic change in the form of a different species name, *Homo helmei*, although that is little used. The 'splitters' strike again: it being an example of appearance requiring a 'taxonomic quick fix' because it is a bit different from others and living some way away from the 'main evolutionary action' in a similar way that Herto has been given a subspecies name. It now fits very well with other transitionary modern humans being found elsewhere. I would suggest that such quick fixing is not a solution to sorting out human variation and evolutionary relationships as much as a 'wait and see what turns up next' approach might be. Being faced with so much temporal and spatial variation in our transition, and thus an equal possibility for species of hominin, has become a nightmare for palaeoanthropology instead of a source of clarification but that nightmare may turn into less of a tangle as more fossils turn up. The contrasting morphologies of the Jebel Irhoud group, Herto, Omo 1 and 2 and Florisbad, show a non-uniform process of modernisation while the basic trend is towards modernisation through gracilisation. It occurred both temporally and spatially with the persistence of older archaic traits coming through in robust cranial features such as browridges. There is also an impressive range of cranial capacity among these examples, from 1300 cc to 1450 cc.

Culture is an additional element that accompanied modern humans but there are many definitions of what defines culture and going into those might muddy the waters further for settling a separation point between our archaic forebears and us. But do we need a point of separation? The truth is, we cannot really find a point where they end and we begin because, as I said previously, there isn't one, morphologically or culturally. Our ancestors built good shelters; made fire and cooked; fashioned a variety of stone artefacts, including retouched points and spears; spoke a language of some kind; organised themselves; successfully hunted and scavenged the biggest animals; used ochre (although we do not really understand what for) and even scratched basic designs on various objects. They were capable of all that up to 400,000 years before Herto or Omo 1 were a twinckle in evolution's eye. This emphasises a gradual transition from archaic to modern humans in different bioclimatic realms both inside and outside Africa. It seems that many skills and capabilities were developed by earlier people, sometimes more than once in different places, while modern morphology took a longer path to shed persistent archaic features that for some may have conferred some sort of advantage.

Like cultural development, the period of slow skeletal development that produced modern humans does not contain neat stepwise movements. Instead, it seems to be a slow, uneven swirl of subtly melded changes moving at different paces, at different times and in different places. No wonder we cannot insert neat taxonomic points of separation into our lineage even if we had many fossils of the right age. However, while we wait for many more fossils to be found we must guess at the length of time our gestation period took although the evidence from Jebel Irhoud certainly extends it wonderfully. Guesses about that change, however, depend on personal perceptions; who you ask where the division should be; the different interpretations of constituent factors, such as new archaeological discoveries and the constant improvements in dating our original ancestors, all of which pushes our emergence ever backwards, as well as into new geographical regions. As I hinted previously, our story *really* began not at 200 ka but at least an order of magnitude earlier and possibly before, depending on where we might recognise *us* as being *us* and whose interpretation one accepts. One beginning, of course, is where *Homo* begins, or where we like to begin calling hominins *Homo*. The difference is in applying the term 'modern'. Hominins 2 million years ago were not modern. Therefore, we walked along a very long and often convoluted path before we emerged as modern humans.

A recent discovery in South Africa has added a strange twist to the picture of emerging modern humans and who they shared their world with. It was the discovery of what was originally claimed to be a modern human ancestor but looked more like a 'habiline' (*Homo habilis*). Named *Homo naledi* after the place where it was found, the Dinaledi Chamber of the Rising Star Cave system near Johannesburg, it was originally undated but on morphological grounds it was thought to be as much as 2 million years old placing it at the time of *H. habilis*. Undated, it could not be properly placed in any temporal or relative context in the pantheon of *Homo* ancestors. However, this hominin is interesting because of the suggestion it may have disposed of its dead, a behaviour normally only linked to Neanderthals and modern humans. Deliberate disposal of dead has never been attributed to early hominins, so the claim was very controversial but an alternative idea of how the remains got there could not be found. Since the first discovery several other 'disposals' have been found in different parts of the cave system. They were in places deep and difficult to access and that underpinned the conclusion of deliberate disposal. Recent publications by the large team of researchers involved, however, state that *naledi* has a mixture of early hominin and derived modern human features (Feuerriegel et al., 2017; Laird et al., 2017; Marchi et al., 2017; Williams et al., 2017). Its hands and feet are almost the same as modern humans complete with an arch in its foot, indicating proper bipedality, yet its fingers are slightly curved, usually a sign of grasping during climbing. This hominin could walk and climb equally with ease. Generally, the rest of the postcranial skeleton is also very like modern humans, but its cranium is not. It has a 450–500 cc cranial capacity, far below that of modern humans and equating it with early hominins like late australopithecines and *H. habilis*, although that does not necessarily make it one of those species Part of the new information includes Naledi's age which has increased the mystery because it has been dated to between 335 and 235 ka, contemporary with early modern humans. That suggests its cranial capacity and other early hominin traits have not altered in 1.5 million years except perhaps its bipedal ability and it is taller than early hominins. But perhaps it is not unique. A similar hominin, with similar brain size, has been found on Flores island in

Indonesia, *Homo floresiensis* and I discuss that later in the chapter. The main difference between the two is height with *naledi* at 152 cm and *floresiensis* 110 cm, with *naledi* shorter than *H. erectus* (185 cm).

Naledi's new age brings it within the transition time of modern humans but while displaying modern cranial attributes and possible behaviour its brain size separates it. I mention hominin brain size throughout this book because generally cranial capacity usually ties a fossil hominin to a one-way enlargement as human evolutionary development continued. Thus, the bigger the brain the more a hominin could do with it including developing cultural advancement. However, organisation of neural networks and associated human capabilities at any given size is not well understood. Size may not always indicate intelligence or capabilities and perhaps the small brains of early hominins may have provided more skills and intellect than we give them credit for, so maybe size does not matter. That leaves us with a caveat that brain size might be useful as a general measure of capability among hominins but it may not be the only yardstick that indicates intelligence or ability. The brain of *Naledi* may be an example of this. Frontal lobe development, for example, can indicate, language, organisational abilities and planning to some degree and this is just what *Naledi*'s frontal lobe development seems to suggest according to those studying this fossil. Not all these traits would necessarily be equally developed but enough in certain combinations to allow behavioural development that lent itself to certain advanced or caring feelings towards others of its type. That would lend support to the possibility of it being able to carry out and organise a basic but deliberate disposal of its dead. These translations of cerebral organisation may also apply to *floresiensis* and is something to keep in mind as cranial capacity is mentioned in further chapters.

The age of *naledi* and the interpretation of what it represents, now makes its disposal behaviour look a little more realistic. Its brain capacity would suggest, however, hominins of small cerebral development would not be thinking about proper disposal of the dead (although that might be prejudice on my part) and there is no real example of such disposal behaviour for at least another 100,000 years anywhere in the world, so that type of activity remains a puzzle. This may be another example of those behavioural traits that we have difficulty in accepting in certain hominins because of how see certain hominin capabilities and thought patterns over time. It is worth considering, however, whether early modern humans took the bodies of *Naledi* into the cave system, there is no real evidence *Naledi* did it. However, hominin capabilities are a subject which I will visit from time to time, particularly their use of watercraft. A last few thoughts or questions immediately arise from this discovery. How did they survive for so long virtually unchanged; how could they have remained on a separate almost non-evolving hominin branch since before the emergence of *H. erectus*; did they ever leave Africa and is it possible others might be found elsewhere in or outside Africa?

CLIMATE AND OUR EMERGENCE

In Chapter 2 I spoke about the role of climate and its influences on early human evolution. The best expression of that would have been made by the onset of strong and long glacial cycling in the Middle and Upper Pleistocene. That must have added various increasing stress loads on our immediate ancestors and made life difficult for modern humans but at the same time it could also have helped them. So, I believe it is fair to say that the transitional stage of our emergence discussed above could have been forged by a cool Earth. It would be surprising if glacial–interglacial cycling did not have an inordinately large impact on the emergence of modern humans and their dispersals out of Africa by presenting both opportunities to move as well as stopping them (Timmermann and Freidrich, 2016). Essentially, we emerged during two very long glacials, Marine Isotope Stage (MIS) 10 (393–336 ka) and MIS 8 (312–245 ka) in which average world temperatures swung between −4 and −9 C. It marked a time of low sea levels, reduced rainforests, expanded deserts, changed and expanded coast lines and caused an overall drier planet. Moreover, fluctuating sea levels and, in some cases, the encroachment of ice produced both corridors and barriers for both humans and animals. Such conditions remodelled biogeography through the redistribution of animal species and herds and changed, shifted and rearranged vegetation mosaics. Some species did not return to their original habitats and some habitats did not recover back to their previous state altering biogeographic profiles and vegetational mosaics around the world. It was a dynamic world indeed.

These altering states brought many challenges and selective pressures for humans although their ancestors had lived through previous cycles. The MIS 9 interglacial (337–312 ka) was generally cool with temperatures higher than today occurring only between 321 and 336 ka which lies well within the time span of the Jebel Irhoud fossils. Those higher temperatures may have initiated a northerly migration of the African monsoon as it did during the high temperatures of MIS 5e. That would have brought a greening in the Sahara enabling early modern people to cross the desert as well as provide good living conditions at places like Jebel Irhoud. The MIS 7 interglacial lasted around 50,000 years (245–195 ka). It was only as warm as our present interglacial for about 2000 years within that time and never rose above it. Looking at it in more detail we can work towards the present through the five substages of MIS 7 that shows it was more a glacial than an interglacial. It consisted of two warm substages (MIS 7e and c) separated by what was a full glacial of 17,000 years (235–218 ka, MIS 7d). Although 7d is not designated a glacial it is reasonable to think that when average world temperatures go below −4°C the earth has entered a glacial phase. Substage 7e was the warmest lasting from 245 to 235 ka with a peak of 2°C hotter than today but it lasted only 2000 years around 242 ka. World temperatures rose again during MIS 7c, from −4C fluctuating between −1°C and −2°C. MIS 7a and b had smaller temperature fluctuations, they were also rapid events that saw world temperatures change by as much as 3 C before they slowly descended into the 60,000-year-long MIS 6 glacial around 195 ka.

The above description is rather dry reading I admit but it demonstrates the topsy-turvey world of a rapid changing climate and the associated environmental extremes that occurred during our transition from archaic hominins: it was our *emergence time*. It is not too imprudent to suggest, therefore, the comparatively rapid fluctuating glacial/interglacial world between 250 and 200 ka demanded appropriate behaviour and survival strategies to be implemented by them. It was a hard but effective method of growing us up, but it is logical to believe those conditions contributed to our emergence as modern humans by confronting us with rapid and repetitive challenges that we needed to respond to.

It was a rapidly changing world that not only came from the cycles but also from strong climatic reversals contained within substages of the MIS 7 interglacial. Glacials and interglacials cannot readily be labelled as good and bad times *per se* because each half of the cycle would have triggered opportunities and setbacks for modern humans. Such changes would have also honed qualities that drove selection pressures (our *9EM*). In geological terms, these changes were comparatively rapid in geological terms, although almost imperceptible if measured in human lifetimes. However, that speed is exactly the pace that genetic selection likes to move at. The length of time involved in a human generation is not counted in evolutionary terms, it is how many generations pass assorting genes as they go. Alterations over generations would have imposed subliminal changes and pressures on human populations.

Environmental change was less pronounced and less stressful overall for humans and animals in lower latitudes. At higher latitudes ice fields and glaciers moved across millions of square kilometres literally burying landscapes, imposing an extremely hostile wilderness and pushing many animal species and plants out. Overcoming and adapting to cold and freezing environments was our biggest challenge and was not successfully overcome till modern humans moved into eastern Siberia and crossed into North America, although this may have taken place much earlier than we have believed, not only among modern humans by archaic people and this is discussed later in the chapter. High latitudes demanded a lifestyle and culture completely unlike anything we had ever been used to. Those challenges did not occur in Africa during our formative growth as modern humans but they did when we ventured into the north.

Nevertheless, while small temperature fluctuations occurred throughout the MIS 6 glacial, as they did in all earlier glacials, they were not of the order of change seen in the MIS 7 interglacial. It is difficult to know what such fluctuations meant for various geographic regions in anything other than the broadest terms. For example, the MIS 8 glacial in Europe must have brought a different degree of impact on Neanderthals than equivalent changes had on emerging modern humans in Africa. It is tempting to invoke these different environments on the two continents as a catalyst in the differential evolutionary pathways modern humans and Neanderthals took and what it meant for us subsequently. We might remember also how modern humans have changed from those who left

Africa into all the many different peoples' that we see across the world today. Those substantial genetic as well as cultural changes occurred in <50,000 years, about the same time it took for modern humans to emerge fully 200,000 years earlier, and both events took place during times of vast climatic and environmental change brought by glacial conditions. As far as we know, Neanderthals and ourselves shared the same *heidelbergensis* ancestor with only our subsequent environmental histories and geographic separation the major factors that separated us. The challenges of glacial cycling almost certainly played a role in biological selection, particularly in the latter stages of the emergence of us both but in different environmental settings. It also provided issues to overcome for those that succeeded and enjoyed success in terms of a continuation of their group.

The first fully modern humans seem to emerge within the MIS 7 (240–200 ka) interglacial and by the end of it there we stood in the form of Omo 1 and 2, an apparently quick transition from the preceding archaic looking *sapiens*. The classic Neanderthal appeared somewhat earlier. For some reason robusticity continued among modern humans probably as a legacy from earlier archaic moderns like Kabwe. Robusticity may have been selected as a way of maintaining cranial strength perhaps conferring an unknown selective advantage useful at that time, we can only speculate. It seems it was less than a smooth skeletal transition from the archaic *sapiens* who preceded us between 450 and 350 ka. They passed on a mixture of features that presented a wide range of morphological variability. Cranial capacity was one of those variables among late archaic and early modern people, and although we need more examples (we always need more examples) it may show that at least robusticity persisted to some extent as a natural part of the archaic–modern transition that lingered many millennia. Skeletal gracility particularly of the cranium, is the symbol of modern humans, robusticity on the other hand, speaks of a brutish or subhuman streak: gracility being the road to a classically modern redemption. Those long-held ideas were passed down through earlier anthropologists, albeit we now know you do not have to have browridges not to care for your fellow humans, Neanderthals displayed that and we do it all the time. Therefore, robusticity, rather than a feature to be selected out as quickly as possible, so divesting ourselves of that brutishness and thus confirming our rise to perfection, may have been one that was often retained as conferring a little understood adaptive advantage under certain environmental and/or climatic circumstances such as glacials.

After 190 ka, the world entered a long, deep glacial with −8 C average world temperatures that remained that way till ~135 ka. Africa's deserts expanded, essentially subdividing the continent into habitable pockets but at the same time preventing people from moving across North Africa and in some instances isolating them. It meant barriers to movement causing 'refuge' isolates trapping small groups. Such demographic outcomes are enough to trigger at least three of the *9EM* group, founder effect, genetic drift and selection, to operate at various levels and to different degrees among those isolates over time. It is possible that it was that robust features were maintained due to these mechanisms in some regions and in some groups more than in others. Small population size and isolation may have also been catalysts for the persistence of robusticity and/or gracility among various populations depending on the genetic predisposition of their founding groups and the type of founder effect that operated. What emerged depended also on the influence of their living conditions and the demographic history of scattered populations as the MIS 6 Ice Age continued and deepened. That might explain the maintenance of robusticity among the Jebel Irhoud fossils. The expanded Sahara during glacials engulfed Chad, Mauritania, Niger and Sudan and moved into Nigeria, Cameroon and northern parts of the Central African Republic. Thus, people were separated from places on the other side of this megadesert and corridors for crossing it were gone. It was now desert from Africa's Mediterranean coast to the Congo. The position of the Jebel Irhoud fossils in Morocco signals the north-western movement of people before ~400 ka that then became a refuge area through glacial onset. However, if that was so it meant migration was in full swing within Africa a long time before the emergence of fully modern humans.

The process of the morphological emergence of modern humans extended from MIS 8–10 that included two glacials and one interglacial. It is on the southern fringe of the South African coast where we see them emerge as physically and culturally modern people. They seemed to have traversed the east coast or perhaps a little inland eventually reaching the southern coastline and they then had nowhere else to go, they had reached a dead end. They set up camps along that coast with the far southerly Pinnacle Point settled by 164 ka, in the middle of the MIS 6 glacial. Many sites dated between 100 and 60 ka pepper the southern coast. They include Langebaan Lagoon (footprints, 117 ka), Klasies River mouth (120–60 ka), Die Kelders, (Klipgat Cave, 80–70 ka), Howeison's Point (65–60 ka), Border Cave (77–70 ka), Hoedjiespunt >74 ka and others. Here are the first signals of modern human cultural complexity such as small tool production, heat treatment of stone in tool manufacturing, fishing, engraving, making jewellery and bone tools and the use of mosquito repellent bedding. In a way, the final polishing of *us* as modern people was accomplished during an Ice Age, it emerged during MIS 5 and it appeared along this coast. It was a time of completion and consolidation of the modern human transition and the final stage in the evolutionary emergence of modern people. Equally, it was the time of the development of what really makes us modern humans: our culture and an encroaching complexity to our society, and most of it occurred during a 60,000-year Ice Age. Considering the vast climate switching during the previous 100,000 years, it seems similar changes in climate and environment probably played their part in the evolutionary making of modern humans. It is also important to reflect on these changes in honing people's ability to move, migrate and live in regions that others had already adapted to and survived for a long time. It is worth thinking about whether the same Ice Ages were having similar effects on archaic groups of which there were a number living during MIS 5 in other parts of the world.

OUT OF EUROPE

While palaeoanthropology has largely been preoccupied with the modern human exit from Africa, something similar had already happened and that was when Neanderthals left Europe and that may have been triggered by the environmental consequences of the MIS 8 glacial. It probably encouraged them to move out through south-eastern Europe and into western Asia as ice began to move south across two-thirds of Europe. Besides ice hiding large swathes of northern Europe, the periglacial regions south of the ice sheet would not have been nice places to live in. Neanderthals had been living in Europe as much as 150,000 years earlier than any modern human exit from Africa. In that time they had spread through Eastern Europe and into the Middle East, western Asia and south to the Levant. They also lived on the western limits of Europe, Portugal and Gibraltar, where, because they were as willing to travel as modern humans, they could have crossed the Straits of Gibraltar to Morocco although there is no evidence for them doing so. The Jebel Irhoud cranium was once considered to be one Neanderthal that did, but this notion has now evaporated for the moment with the recent discoveries in Morocco. Neanderthals lived in Gorham's and Devil's Tower Caves, but 100,000 years after Jebel Irhoud wandered into Morocco. Neanderthals also moved into the Zagros Mountains of Iraq (Shanidar Cave, 45–35 ka), and on to Uzbekistan (Teshik Tash ~70 ka) the Altai (Oklandikov 44–38 ka) and may even have penetrated north to the Russian Arctic (Byzovaya, 33–28 ka) (Slimak et al., 2011).

The extinction time of Neanderthals is now thought to have been around 40 ka, earlier in some areas and perhaps later in others. Their patchwork pattern of extinction is not surprising given an assumed small population overall and its fractionalised and scattered pattern across a large area. Differential extinction events among the megafauna they lived with were not an unusual occurrence and they were part of that ecosystem. Among mammal populations extinction is erratic not uniform, leaving isolates where some survive longer than others depending on their ability to survive in their particular niche and their population size, particularly the number of female animals. Widely separated groups unable to sustain their populations is another reason for extinction as backup personnel and lowering birth rates eliminate the group, and so it probably was among Neanderthals. It also seems likely that some did not go extinct in the true sense of the word but by mixing with modern humans. There are many reasons for such extinction patterns and no doubt Neanderthals were as subject to those as the megafauna and any other animal (Webb, 2013). Nevertheless, Neanderthal migrations are mute testimony to their success as a human species over a long time. It may have been the case that modern human dispersals across southern Asia initiated changes in Neanderthal demography as well as breeding and settlement patterns even pushing the last few populations into western and central Asia, northern Russia and Siberia. Perhaps they just kept ahead of those they wanted to avoid and it may have been at that time they moved to eastern

Asia using their cold-adapted physiology, morphology and technology to cross north of the Himalayas through southern Mongolia or the Altai. Even moving into central and northern China may not have been beyond them as well as breeding with their Denisovan cousins who lived in the region.

Language provides many cultural and survival benefits. It almost surely developed very slowly over a very long time with an exponential flurry in the Late Pleistocene. It might be thought of as emerging when a certain level of higher intellectual development took place, perhaps one feeding off the other. The advent of language enabled the growth of social complexity providing an enormous edge to gaining adaptive strengths through traditions, education, organisation, instructing, comparing the world with others and constructing shelters and other constructions. The finding of two semicircular, low-walled constructions made of stalactites and stalagmites 300 m inside a cave in south-west France dated to 175 ka, is the first evidence Neanderthals built any sort of construction (Jaubert et al., 2016). Traces of fire were found within the semi-circle but the purpose of the work, not surprisingly, is unknown, although dated to a time deep in the MIS 6 glacial might suggest the construction was meant to be shelter within a shelter, for two families, against the bitterly cold conditions outside. Making and building large constructions requires organisation and linguistic direction to follow a plan and teach and reason with others why the work is being undertaken. Moreover, if ceremony and ritual are involved they too require directions and explanation about what is required from participants, its meaning and why it is necessary. Language for the two branches of humanity living in the West at that time, was not only being spoken it was essential to the way of life both for modern humans and Neanderthals.

But the debate over whether Neanderthals had language, particularly complex language, continues. Genetic evidence in the form of the *FOX-P2* gene, implicated in the ability to acquire language and which is found in modern humans today, is found in the Neanderthal genome. Neanderthals also had a normal hyoid bone, the small crescent shaped bone found high in the throat that acts as a hanger on which the small pharyngeal muscles enabling speech are suspended. A hyoid bone has been found at Kebara Cave (60–48 ka) Israel, a site occupied by both Neanderthals and modern humans. Modern humans had language at least from the time they left Africa and probably well before. But it has been suggested that Neanderthals may not have had the *FOX-P2* gene before they mixed with modern humans and received it from them, they had a hyoid bone, however, was that waiting for a time when a *FOX-P2* gene might, by chance, come along! One argument is that while the hyoid bone belongs to a Neanderthal, mixing between the two groups may have conferred a hyoid on later descendants. However, the combination of *FOX-P2* and the hyoid make a compelling case for the existence of language among Neanderthals. Moreover, the idea that only modern humans could speak with any sort of complexity once again smacks of the superiority of modern humans who while they look like the winners may have come close to being losers later.

Contemplating the complexity of language 100,000 years ago and even earlier, and the difference between the language of modern humans and Neanderthals and the modern languages of today, is intriguing. But we cannot do more than wildly speculate about it or what it sounded like, although I favour a type of click language, like that of the Khoisan, which may go back a very long way. Nor can we know whether there were dialects and language variation between groups living in different regions of Europe or parts of Africa such as those along the South African coast. Perhaps there were different dialects or languages among modern humans or between Neanderthal groups living in Europe, the Middle East and western Asia. Recognition of a mutual, although largely unintelligible, means of communication between modern humans and Neanderthals may have helped in their first face-to-face contact. It is not fanciful to suggest that sign language may have been beneficial for emphasising certain words or requirements and no doubt that is what they used when hunting, clicks would also be useful.

MODERN PEOPLE PRACTISE LEAVING HOME

Various fossil hominins and archaeological sites across Africa indicate purposeful movement to occupy regions in most directions. Jebel Irhoud is only one example, but how did it get to Morocco? We can only presume that when they could they did, possibly following riverine or green corridors that crossed the Sahara south to north. So too, modern humans explored within the continent before they left it. Therefore, modern humans of different morphologies appear as widely spread, or widely separated, seen from another perspective. Separation and isolation were real possibilities particularly when periodic glacial cycling turned savannah into desert barriers of separation and both factors played their part in how respective morphologies turned out. Some groups found themselves isolated, perhaps for some considerable time, allowing genetic drift and founder effect to enforce directional change in cranial form and in other ways. That may have been enough to provide a range of variation between isolates. Such processes are important when sequencing fossil crania and when viewing differences between spatially separated populations. They become indicators of their relationship to considered norms for the species at any one time and the environmental circumstances they found themselves in. The mix and division factors of human migration, that allowed different groups to meet and discover others during pulses of movement, operated for a long time. Those mechanisms are also important to consider when assessing morphological differentiation between fossil hominins that were widely separated in time and space and particularly when considering the sequence of repetitive climatic shifting that took place during the MIS 6 glacial alone: that vital stage in the story of early modern human development. Early modern humans were spread across many hundreds of thousands of kilometres, so when we talk about single avenues of descent and then consider genetic inheritance and climatic and environmental change we cannot avoid seeing differences between fossil humans from different regions as anything other than complicated but almost certainly a product of their enviro-climatic times and demography rather than separate lineages.

Some hominin examples are considered modern transitional forms. In Africa, they possessed archaic skeletal robustness, like Florisbad and Ngaloba (aka Laetoli 18, 120 ka, 1367 cc). The latter retained its robust appearance till rather late possibly due to processes of adaptive selection as described above. The likely reason for Jebel Irhoud's presence in Morocco is that it signals a movement of people from northeast Africa, eg. the Ethiopian/Kenya region and then their journey was either 5000 km, as a direct crossing to Morocco, or 6500 km via the Nile Valley and along the North African coast. It was an 8000-km journey across sub-Saharan Africa, along the edge of the Equatorial belt, to the Atlantic coast and then north to Morocco. But as I mentioned before, where hominid fossils are found may not mean anything about their geographic origin or how far they travelled, it only means they died where they were found. If climatic and environmental factors played a part in such movement, and I can only assume they did, a Saharan crossing happened at the end of MIS 9 for the best crossing time.

The widespread distribution of archaic features among what are ostensibly early modern humans occurs among temporally and spatially separated African fossils. Some are separated by 150,000 years and 3000 km, like Florisbad and Ngaloba, while Jebel Irhoud and Florisbad are separated by 50,000 years and 7500 km, as the crow flies. Among the earliest moderns is a mix of features that range from robust to gracile with no apparent reason for their geographic dispersal unless selection played a part in different regions and at different times because of world climate change to glacial conditions that prevailed for 60,000 years.

It was during the AE3 stage in Table 3.1 that modern humans changed skeletally from a largely rugged and archaic looking individual to a more gracile type. Archaic features persisted among some including prominent brows, facial buttressing and marked areas of muscle attachment, particularly around the cranial base. Today, we would notice Kabwe in a crowd and he would definitely get a stare, but not necessarily Herto. He would fit nicely on a rugby field although probably get a second glance if he entered a ballet class. It was in the AE4 *Early* and *Middle* stages that modern humans went to southern Africa whwre they set up their coastal camps. While they were doing that, on the other end of Africa they were leaving the continent. And when we took our first steps out of continental Africa into the outside world we embarked on on a very steep learning curve.

MODERN HUMANS WITHIN AFRICA

Table 3.1 shows AE3 spanning the modern human *transition phase* (350–130 ka), when archaic and earlier modern humans were travelling within Africa. Once again, placing a specific demarcation point is fatal, so it is likely some movement out of Africa could have taken place well before 200 ka and within a few years I am sure we will see evidence for it in the Middle East or beyond and I would not be surprised if it was at least between 300–200 ka. However, modern humans were roaming widely within the African continent (Rito et al., 2013). Their spread south to the limits of the continent is widely recorded in the many sites discovered around the coasts of southern Africa in a wide arc from the South African–Swaziland border round to the Atlantic coast. The oldest is Border Cave (227–164 ka) (Watts 2009), but there are other early modern sites on the way south, including Gnjh-15 Kapthurin (Kenya) and Twin Rivers >200 ka (Zambia) (McBrearty and Brooks, 2000; Barham, 2002; Lombard, 2012; Leslie et al., 2016). Their MIS 7 interglacial dates indicate they enjoyed good conditions but they may indicate a Rift Valley corridor was used where, during MIS 6 dispersals, resources may have been maintained even when areas either side were suffering reduced resources due to environmental downturn and arid conditions. It is worth noting, however, that while glacial conditions generally create generally difficult or impossible conditions inland, it was also a time of continental shelf exposure in various places along the East African coast which meant new environments to explore, hunt and forage and possibly present an alternative corridor south to that of the Rift Valley.

Ochre pigments found in an archaeological site are normally associated with body decoration, artistic work and ceremony and ritual, all cultural markers usually associated with modern humans. This may not have necessarily been the case, however. Rather than indicating an association with ceremony and art, ochre could have been just a novel material: fascinating to hold, crush, play with and just rub on the body in a secular or playful manner under the fascination of its colour and texture as children might do. Novel material in ancient sites, such as Acheulean crystals at Singi Talat, India, and natural glass at other sites were apparently there without any other indication of artistic or ceremonial use (Clottes, 2016). Nevertheless, the consensus is that ochre in an archaeological context does indicate ceremonial/ritual use and/or body adornment; mainly because it is not particularly useful for anything else and ethnographically hunter-gatherers use it in that way. Perhaps ochre use should accompany the making of jewellery, bone points, heat treatment of stone and other complex material culture. It then makes more sense as something more than just a plaything. Its presence should be parsimoniously taken as an indication of a new cognitive stage in the transition of modern humans from the lower to the upper substage in their late transition not immediately apparent from other cultural remains. However, it still depends on the context in which ochre is found in the wider archaeological picture.

Early modern humans reached the southern extremities of Africa by 164 ka. Excavations at Pinnacle Point on the southern tip of Africa have revealed several significant aspects to modern human activity and culture there. Settlement at 164 ka means modern humans lived there deep in the MIS 6 glacial when seas were at their lowest and it was a very cool environment. They could have moved along the coast foraging on the exposed South African continental shelf

(Agulhas Plain) that extended out as much as 250 km from Pinnacle Point during maximum low stand (Fisher et al., 2010). Apart from modern humans reaching this southerly tip of the continent, it is worth remembering that at the same time Herto lived almost 5500 km away in Ethiopia. Pinnacle Point shows people harvesting sea food, making small bladelets and using modified ochre pigment (Marean et al., 2007). The discovery of small tool technology (bladelets) and ochre in that site put the use of these items back 90,000 years. Their discovery not only indicates developing cultural practices and technological skill but also they show how our view of society at that time can change with one discovery, the sudden thrust of culture back in time, and that we can always be surprised what we discover. Although widely separated, Pinnacle Point and Herto shared the same time frame deep in a 60,000-year Ice Age with all the massive environmental and climatic changes that went with it. But those changes were different in different areas (Fig 4.2).

Intracontinental movement would have been difficult during some stages of MIS 6 as the Ice Age divided northern Africa from sub-Saharan Africa, the south from the northeast and, next door, rendered crossing the Arabian desert impossible. So, with these barriers how did people reach Pinnacle Point?

Their arrival could suggest modern humans were testing their skills of survival and exploration by turning inwards on the continent and the glacial conditions met with tested and honed those skills in the *apprenticeship* I spoke of before. Those skills together with behavioural, cultural and biological adaptation that occurred could have laid the foundation required for leaving Africa. Ice Ages may always have forced an acceleration of skills, strategies and adaptive knowhow among all hominins but was, perhaps, greater among modern humans. They require adaptive behaviour at a level not normally required in a tepid, quiescent interglacial. Consider, hominins of all persuasions had to live in vastly changing climates and environments, these were ideal times to test their

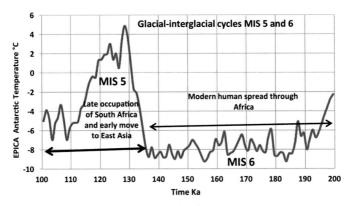

FIGURE 4.2   Glacial cycling from MIS 7 through MIS 5 showing the deep, long MIS 6 glacial and two phases of modern human occupation in Southern Africa (After EPICA Community Members, 2006).

metal and see how they could cope and adapt to changing conditions, opportunities and downturns in previously advantaged circumstances. They did not hide or hibernate during glacials they lived through them and in most cases survived by learning new skills, survival techniques and using adaptive behaviour like moving to better places and thinking about how to protect themselves from environmental stress. Those things challenge the brain and sparked ideas and ingenuity emerges but in so doing it adds just a little more to the capabilities of those undergoing a baptism of radical climate change. Those that did not do this went locally extinct particularly in remote or isolated places.

Modern humans at Pinnacle Point signal a journey to the southern limits of the African continent. While those travellers were contemporary with Herto, would it have crossed Herto's mind to move the 400 km to the Red Sea and cross the Bab al Mandab Strait? Did the mind within that 1450 cc cranium think in the same way as those living 5500 km to the south as they gathered marine resources at Pinnacle Point? I believe it is likely they did think differently from Herto, mainly because of their experience, although it probably took generations to make the journey south. But it makes one think that people that take off and begin a long journey south could equally have reached the Bab al Mandab Strait and would have crossed it. The evidence for the taking of shellfish in the south shows people had the ability to live off coastal resources just as well as they could off inland resources; a non-specialised approach to food gathering was an adaptive strategy and the key to survival. The coast of the Red Sea, therefore, could have been exploited just as well as it was when those people made it to the South African coast. However, as we saw in Chapter 2, they were not the first to forage for shellfish, *H. erectus* gathered freshwater mussels 700 ka ago, and we regard them as the ultimate non-specialised foragers.

It is tempting to suggest that while African inland migration routes were cut or difficult to traverse, a coastal path or beneficent corridor linking northeast and southern Africa was used during the MIS 6 glacial. Lowered sea levels exposed the Agulhas Plain to various widths along the coast, although its widest was at the tip of South Africa and along the Atlantic coast. For example, it was narrowest around Klasies River and broader near Blombos Cave and Pinnacle Point, narrowing again at Die Kelders (Fisher et al., 2010). Klasies River (125 ka), Deipkloof (115–60 ka) and Blombos (110–100 ka) are other significant modern human sites but they are much later, between 60,000 and 30,000 years younger than Pinnacle Point. The many occupation sites and their placement suggest later pulses of modern humans arrived after those at Pinnacle Point, or was it Pinnacle Point people moving? Alternatively, many sites were probably out on the Agulhas Plain and were later drowned after sea level rise which began around 135 ka pushing those on the plain slowly inland as shorelines moved in. The late occupation phase shown in Fig 4.2 is, no doubt, an artefact of the drowning of earlier sites on the Agulhas Plain that would show a continuum of occupation along this part of Africa if only we could see them.

I briefly explained earlier that the modern human journey could have brought about changes in morphology and culture that altered people, after several generations, and removing them somewhat from their ancestor's culture and appearance. Generations must have passed on those journeys inside Africa, although how many is only a guestimate as it is for the later journeys outside Africa. However, it seems only logical that inevitable changes in morphology and behaviour occurred during dispersal histories wherever they took place. Naturally, the earlier modern humans left Africa the more likely it was that they met and were possibly influenced by archaic groups inhabiting regions they passed through. Evidence from Eastern Europe provides a glimpse of such a biological influence that took place there when modern humans and Neanderthals mixed. Cranial traits observed in several eastern European and Levantine fossils include Mladec´, (Czech Republic, 31 ka), Skhul (90 ka) and Manot Cave (Israel, 52 ka), the Oase 2 (30 ka) adult and the Muirelli child cranium (30 ka) both from Rumania. Their obvious morphological differences from other modern humans together with genetic sampling of Oase 2 showing that it carried 6%–9% Neanderthal DNA, indicates modern human ancestors had mixed with Neanderthals possibly eight generations before (Fu et al., 2015). These examples are testimony to the influence modern humans and indigenous populations coming together especially when we realise how widespread that could have been when contemplating the odds of finding all these examples from their time frames. But I am getting ahead of myself here.

OUR JOURNEY

Assembling a picture of modern human movement out of Africa and the subsequent dispersal patterns is an extremely difficult task. An interdisciplinary approach is required as well as using a variety of archaeological, palaeobiological and genetic evidence and several papers written covering the topic (Beyin 2011; Oppenheimer 2012; Garcia 2012; Boivin et al., 2013; Reyes-Centeno et al., 2015; Groucutt et al., 2015a,b; Reyes-Centeno 2016). The task is difficult because we will always lack vital details to make firm conclusions. But we should start somewhere and the type of movement that was undertaken during these dispersals is one place to start.

Essentially, modern humans were not really doing anything new, they were just following in the footsteps of early hominins like *H. erectus* or even *H. habilis*. It is not unexpected, therefore, that that there are many parallels between the two journeys. Both groups emerged in Africa and both took advantage of climate change to provide passage across and between continents using changed environmental conditions and lowered sea levels that formed land bridges is some places. Although we are not so sure about the routes taken by *H. erectus*, modern people probably took roughly the same routes through the Middle East and across the Saharan, Sinai and Arabian deserts. They could do that by following palaeoriver systems, the mercurial streams of interglacials, that brought fresh water to previously arid regions, changing deserts to habitable places at least along corridors either aide (Crassard et al., 2013). These places were impossible to cross during glacials. Interglacials made silent and often buried river beds spark into frenzied flows, but even during some parts of interglacials that did not always happen.

One of the problems for both *H. erectus* and modern humans, was that they required 'optimum conditions' to migrate particularly if they chose the southern exit point. But that combination occurred only rarely because the two conditions usually acted out of phase. Water was required in desert regions in the form of rivers and, in some places, lakes. Freestanding fresh water opened migratory corridors enabling the initial movement out of Africa and crossing of the Middle East and substantial areas of southwestern Asia. Those corridors opened only when climatic conditions brought regular and sufficient moisture to change desert into savannah or another benign environment that would support migration. The problem was, such conditions did not occur very regularly and when they did there could be tens of thousands of years between them. It could be argued that conditions good enough to exit Africa on foot with optimised environmental conditions happened only rarely. For modern human exits, they occurred during MIS 5, particularly in substage 5e, and for earlier exits MIS 12 around 400 ka, although interglacial conditions before 800 ka are not clear. To cross water gaps people needed an Ice Age to lower sea levels to cross straits, move from island to island in simple craft and cross narrow water gaps.

There is little to suggest, modern humans moved any more logically than early hominins. I find it difficult to believe that they had some sort of planned route to follow or deliberately tried to achieve special over-the-horizon geographical goals that they could not possibly be aware of. Knowing the exact routes taken by AE1 hominins is not possible beyond an educated guess based on where their fossils and camp sites are found. The same can be said for the routes taken by the AE4 humans. To know them exactly would require knowing behaviour and how it changed across time and that is beyond difficult to know with any certainty. Human behaviour changed over time making it difficult to use the behaviour or cognitive processes of modern hunter-gatherers as a proxy for those of AE4 people living 120,000 years ago. Moreover, modern hunter-gatherers stick to their 'country', they are not travelling across the planet. Today there are few hunter-gatherers living a traditional life and most are influenced in one way or another by modern industrialised social systems. Time passes and the construct and concepts of the human mind moves on also.

We cannot readily appreciate the logic modern humans applied to move and the direction they chose to move in. Neither can we understand what was important to them other than to eat, find water, reproduce and survive, even though they were modern humans like us. We can make educated guesses about these things and we can argue: is that useful or worth anything? The parameters involved in all AE movements are a source of ongoing confusion because of the many possibilities involved. In the AE4 case, as in any other, we need to know many African exits there were, then what the timing of those exits and how far they travelled after that. However, exit timing using two exits coupled with the variety of possible dispersal patterns, produces many combinations. Adding the fickle nature of humans really make it complicated. Because of these, it is extremely difficult to make an estimate of the length of time any individual journey all the way to Australia may have taken. Changing climatic and environmental conditions at the time of leaving Africa as well as afterwards is another factor that would influence decisions about the route taken. All these issues have become central to knowing how long journeys took and we have little chance of knowing about any of them. They become more important when those who reached Australia are considered because of the 'snowball effect' of accumulated setbacks, difficulties and route changes along that longest of journeys. We do know, however, those who were Australia bound were the first to undertake a proper ocean crossing. So, we return to capabilities. They must have increased with experience proven by the fact they must have had proper watercraft to make the final crossing—probably no elephants on that one. Therefore, I want to go back to the beginning of modern humans and look at the story of our emergence. Perhaps, by examining what we know and using some logical propositions some details, however basic, regarding the issues listed above can be put together. The story is very convoluted, complicated, somewhat controversial and much of the information we would like to have we do not have. Our story includes others who came before us; those we met on our travels and lived beside in different places as we roamed the world. Because of that, their origins and stories are as important as ours because our parallel worlds became intimately intertwined and, as I mentioned before, we probably influenced each other biologically and culturally to some extent and that included those who became the first Australians.

HOW LONG IS A PIECE OF STRING?

When modern humans left Africa, they faced the challenge of maintaining population numbers set against the possibility of their band going extinct. They might not have contemplated that but with hindsight we can. Band size was critical and a certain minimum size had to be maintained. Without retaining certain numbers in bands, continuing their journey would be difficult and they certainly would not have reached Australia. The smaller the band size the more vulnerable they were, so reduction in band size or numbers across several bands could result in them vanishing. Apart from reducing their ability to hunt and gather food, a small group was weakened for any response they might need against predators or other humans. Maintaining a particular size may not have been a conscious activity but they would have been conscious of losing members and the inherent weakness in their group that that brought. The basic factor in band viability was maintaining a required number of people to be safe and provide food. Combining or travelling with other bands may have been one strategy to strengthen numbers and combat that situation. The reduction in the numbers of bands and the melding of bands would reduce overall the people making the journey but it was a strategy for survival and almost certainly must have taken place regularly.

That brings us to another issue, exploration by single bands or groups of bands. The original exit must have involved more than one band. If one band had struck out alone it was immediately vulnerable and became open to extinction far quicker than several staying together. Moving forward was a matter of safety more than anything else, but moving was forged by accessing ample food and water and somehow trying to maintain numbers. Procurement of food and finding water were primary movement drivers and capable and strong hunters were essential for the process. Therefore, the environment they passed through was itself important and sometimes they had to learn fast because it restricted them to a safety-first principle in terms of resources for survival. To a certain extent, the fewer the resources the faster onward movement proceeded. Having enough water to drink also meant there was probably enough to sustain animals and that meant food but they had to catch them and the behaviour and habits of new species encountered had to be interpreted and understood.

Preagricultural human populations came and went, grew and collapsed and occupied and abandoned many areas across the globe. One reason for that was enviro-climatic change, a formidable combination that always forged human demographic patterns. From time to time these factors could form human-empty regions and halt dispersals altogether. People vacated very large areas during glacial events where land disappeared under ice or it became denuded of vegetation as desertification and the drying of water resources occurred. Equally, environmental change altered the biological and social composition of exploring groups. It did so by dividing and isolating them and at other times driving them onwards, perhaps quicker than was normally safe to do. Environmental change also altered living conditions and food supplies variously over the short, medium and long term. We might never know what else caused the ebb and flow of some groups or people and what guided them along the routes they took. Poor health, in the form of tropical diseases, is one that comes to mind, however. The presence of Yellow Fever and Malaria both of which are endemic in primate populations could certainly have existed along some migratory paths. All these are some of the issues affecting migratory progress that determined success or failure.

In our times, it is difficult to imagine going nowhere, slowly, and entering vast expanses of human-empty wilderness. For modern human dispersals that was generally the case. Forward progress may have been through range expansion but that was slow and more likely to have been something confined to earlier Plio-Pleistocene hominins who were possibly more directionally challenged. For modern humans, there was no special goal either, except perhaps being more inquisitive and with somewhat greater capabilities that included navigation by the stars and other natural phenomena. Those abilities would help them move more easily through the next valley or across the next stream or river and even eliminate some of the randomness of dispersal that we might expect in those earlier hominin explorations. Movement may have stopped when good conditions were right, particularly safety. We cannot know exactly how or even if a pattern of stop–start, forward–sideways–backwards movement took place or what forward momentum was achieved in a given amount of time. It could be suggested that AE4 journeys were not particularly straight or as logical as we might think. Even modern human movements are not easily understood or estimated, especially by a 21st century palaeoanthropologist making measurements on a map trying to guess how long it might take to move along an imaginary line between geographical points A and B. Therefore, what the 20,000 km plus route taken by people that eventually arrived in Australia was, or what time that took to cover, is largely impossible to know, except there was an ocean crossing at the end and no other modern human dispersal took that step. Like Lower Pleistocene hominin journeys, we see modern humans at the beginning of their journey in Saudi Arabia but then the view is very misty till they emerge in South Asia, more mist, then China and eventually Australia. Although that statement is not strictly true, as we shall see, we do have large gaps in those journeys that need filling in particularly in the approaches to South Asia and from Southeast Asia to Australia. However, from what has been discovered so far, there is a little better understanding of modern human journeys than those of early hominins as we see glimpses of those journeys in several places between their start and finish. So, what do we know?

Band movement must have been sporadic and composition variable. They must have ranged from small to medium in size, perhaps they kept in contact because travelling in isolation was an invitation to group extinction. There is security having neighbours, hunting is easier and more successful and the group has a better chance of replacing lost individuals. Contact between bands could have been constant, intermittent or non-existent but each of these implies an increasing element of danger, respectively. Although they took somewhat longer journeys, essentially modern humans were not really doing anything new, they were just following in the footsteps of more ancient hominins like *H. erectus* or perhaps *H. habilis*. But there are also many parallels between the two journeys. Both groups emerged in Africa and both took advantage, although unaware, of vast climate cycling that sometimes provided a way for them to cross over continents and other land masses and connected them with places they could never have visited via lowered sea levels and land bridges. Therefore, those same changing environmental conditions also opened and closed migration corridors. Although we are not so sure about the routes taken by *H. erectus*, modern people probably took the same routes, generally through the Middle East and across the Saharan, Sinai and Arabian deserts following rivers that flowed across them at those 'sometimes'. Those places were impossible to cross during glacials as well as in some interglacials, they did not always spark silent and often buried river beds into frenzied flows as they did during time of greater rainfall and that did not always take place in an interglacial. One of the problems for *anybody* making the journey within the last 2 million years was that they required 'optimum conditions'. They included water in deserts and lowered sea levels on coasts, but that combination happened only rarely because they are normally out of phase: in interglacials and glacials, respectively. Passage across what we now see as the vast regions of the Sahara and Arabian dessert required freestanding water and rivers and lakes. When they appeared so did vegetation and game. That combination opened corridors for migrant hunters, but they were irregular and when they did happen they were separated by tens of thousands of years. Therefore, successful African exits only happened infrequently. There is also the question of not just when the first African exit took place but when the first *successful* exit took place? Just as for the earlier hominins, it seems more than one exit of modern humans occurred and it is likely many of those did not make it very far, although eventually some made it all the way to Australia.

Besides the possibility of a very early modern human exit, the three sub-phases b, c and d in Table 3.1 cover all exit times. Those times depend on the DNA and archaeological evidence which has been somewhat out of phase together with the number of exits made, which not all researchers agree on, and the subsequent direction taken. At first glance, it seems early hominins and modern humans had only one way to go—north along the Nile river corridor then east. Moving to Africa's extreme northwest and crossing the Straits of Gibraltar was possible placing them directly into western Europe but like Neanderthals passing in the opposite direction, there is no evidence for it. Therefore, there are two possible exits, one in the north and another in the south and both were taken at different times. However, the assumed origin of the earliest modern humans is in Ethiopia placing them next door to the Bab al Mandab Strait. Lowered sea levels during glacials would have helped them cross there but environmental conditions in the receiving area would have been poor. In fact, during glacials the Arabian Peninsula can be expected to have been more arid and often more expanded, if that is possible, than they are at present. Nevertheless, the most parsimonious idea is that the first modern human exit took place right there, in an interglacial and there is some evidence for it.

In MIS 5e world temperatures were as much as 5 C higher than at present changing climate across the region and altering monsoon behaviour. Sea levels were also higher than at present. The timing of and required conditions needed to exit Africa and reach Asia are confusing. Present evidence points to an earlier exit at around 120–110,000 years ago which, for reasons that remain unclear, did not go farther than the Levant and can be called the first exit. That did not require a sea crossing but a second exit then seems to have taken place 70–60,000 years ago and that probably did, both at the beginning and at the end of the journey. But on either occasion optimum conditions of environmental suitability and low sea levels were required. Using the three major river systems that crossed North Africa and several others that flowed across Arabia west to east enabled modern humans to journey north and northwest across the Sahara to the Mediterranean Sea coast and cross the Arabian Peninsula.

I might suggest at this point, the first modern humans to leave Africa almost surely did not continue all the way to Australia. In fact, it could have taken so long to reach there that it seems extremely unlikely any of those first groups to do so were the same people that left Africa. It depended on their speed of travel that in turn depended on how many stops and detours they took. It is more likely that the journey to Australia comprised a series of movements over an extremely long time involving many generations, meaning that the first Australian arrivals had been out of Africa for a very long time. Generation after generation could have gone by as they slowly and convolutedly moved through the Middle East, into South Asia, down into Southeast Asia and then through the many islands of the Indonesian archipelago.

Northern Exit Route

The precise timing of the first modern human exit from Africa is still debated, but assuming modern humans began their out of Africa journey in the Ethiopian region, the only exit points are north via the Nile Valley or perhaps moving farther west to take riverine corridors that flowed across the Sahara at that time to the Mediterranean and then moving east along that coast. The other is a direct southern crossing to the Arabian Peninsula from Ethiopia to Yemen via the Bab al Mandab Strait. The way north could follow either side of the Nile river towards the Mediterranean 2000 km away. Travelling on the eastern side avoided crossing the Nile's broad delta, but that obstacle was unknown to them. Crossing it, they faced the twin dangers of strong currents in the wide river and large Nile crocodiles (*Crocodylus niloticus*), that infested the river and its delta. Reaching the Mediterranean coast, they could move east or west, depending on which side of the Nile delta they found themselves. Travelling across the Sahara from Ethiopia between 190 and 135 ka (MIS 6) was impossible in its hyper-arid condition. An earlier crossing could take advantage of much better environmental conditions, just one example of how exit timing is controlled by environmental circumstances. Moving east from the Nile led to the Levant by following the eastern Mediterranean coast keeping close to the coast where fresh water and food were more likely to be found. Eventually they reached the Levant and that is where some of the earliest skeletal evidence for modern humans outside Africa at Misliya Cave in Israel (177–194 ka) has been found.

Another route that could be taken after moving through the Nile corridor was a direct crossing of the Gulf of Suez during times of ocean low stand. Any sea level drop of more than −70 m turned the gulf into a dry flat plain 16–25 km across even at its southern end. That crossing would mean our travellers would bypass the Levant and put them in the Sinai Desert. They could then move north following the western side of the Gulf of Aqaba and it was possible to make another 6-km dry-foot crossing to Titran Island at the southern end of that gulf and open the way into the Saudi Arabian Peninsula that stretched southeast and south. These crossings were possible only during glacials but desert areas were not necessarily welcoming at those times. However, that probably did not deter modern humans from undertaking a similar route far to the south.

Southern Exit Route

An alternative exit from Ethiopia was a more direct route to Asia through the Bab al Mandab Strait. During low stands, this narrow neck at the southern end of the Red Sea almost closed. Today, during an interglacial, the strait is 29 km across but glacial sea levels closed opposing coasts to within 4 km and a 15 m maximum water depth. There are contrasting views about the presence of a complete land bridge, however, with some suggesting there has not been a connection since the Miocene (Fernandes et al., 2006). Others propose one must have occurred from time to time for very short periods and probably played a role in the dispersal of Hamadryas baboons (*Papio hamadryas hamadryas*) from northeast Africa to the Arabian Peninsula, sometime between 220 and 86 ka (Beyin, 2011). Lacking any rafting skills, the baboons either made a dry-foot crossing or swam. Either way, that date range encompasses most of the time when modern humans left Africa and if baboons had no difficulty crossing neither would they. Coincidentally, a large section of that baboon migration window spans the entire 60,000 years of the MIS 6 glaciation (190–130 ka) when seas were at their lowest and so there is no reason humans could not have made an earlier crossing than 130 ka as they seem to have done in the north.

Further Along the Southern Route

There is a small lag time between the end of a glacial when polar ice melts and seas begin to rise right round the world as the earth warms and becomes wetter. That would have presented an ideal window in which to cross to and continue east along the southern Arabian continental shelf. The ameliorating climate possibly formed narrow corridors that contained enough resources to offer bands opportunistic passage along the Yemen–Oman coastline. The width of coastal shelf exposure was not even, however, there being only three places that gained substantial shelf exposure along that coastline. That exposure for each measured about 1540 km long with varying widths of between 30 and 70 km maximum at −100 m low stands. They existed along the first 250 km of the Yemeni coastline and along the Oman coast. Freshwater soaks could be found along the exposed southern Arabian shelf, forming refuges with localised pockets of vegetation making very suitable camping places (Faure et al., 2002). They would attract animals with humans following and act as staging posts forcing a slow forward movement and temporary settlement as travellers moved from one to another. Behind the travellers rising sea levels from a strengthening interglacial made any thoughts of a return to their African homeland (if they had any) increasingly difficult without suitable watercraft as the Bab al Mandab Strait slowly widened. Besides watercraft, one thing not found in southern Arabian archaeological sites, however, is the skeletal remains of the travellers unlike much farther north in the Levant.

Stone tools reflecting a north east African typology, and associated with modern humans, have been found by Hans Peter Uerpmann at Jebel Faya on the eastern end of the southern Arabian Peninsula dated to 128 ka (Armitage et al., 2010). It lies in Sharjah, UAE, 52 km from the Persian Gulf and close to the straits of Hormuz and other points on the coast of the Gulf of Oman. At that point people were 2000 km from the Red Sea and 100 km from South Asia, and when people lived at Jebel Faya, in the MIS 5–6 transition, sea levels were low enough for a walk across the Persian Gulf (EPICA 2004, 2006). There is no firm indication who lived at Jebel Faya, nevertheless, the date and artefacts means many things. Firstly, people left Africa before 128 ka, possibly hundreds or even thousands of years before, and at that date the likelihood is they were modern humans. Secondly, the site shows they were not only well out of Africa by that time, by both northern and southern routes, but had moved on and were poised to cross into South Asia. Drying of the Persian Gulf to a point just south of the Straits of Hormuz left the way open for people living in the Jebel Faya region to walk to the Makran Coast (today's Iran) from south of the Musandam Peninsula. After a full rise to slightly above modern levels, around 114 ka, another high stand began in the very cool MIS 5d stadial (114–102 ka) allowing those at Jebel Faya and other sites on that side, another opportunity for a dry-foot walk to South Asia. A flat plain lay before them beckoning them into lands that have been termed 'savannahstan' (Dennell and Roebroeks 2005; Dennell 2010). What lay ahead was indeed a savannah with coastal mountains inland and aridity. This was the path to India but it could have been a very harsh pathway when it changed into a full desert landscape as moisture levels dropped and that might be expected during a glacial when sea levels were low. The way forward was land only, except for crossing, or trying to move around, large rivers. The next sea crossing would not be encountered till the final leg of their journey to Australia. That crossing would require all the ingenuity the travellers possessed including water craft, it was no dry-foot crossing.

Jebel Faya's age places it close to the beginning of the MIS 5 interglacial. That was the warmest interglacial since MIS 11 (429–393 ka) and seas rose up to six meters above those of today. MIS 5 is divided into five substages (a–e) consisting of a series of highly contrasting climatic phases from very warm to almost full glacial conditions resulting in very variable temperature swings and moisture levels across the region as well as accompanying sea level fluctuations. MIS 5e was the warmest when coastal as well as inland enviro-climatic conditions were favourable for moving through the region. Speleothem evidence from southern Arabian caves has joined results from megalake studies in western Egypt and the Negev and Nafud deserts showing high rainfall across North Africa and the Middle East during MIS 5e (Table 4.1). This signaled opportunities to moved inland and across the Arabian Peninsula particularly

TABLE 4.1 Palaeolakes and Their Dates from Africa and the Middle East

Speleothem and Megalake sites	Marine isotope stages						
Area	Lat°	9	9–8	7	5	5–4	References
Southern Negev Desert Israel	30°N	v	350–290	220–190	142–109	–	Vaks et al. (2010)
Northern Negev Desert, Israel	31.5°N	–	–	200–150	137–123	76–25	–
Nafud Desert, Saudi Arabia	28°N	410	320	200	125	100	Rosenberg et al. (2013)
Mukalla Cave, Yemen	18°N	–	330–300	245–230 209–195	130–123	105–100	Fleitman et al. (2011)
Mudawarra, Jordan	29°N	–	–	7a–6e	5e, 5c–a	–	Petit-Maire et al. (2010)
Hoti Cave, North Oman	23°N	–	330–300	v	130–120	82–78	Fleitman et al. (2003)
Murzuq Basin, Libya	25°N	–	380–290	220–190	142–109	–	Geyh and Theidig (2008)
Rossing Cave, Namib Desert	20°S	420–385	–	230–207	120–117	–	Geyh and Heine (2014)
L. Tushka Western Egypt	21–26°N	–	320–250	290–140	155–120 110	90–65	Szabo et al. (1995)
Kharga Oasis, Egypt	–	–		Speleothem formation	Speleothem formation	–	–

in the Hadramaut Mountains of Yemen and inland along the coastal regions of Oman. Higher moisture levels resulted from changes to the Indo-Arabian and East African Monsoons that brought rain into and across the Saudi Arabian Peninsula. The result was a series of fresh water streams and lakes forming 'freshwater highways' (wadis) enabling a movement across the centre of the Peninsula that is today arid desert. Settlement sites discovered around palaeo-lakes, such as Jubbah palaeolake in the Nefud desert, provide stark testimony to penetration of the interior by modern humans during MIS 5 and earlier by their direct forbears during MIS 7 (Petraglia et al., 2012; Crassard et al., 2013). Much wetter times had occurred hundreds of thousands of years before when earlier hominins crossed these same regions (see also Chapter 3). The demographic consequences for different hominins groups over time, directed by glacial–interglacial cycling, have left a confused and complex picture of hominin history across the peninsula to sort out. For many years Arabia was not considered part of the discussion of human evolution and spread of humanity but that has changed. It is now obvious that it was a crossroads for dispersing modern humans as it was for earlier hominins.

Northern Route from a Southern Crossing

Besides moving east after crossing at Bab al Mandab, it was also possible to go north following the Red Sea coast to the Gulf of Aqaba. Continental shelves along that coast extended out up to 250 km either side of the Red Sea when seas were below 80–100 m drowning camp sites used as that time. The problem of camp sites below what is now ocean, has made tracing and sequencing modern human journeys very difficult in places where they and others before them moved along coasts or crossed dried out sea beds now flooded. During MIS 5, continental shelves were generally under water but there were several places along the eastern side of the Red Sea where it was possible to turn east into the Saudi Arabian hinterland following rivers flowing across the Peninsula. A complete crossing was possible if those rivers joined others flowing to the Persian Gulf allowing a crossing of the Arabian Peninsula. That route also took people far from the Levant, drawing them towards South Asia and removing the choice to move east or west. The most likely route that would take people to the Levant was by sticking to the banks of the Nile or the western side of the Red Sea and ending up on the Mediterranean coast.

That seems to have been the route taken by the modern humans at least 180 ka also others found at Mount Carmel in Israel. Investigations there began between 1923 and 1926 and excavations in the Qafzeh and Skhul Caves revealed a series of skeletons, the oldest dated to around 110 ka. Other Mount Carmel caves such as Tabun, Kebara and Amud were occupied by Neanderthals much later around 70–50 ka, but they had been living in the region on and off since 150 ka at the height of the MIS 6 glaciation. Moreover, it may have been that very long, cold glacial that sent them south out of Europe to find better conditions, but why they had not done that in a previous glacial is baffling. Moving south, they may then have become isolated from other Neanderthal bands in Eastern Europe, particularly if their population was a series of islands separated from each other. The wider European population may have been sparsely distributed with slowly reducing band sizes and a rising frequency of isolated bands. We assume the Levant was empty of humans when Neanderthals arrived and occupied the variety of caves they found but the evidence now seems to suggest that they and modern people arrived then left at different times inhabiting the area for different lengths of time and possibly missing each other in most of cases.

One Qafzeh skeleton is a well-preserved, fully modern female who possibly died in childbirth, not an uncommon occurrence in those times one would think. In contrast, the cranial remains from Skhul Cave present a variety of features, particularly those of Skhul 4 and 5. They include a long low head (Skhul 4) and both have prominent brows, features reminiscent of Neanderthals. However, the large fully oval cranium and prominent chin of Skhul 5 suggests a modern human with rugged features. That confusing morphology has led to suggestions that upon reaching the Levant modern humans were not moving into a human-empty world as their ancient hominin predecessors had done 1–2 million years earlier. Neanderthals possibly occupied the Levantine caves sporadically and this is where they and modern humans met up for the first time but Misliya Cave evidence suggests modern human reached the area before Neanderthals did. The region then presents a very confusing but interesting picture. One present evidence it would seem first came modern humans, then Neanderthals then more modern humans. So, it is not surprising people with a mixture of features from both peoples have been found at Mt Carmel. The exact timing of the meetings is unknown and subsequent occupation of the area by the two groups presents a complex picture, because modern humans seem to then disappear for about 20,000 years. Whether they were outcompeted, killed or genetically assimilated by Neanderthals or descendants of the mixed population is unclear but the mixture of robust and modern cranial features on some individuals suggests the latter may not be fanciful.

Meeting the Relatives

Can we imagine the first meeting between modern humans and Neanderthals? Did modern humans move in silence as they slowly approached, both peoples stopping within metres of each other or did they begin an immediate skirmish perhaps sparked by surprise at the sudden encounter and who was shocked more? One or other or both may have observed each other secretly for many days from far off. Listening and watching each other's fire smoke and habits before precipitating the final meeting, knowing that this moment would come at some time. Perhaps a signal fire was deliberately lit to reduce the possibility of the sudden, unexpected meeting. I can only think that it was a shock to whomever saw the others first whether they were previously observed secretly or not.

The hunting skills of either party may have played a large part in any prior observation or tactics used to reduce the disquiet and that required the upmost secrecy and stealth—which of them won this quiet battle of skills? Probably neither; the highly honed senses of sight, hearing and smell of both parties make it impossible to believe that they did not know of each other's presence days, even weeks prior to a final meeting.

Neanderthals were generally much like modern humans but different enough to make them look different from the new arrivals. It was a meeting between tall, slim people and stocky, short, and powerfully built people. It was a facially gracile people with oval heads facing those with robustly constructed broad projecting faces, prominent browridges, large noses and long large heads containing, on average, a somewhat larger brain. Was the larger bodied, big limbed and powerful Neanderthals seen by the more delicately built moderns as terrifying or clumsy when they walked or hunted, who knows. It was also the first meeting between people of different colour: when dark or black-skinned moderns met lighter skinned people with different hair type and colour. While Neanderthals were lighter skinned, it was a skin that was very weather beaten, leathery and tanned and that would have toned down the difference in appearance. Moreover, their skin was covered in years of grime, smoke stain, dried animal blood and grease that camouflaged its colour further. And on a more down-to-earth note, it is more than likely that their keen sense of smell would have detected the other's odour well before either group spotted the other. They would have dressed differently, and there were differences in the weapons they carried but smaller differences in the stone artefacts they made. Children in both groups would have been frightened as well as curious seeing people and children of the other group and perhaps some adults were nervous also. Their language was certainly different which always adds to the suspicions of one group about another. Not only was it a meeting of physically different people, it was an historical meeting because for the first time they must have both realised they were not alone after all, the new people were not like them and to make it worse they could not communicate properly.

Neanderthals and modern humans shared a common ancestor possibly 350–450,000 years ago and now they were meeting up for the first time since then. Whatever form the meeting took both groups occupied the same Levantine cave systems for many years and skeletal evidence from the area suggests they became quite friendly. There were certainly Neanderthal females living there at the time the first modern humans arrived because one has been found in the nearby Tabun Cave (120–100 ka). The fact we carry Neanderthal genes shows we did mix with them in some place at some time. It shows also that even after hundreds of thousands of years of separation we had not become strictly speciated. Apart from some suggestion of male infertility that was possible but unproven, we seem to have apparently benefitted from some aspects of their genetic legacy particularly our immune system. It would indicate also that meetings and matings between modern humans and other archaic lineages may occurred elsewhere without any or very few negative consequences. Nevertheless, the meeting begs the question of how far back a division between two hominin lineages can go before speciation takes place.

A Choice of Routes

Modern human exits beginning in central Ethiopia had two possible routes to follow (Fig 4.3, *large black dot*). One was across the south end of the Red Sea (Fig 4.3A), which is most likely way the Jebel Faya people as well as later groups found in Dhofar went (see below). The other was in the north after traversing the Nile corridor (B). The southern crossing allowed movement directly east, either following the southern Arabian coast (C) and skirting east of the Hajar Mountains (D) or moving inland during wetter periods and following tributaries to Wadi Hadramaut (C) that led east towards Aybut al Auwaj nearer the southern coast.

Meantime, the southern crossing at Bab al Mandab not only allowed a direct move east, it offered the choice of a second route, leading north following the

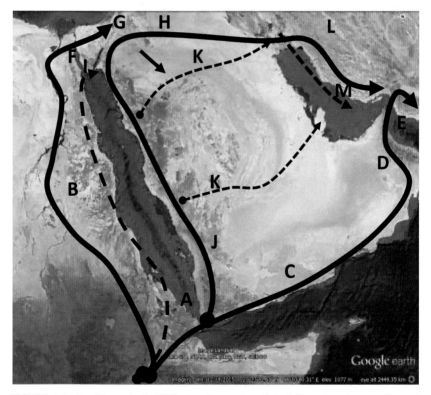

FIGURE 4.3 Possible main (*solid line*) and secondary (*stippled*) modern human dispersal routes out of Africa and through the Middle East.

eastern side of the Red Sea coast (J). Continued eastward movement eventually led to a dry Persian Gulf during glacials and an easy crossing to the Makran coast of Iran marking entry to Asia (E). Following the Nile corridor or Red Sea coast to its end, during raised sea levels, meant continuing to the Gulf of Suez (F) and the Mediterranean coast before a right turn led them towards the Levant (G). However, a dry Gulf of Suez and Red Sea contraction could have meant a crossing at the north end of the Red Sea to the southern tip of Sinai (I). It was then possible to cross the mouth of the Gulf of Aqaba opening the way into Iran by crossing to the north of Arabia (H–L), or an entry into central Arabia (H *arrow*). The (H–L) routes also mean tackling the Tigris–Euphrates river and that may have required some further directional change possibly towards the northeast if they needed to avoid the river. Alternatively, route H–L took humans through the Tigris–Euphrates valley north of the Persian Gulf and funnelled them south between the Zagros Mountains on the left and Persian Gulf on the right (M, *dotted line*).

During wet times, rivers, lakes and marshlands formed across many areas of the Saudi Arabian Peninsula. Moving east was possible from places along the northern route following streams that ran during the MIS 5 interglacial and several times since then (K). Principle among these streams, north to south, were Wadi al Hamd, Wadi ad Dawasir and Wadi Hadramaut, all draining roughly west to east. Wadis Sahaba and ad Dawasir drained from the mountainous western Arabian Peninsula and flowed towards the Persian Gulf and after traversing the Rub Al Khali desert it exited where the present border of the United Arab Emirates lies and flowed into the Persian Gulf. Likewise, to the North, the Wadi al Batin drained from the mountains in the west and flowed to an exit point where Kuwait City now lies, contributing to any water from the Tigris–Euphrates river (Breeze et al., 2015).

Continuing north and crossing the Assir Mountains running parallel to the eastern Red Sea would bring people into contact with these handy highways and palaeolake stopovers. The channels and lakes formed stepping stones and 'roads' for dispersals during incursions of the Indian Ocean summer monsoon into southern Arabia and when the East African monsoon was activated, both bringing wetter times to west and north-western Arabia. Journeying would always depend on environmental and climatic conditions at the time. Of course, it is possible to move north of the Gulf into Iraq even during high sea levels (L), or south along its dry bed (M) at other times. The main Tigris–Euphrates river in full flow might have proven an obstacle depending on catchment rainfall but to cross it journeying up the waterway to find the headwaters could have presented a very circular route for people to move into Central Asia and, indeed, would have placed them in close to Turkey.

Routes shown in Fig 4.3 show only a few possibilities, if all possibilities were shown the map would be covered in lines. Routes were controlled by environmental conditions that fluctuated constantly in cycles measured only in decades or a couple of centuries at most as well as the very long term glacial cycling. That presented opportunistic pulses of travel for onward dispersal. But Fig 4.3 does demonstrate that the Levant need not necessarily be on the itinerary of travellers taking eastbound routes and could be easily bypassed. It could be avoided just by choosing directions H and L and that also avoided entry into Europe. People could and did move in many directions without ever entering the Levant and the perceived gap in modern human movement in that direction may be a figment of enviro-climatic opportunities that guided modern humans elsewhere and into a different outcome for their dispersal pattern. Moving along the Nile corridor during high sea levels was the only route/time that was likely to place people in the Levant without much choice.

There were many times when the Bab al Mandab strait was difficult to cross, and so were the Gulfs of Suez and Aqaba. Any crossings earlier than 132 ka could only do so at the tail end of the MIS 6 glacial, although internal travel through the Arabian Peninsula would have been difficult with harsh conditions prevailing. However, while sea crossings were difficult during high stands, internal environments were generally more habitable thus producing a series of stop–start dispersal movements. Using these conditions to predict movement opportunities, however, is not always possible. Detailed knowledge of the environmental impacts of glacials is difficult and complex because of many unknowns including band survival strategies and exact interglacial enviro-climatic patterns. The AE4 *late* dispersals would, almost certainly have been caught up in these. For example, the MIS 5 interglacial should have been an optimum time for modern human dispersals but it is not till its component stadials and interstadials are examined that we see opportunities for and barriers to movement even at a coarse level. The patterns and reasons for the exact way modern humans dispersed can only be speculated upon. However, the dispersal routes shown at H, L and E in Fig 4.3, might offer a tentative explanation of a couple of mysteries at least. The first concerns why they arrived in one place but not another. For example, why did modern humans arrive in Europe so late? The other is how did they reach East Asia so much earlier than Europe when Europe was on Africa's doorstep metaphorically speaking. It seems likely that what is shown here might indicate that modern humans may not have always 'chosen' a route but the route 'chose' them, directed by what was opportunistically available and that was a southern route that moved them east.

The answer to the apparent disappearance of the AE4 *early* dispersal could be a lack of evidence for them, but another may be that their choice of a route out of Africa was to blame, particularly if that choice was a southern route and we are looking in the wrong place. Another possibility is that they just disappeared, perhaps through encountering Neanderthals at Mount Carmel with unfavourable results. Perhaps their band size(s) was small hampering their ability to survive natural vicissitudes as well as large and/or more capable Neanderthal bands better adapted to a region where they had lived for some time and who were tougher and more numerous. The Neanderthals were semi-sedentary, in the loosest terms, as localised sitting tenants in the region. The original arrivals may have had larger bands, and without backup dispersals of modern humans a situation could have arisen whereby modern humans could be left outnumbered and vulnerable, either through mixing with indigenous Neanderthals, being outcompeted or murdered by them or a combination of all those possibilities. However, it is tempting to suggest that Skhul 5 may indicate that not all contacts were aggressive. If correct, it indicates modern humans did not keep to themselves and they had no aversion to closely associating with others that did not look exactly like them unless women were taken by force by either group. Whatever the reason, modern humans were stopped in their tracks.

Modern humans would venture no further at that time except as a mixed modern-Neanderthal people with their genetic component contributing less and less each generation. If we assume they became part of the Levant Neanderthal population, then it could look as though they disappeared particularly with the possibility that Neanderthals or Neanderthal–modern human mixed people, retreated north away from the Levant in response to the warming that MIS 5e (133–114 ka) brought. However, it is possible that if modern humans did not enter the Levant for so long, they were choosing another route out of Africa, perhaps the southern route. That pointed them towards Asia and could have put them in China as early as 120 ka or even earlier and, as we will see later in the chapter, it seems that is what might have happened.

We cannot ignore later travellers of the southern route that suggest it was a viable route of travel and one favoured by those move out of Africa. Archaeological evidence from Dhofar in Oman (mentioned earlier) dates to 107 ka. This evidence suggests later dispersals took place along the southern exit route replicating the journey of those that reached Jebel Faya (Rose et al., 2011). However, they were following on 20,000 years later and doing it during the cold MIS 5d stadial, a time of drier conditions and somewhat lowered sea levels. A stadial is usually a cooler phase within an interglacial, but the MIS 5d stadial was more than that, it was severe. It contained temperature reversals ranging from just above to full glacial conditions (taken here at an average world temperature -4 C below today's world average), and the environment no doubt reflected that. EPICA data from Antarctica ice core records show that over 15,000 years (123–108 ka) average world temperatures plunged 10 degrees from $+3$ C in MIS 5e to -7 C in MIS 5d, passing into full Ice Age conditions around 113 ka (EPICA Community 2004, 2006). Maximum ocean low stand was reached around 108 ka providing a chance for dispersals to again cross the Bab al Mandab strait and perhaps that is what the 107 ka Dhofar date reflects. A 4-degree warming occurred by 106 ka before another cooling around 103 ka took temperatures again to -7 C before rising to -4 C again around 101 ka. These are comparatively overnight changes in world climate, over 8000 years and they must have brought equally radical, rapid environmental changes and challenging times for travellers reinforcing their adaptive strengths and broadening their generalist survival strategies. It might have been so that such an enviro-climatic roller coaster was also a catalyst for people choosing one way out of Africa over another; forcing or constraining exit timings and determining subsequent dispersal routes.

The Dhofar evidence indicates a southern route out of Africa was not only favoured by then, it was the main exit choice. Again, fluctuating conditions make human migration patterns difficult to predict. The Qafzeh Cave evidence suggests a modern human dispersal took place between 95 and 90 ka following a northern exit either through the Nile corridor or along the Red Sea's west coast. From 110 ka until the Holocene, sea levels were always lower than today fluctuating between −30 and −80 m, although trending below −50 most of the time after 80 ka, meaning humans would be greatly assisted to cross the Bab al Mandab during that window of opportunity. Qafzeh people left a little later than those camped at Dhofar but both could have crossed the strait when sea levels were around −60 m. The evidence from these sites might suggest modern human bands were taking northern and southern routes out of Africa almost simultaneously with perhaps only sea levels forcing the choice. But it is interesting to think those going north would have been positioned towards the Caucasus and Europe while those in the south were headed for South Asia and the Far East. However, while we see only these two routes, moisture levels probably determined other possible routes forging random dispersals, some successful others not. At present, it seems the earliest modern humans attempting a northerly route out of Africa were successful to a degree, but only as far as the Levant. The perceived disappearance of modern humans between 45 ka when we know they entered Europe, and 100 ka, may indicate either failure to continue, for reasons given above, or, for some reason, they just favoured a southern route along the southern Arabian coast. Perhaps it was not till later, when more or larger bands of modern humans moved through the Levant, that they could overcome any obstacles they met there such as the dwindling populations of Neanderthals. The apparent late entry of modern humans into Europe almost 90,000 years after they reached the Persian Gulf is another puzzle that lingers over the use of the Levant corridor as a migration route. Just a last thought about this and the Neanderthals. They may have stopped modern humans entering Europe till around 45 ka but they were gradually disappearing and were slowly unable to present any sort of defensive force. Like the megafauna they knew so well, they were slowly becoming extinct over several glacial cycles, but in the end the last few rapidly succumbed because they had crossed a threshold of survival in terms of dwindling populations size and viability against the hardship of survival as a new Ice began. Is it possible that the late entry into Europe by modern humans was a factor of Neanderthals presenting a barrier until they began to seriously lose numbers as they slowly went extinct, like the megafauna and for the same reasons: a small overall population that slowly fragmented into inefficient, unviable and non-sustainable groups, scattered and losing touch with each other; inbreeding and losing those who knew how to survive. Did modern humans carry diseases Neanderthals had never experienced before?

'SAVANNAHSTAN' AND SOUTH ASIA

Crossing the straits of Hormuz was not the only way into Iran. Under the right environmental conditions and during times of lowered sea levels, a route through northern Saudi Arabia, Iraq and Iran was also possible, particularly for those moving southeast from the Nile corridor or travelling along the eastern edge of the Saudi Peninsula. Once in Iran they could follow the eastern Persian Gulf and Makran coasts at times of high stand seas, or along the emergent Persian Gulf when they were lowered. Both routes offered a way into 'Savannahstan' and east. Sea level timing was the prime determinant for the initial routes taken at the exit from Africa and for other steps along the way to South Asia but the next stage did not rely on them so much. We have little evidence for modern humans between the Gulf of Hormuz and South Asia, they disappear. We do not know where or when they are there and that makes it difficult to place or track them and the environmental conditions they faced crossing the region. Although we do not see them, some must have traversed the region during MIS 5. The radical environmental changes during the five substages of this interglacial could turn dispersals into pulses governed by enviro-climatic switching that could have brought progress to a halt. Modern humans crossing north of the Persian Gulf and continuing east and inland would reach central Iran. It meant a slim possibility of meeting Neanderthals especially if modern humans moved northeast which brought them into contact with those already living in eastern Uzbekistan. But following the Euphrates river to its headwaters increased the chances of a meeting with their heavily built cousins.

Crossing the straits of Hormuz and moving along the Makran coast is the most likely route modern humans took to reach South Asia but there is no real evidence they did at present. The Makran coast presented a similar environment to that of coastal southern Arabia. Whether fresh water springs were present as along the southern Arabian coast is unknown. But the modern Iranian and Pakistan coasts constituted a better corridor to follow than chancing their luck moving across the harsh, mountainous inland, particularly under hyper-arid glacial or stadial conditions. An exposed continental shelf and the possibility of refuge areas for limited settlement may have existed along the Makran coast as they did along the exposed Yemen shelf. As discussed in the previous chapter, meeting the Indus brought two choices, cross it or move up its valley. It is possible that modern humans entered India in their earliest dispersal during MIS 5e between 128–108 ka possibly at a time when archaic hominins lived there (Petraglia et al., 2012). Evidence for that comes from 16R dune near Katoati in India's central northwest. There has been some debate whether the Katoati site dated at almost 130 ka is a modern human camp or that of a previous group of (archaic) hominins. Now, with the recent discoveries in China of the presence of modern humans at least 120 ka, the Katoati date of almost 130 ka looks more like a modern human campsite and a stage in their successful move further into Eastern Asia.

DNA evidence shows Neanderthals may have begun their spread east only around 150,000 years ago but it seems unlikely that it was that late. There is reason to believe they could have had moved east at any time in the previous 100 or 150,000 years. Others were certainly making their way east a lot earlier. What the discovery of the Denisovans and the enigmatic *Homo floresiensis* has done is to highlight two very important questions: who else is out there to be discovered, and what else has human evolution in store that will alter our way of thinking about human evolution? The 'Hobbit' discoveries show the uncertainty surrounding human evolution at the time of modern human dispersals particularly during the earliest phase. The earlier modern humans dispersed the more likely it was that archaic people lived somewhere along their pathway who they could have met and interacted with, what that might have meant for them and their onward journey is unknown but contacting people who were totally unknown to them must have had a similar impact to the Levant contact, and that could have presented consequences for their ongoing journey. It was not so long ago that we considered that 100,000 years ago it was just Neanderthals and us but this is no longer the case. We know about the Denisovans, their distribution and connections with other people, archaic and modern. We can add to that our ignorance of what Maba and Dali might represent and their relationship to earlier Chinese *H. erectus* fossils (if any) and what archaic *sapien* forms played in the complex tapestry of human evolution in China. Who, for example, is Penghu man? Moreover, it seems we are still in the world of guessing when putting our story together, because of the many possible outcomes when and if we met up with these archaic populations. Archaic people are very likely to have lived in India when modern humans arrived and that would be another archaic group to factor into the equation of modern human dispersals and their outcomes in the region and, as usual, we know almost nothing about them. And what of the possible combinations of interaction between archaic hominins alone? The story of modern human dispersals through South Asia is still hardly known although there are tantalising clues and exciting prospects. There are many Middle Palaeolithic sites across India but most remain undated, although some of the stone technology found on them has affinities with that in the Arabian Peninsula and even North Africa (Blinkhorn et al., 2013).

A somewhat later site than 16R is found at Jwalapurum in the southeast of the subcontinent, that shows at least two occupation phases (80–74 and 50–38 ka). It is a modern human site but importantly its phases bracket the eruption time of the Toba supervolcano showing modern humans neither suffer the devastation of the eruption to the extent that has been claimed nor did it cause a population bottleneck that was also claimed to have brought about drastically reduced animal and human populations (Ambrose 1998; Haslam et al., 2010, 2012). The geographical position of Jwalapurum also indicates modern humans moved well down into the subcontinent, not necessarily taking a coastal route around the subcontinent but exploring within it. Again, it demonstrates the opportunistic strategies employed by modern humans: if it's possible to go exploring inland they did and why not. That issue is important when we come to the exploration of Australia by its first people. It is interesting to speculate that this type of exploration pattern may also have occurred in Australia adding another explorative strategy to that of the favoured coastal colonisation model that has been widely proposed for years (Bowdler, 1977). On the other hand, central India is much more inviting than central Australia which was harsh and arid most of the time. However, could such behaviour have implications for other areas, especially when considering direct coastal routes and the often-proposed avoidance of inland regions? Late Achuelian tools found in the Jurreru and Son valleys in India shows that earlier archaic populations had also carried out the same type of inland exploration pattern: going inland when conditions were good enough.

Penetration of India's Thar desert by modern humans is another intriguing prospect of their story in South Asia. Intermittent use of the desert occurred when climatic conditions were reasonable between 95–40 ka (Blinkhorn et al., 2013). Desert travel is similarly reflected among modern humans in the Arabian Peninsula and in North Africa under moderate conditions. There is a clear behavioural adaptation of being able to move into and take advantage of generally poor environments as they improve even for short periods by opportunistic forays, a tactic that would have been advantageous in their Australian explorations. There is also clear evidence that Middle Palaeolithic stone typology was used well into the Late Pleistocene when microblade technology is then developed/employed locally (Mishra et al., 2013; Petraglia et al., 2012). But rather than bringing such technology with them as they moved into South Asia they seem to have produced their own and that included smaller more delicately fashioned tools perhaps indicating an adaptive ability in style and the use of local materials, which helps to explain their dispersal success as well as their later explorations through South East Asia and across Sunda.

JOURNEYING ON

The 128 ka evidence from Jebel Faya is perched on the brink of Asia. If those people had continued their journey successfully we could expect to see them in South and even later in East Asia by at least 120 ka and, as indicated previously, it seems that is what may have happened. The 16R dune in India (128–108 ka) and several sites in China dated to MIS 5 suggests this, with the Fuyan Cave site expected to be closer to the 120 ka upper date in its range (Liu et al., 2015; Petraglia et al., 2013). That discovery might signal success for the earliest known modern human dispersal across the Old World and even into the New World (Holen et al., 2017). If such an early dispersal was successful enough to take people that far, I see no reason modern humans (unmixed or mixed with archaic people) could not have reached Australia soon after that, perhaps within the 100–80 ka range. However, that might depend on whether they survived as a dispersing group after meeting endemic archaics of which there were several types around at that time. If a later group of modern humans left Africa around 70 ka, as the DNA has suggested, and they took a route down the Malaysian Peninsular, crossed Sunda and directly crossed into Sahul, we might not expect to find the earliest arrivals in Australia till around 60 ka or a little before but that is not the case as the >65 ka date from Madjedbebe in Arnhem Land, at least for the earliest arrival on this continent (Clarkson et al., 2017). Other evidence from South and East Asia and perhaps North America supports that scenario. It also makes the genetic timing of a 70 ka Out of Africa likely to bring unmixed modern human to Australia because most of those were either much less frequent or extinct in the region by then.

Another dispersal into the Levant took place around 100 ka, with modern humans moving essentially north contrasting with the direction the Indian and Jebel Faya evidence tells us. The latter looks like people were poised to move into southern Iran and that was the migratory route taken by those bound for Australia. Several scenarios have been suggested above to reach the Far East and for now we must take our pick of them. But before leaving this section there is a need to mention the fully modern cranium found at Liujiang in China's north central Quangxi Provence in 1958 and dated to 68 ka. What might this person represent? There are only two possibilities for its origin; it could be a modern human descended from previous generations of modern indigenous people that remained unmixed after their arrival, or it might represent a new modern human dispersal reaching China around the time of its death. If Liujiang was a descendent of earlier moderns, an indigenous modern in China and perhaps a descendent of those represented at Fuyan Cave, we must contemplate its ancestors lived beside archaics for some time. Therefore, its very modern appearance could better suggest it represents a more recent arrival when fewer or no archaics were around and it would better match the DNA evidence as part of an out of Africa dispersal happening around 70 ka or a little earlier. Either way Liujiang is living in southern China in the middle of the Late Pleistocene, but is it on its way somewhere or just enjoying its long Chinese heritage?

TO AUSTRALIA

Simply saying people 'came to Australia through the Middle East and South Asia' hides a great many issues and makes the journey farcically simple. It also avoids the many difficulties, issues and possible failures met with by those who accomplished it. The subject of the first Australian arrivals, when they landed and the routes they took to get here is a discussion I have had before both with colleagues and in print, but issues, evidence and minds change in this business as time goes by and fresh evidence emerges. Alternatively, it is easy to become embroiled and lost trying to work in the complexity and minutiae of possibilities, predictions, events and myriad of choices of routes taken in such a journey particularly when it comes to which way humans moved through Asia, Sunda

(Papua New Guinea, Australia and Tasmania) and the myriad of islands that lie between South East Asia and Australia. Moreover, in some respects, without direct evidence, it is only my guess or choice of circumstance versus that of others. The many routes that could be taken as well as their combinations and alternatives as well as the fickle nature of humans making quite different choices from those predicted by palaeoanthropologists and archaeologists, produces an array of confusing possibilities that can confound any discussion of *the* route taken. It is my experience in thinking about such a journey that it becomes almost impossible to discuss it in any meaningful way without evidence and the route can be anybody's guess. So, we either rattle on about things we have no real knowledge of or forget trying to describe *the* route in any other way than in general terms and by just the best we have using the archaeological signposts we have that suggest *the* route taken.

Perhaps then, we should examine the way we think about the issue and there are a few clues we might use. One reason for the complexity is the problem is in our own heads. We can see the many possible ways from our *satellite view* of the world via electronic maps and images and we can measure any route we like willy-nilly and make the most parsimonious predictions for travel. Those first travellers did not have such gadgets so they would do what was within their capabilities and from what they could make out of the world at ground level in front and around them, and make decisions based on what was the most logical next move from that standpoint.

Glacials brought a wide range of changing opportunities for land-based journeys and island-hopping, each bringing environmental change in different ways in different places. The result was that migratory possibilities changed as possible routes opened and closed at different times and in different places. Glacial sea level change formed narrower and shallower sea crossings and joined islands or larger landmasses. Terrestrially, landscape changed as deserts expanded. Smaller deserts joined up; some contracted to become savannahs and then reversed again; rainforests retreated and expanded; cold, high country became glacial; forests came and went; tree lines moved up and down and a massive rainforest grew across the once flooded Sunda shelf that would totally disappear again under rising sea levels but probably existed when people crossed to Australia. All of which could make journeys shorter or longer depending on glacial/interglacial conditions. We have already seen how these alterations provided natural valves, at times allowing travel at other times isolating people and stopping them, it was the same with interisland travel. This type of travel was probably most successful when sea levels were low allowing direct movement south as islands joined up or crossings were shortened, shunting people in different directions. But it may not have proceeded in that way at all. Sometimes travel sped up at other times it slowed down, at other times it stopped. People coped, learnt, adapted, managed and improvised, moreover, it all happened slowly, imperceptibly at its quickest changes took place at an intergenerational pace. Random movement was always a factor where no particular direction was taken except onward in one direction or another, whenever it was thought possible and in a direction that whatever people thought fit or easiest or what suited their purposes and circumstances at the time. Out of such a turmoil of factors and whichever route they chose some ended up in Australia.

Australia and eventually South America were end points in the movement of modern humans out of Africa. Migration through island South East Asia to reach Australia was somewhat different from the land-based movement previously undertaken, and those in the north continued to some extent all the way into the Americas, probably skimming or walking the coast close to the mainland to reach there. For those going south through the tropics, moving great distances across land was now replaced with island-hopping where the use of water craft was at some time essential, the only way forward and that was different again in the repertoire of modern human migration. By the time they were ready to make a larger sea crossing they were probably used to such craft and ready for an ocean crossing. Generation by generation they subliminally took in the skills of operating and building a suitable craft for the crossing as they also subliminally learned the ways of the sea, its currents and how to live off the food available in it. They are likely to have cut their teeth on smaller coastal and interisland journeys through the Philippines or among other islands of the region. They had accomplished the required skills and marine knowledge to use water craft of some kind to reach the northern Philippines by at least 67 ka as the Callao Cave evidence seems to suggest. But it was not very different from the strategy of their kinfolk roaming close to the Arctic possibly with others.

THE LAST CONTINENTS

The arrival of people in Australia and the Americas brought to a halt the first of three major phases of human exploration and discovery. The first was moving across the Old and New Worlds. The second was oceanic exploration particularly of the Pacific Ocean, and the third, is now our move into space and that has hardly begun. The first phase moved humans across the Old World, took them to the other side of the planet and required them to explore in a moderate way with watercraft. After tackling the wide open spaces of Eurasia, islands of various sizes and crossing small ocean gaps and large rivers they arrived on another continent, an island continent, and life for them would be different in many ways. It contained a microcosm of virtually all the world's environments and new animals, but it was also the first of two dead ends for the first phase of modern human dispersals; the other lay at the tip of South America.

Like crossing to the Americas, stepping ashore in northern Australia meant there was no onward journey beyond exploring the continent they had landed on. Indeed, there was much left to cross although not as much as the Americas. Australia was the last island to hop on to, there were no more. The two either side, Papua New Guinea and Tasmania were joined during low sea levels. Indeed, there is the very real possibility some people entered Australia from some point along the New Guinea landmass without a boat. Australia was the biggest island they had hopped on as they crossed from Sunda to Sahul, they were at a dead end. When they moved inland they met a vast, empty open space taken up by a fearful amount of desert and depending when they arrived, there was more then than there is now. None of those intrepid explorers had ever experienced desert and aridity, not unless they had rushed from the Thar desert in India to Australia in a few years. Those that had experienced desert were long dead probably many generations before. They now had to explore a giant land far bigger than the one we know today: 2,500,000 km^2 bigger. There were many new challenges; the biggest was environments ranging from Alpine to rainforest. When they eventually reached the southern and eastern coasts, one look at the high-energy breakers and no land on the horizon, surely convinced them that this was water that could not be entered or crossed. This was vastly different from the comparatively quiet ocean they knew in the tropical north and their skills were not a match for this sea. There were also very different animals, some had two heads; one sticking out of its stomach! Whether they knew it or not, they were confined on a very new continent; this was the end of the line. They perhaps did not realise it but they were setting up one of two of the most isolated outposts of modern human settlement so far, and far from their original homeland. They must have eventually realised that they could go no further. They were not used to dead ends so they kept moving ever onward across and/ or round Australia, or Sahul as it was during times of oceanic low stand. They preceded to the southern most parts of the new continent where, at present, we find many old campsites. Now they could only turn back on themselves or continue along the coastline.

The difference between going to Australia and crossing to the Americas was that it was punctuated with a warning, although that may not have been picked up at the time. To reach Australia required a decent sized sea crossing and that was not necessary when crossing to the Americas, although close coastal voyaging may have been used following the Bering bridge and the Alaskan and Canadian coastlines which has been suggested as an alternative to walking and offering a quicker entry to the Americas. But that was at ocean low stand during an Ice Age. So, they crossed during low stand in freezing conditions, or crossed in watercraft in somewhat warmer times. To reach Australia, however, there was always the need for a deliberate and comparatively long sea journey of perhaps several days or even a week or so, depending on currents, weather and where the take off and stepping off point was. But it was always warm. There was also a time on that voyage when the horizon in front and behind did not contain an image of land that was not the case crossing the Bering Strait.

Groups that were eventually to move to the Americas and Australia parted company somewhere in Asia. Until now, the evidence suggested that the southern migration moved away, leaving others in central Asia possibly around Mal'ta in the Caspian sea region (Fu et al., 2014). Recent DNA research points to a common stock that began there for the later First Nations people of the

Americas who are all related to the oldest entrants. The process must have been dictated by the initial decisions taken, as they entered Eurasia that found themselves moving across Central Asia, and their subsequent life history took them in one direction or another. But what was the process of the conquest of these final two continents?

The journey northeast and into the icy wastes of the Beringia may have taken some time mainly because of behavioural, cultural and biological adaptations they had to make before being able to cope with the oncoming environment and thrive in it. The widely accepted time of people arriving in North America is somewhere between 30–20 ka, suggesting a slow push through northeastern Siberia gathering their adaptive strengths required for high latitude migration. As a contrast, the Australian date of >65 ka would suggest they struck out earlier through southeast Asia towards the warm tropics. Of course, this simplifies the undoubted complexity of the process with its inevitable twists and turns as people moved this way and that and made decisions. Individuals and, indeed, whole groups may have been lost through accidents, carnivores and even disease, but all the time the gap between the two main groups was widening as they moved away from each other on different paths. Again, this is a time to remember the the operation of the *9EM* that were, once again, slowly moulding the two exploring parties to suite their changing circumstances and environments and honing separate, although subtle, morphological appearances. It meant a shorter person to retain body heat, and a lighter skin to absorb minimal Vitamin D from the sparse sunlight and long, dark winter nights of the high latitudes for those travelling northeast. Also, an epicanthic eye fold protecting eyes against driven snow, reflected glare and ice flakes. Going south moulded a slimmer, tall person to lose body heat for cooling, with a darker skin for combating the damaging effects of ultraviolet radiation on the skin in the tropics, but robusticity persisted.

As usual, their chosen paths were almost certainly not deliberately planned because while we can see the result the people did not conceive of an *end* or a 'right way' to go. The northern dispersal would also require more time for the adaptations required than that of people moving towards and crossing the equator. One important contrast between these ventures is that moving south modern humans did not confront cold and the particularly severe hardships associated with a very cold climate. The southern movement took modern humans into the tropics for the first time since leaving Africa, although any similarity with Africa would probably not have lived on in their collective memory, separated as they were by many generations from those who first left Africa. Their collective journeys had included biological and cultural adaptations and possibly they even required a different material culture, tool design and manufacturing forged by regional differences and available raw materials but all these were not very different from what they were used to back in their geographical origin. They may also have lost some skills in artefact production as mentioned above and I propose to outline why below.

Southbound groups gradually and easily adapted to the tropical environment on the way to Australia. No longer were there arid and semiarid landscapes and mountainous regions to cross or skirt around. The eternal search for fresh water became a thing of the past. It was now all around them in the form of tropical rain. Rainforest and tropical savannah were dominant with their abundant fruits, vegetable foods and small deer, rodents, pigs, monkeys and other small to medium-sized game, all plentiful and easy to hunt. During low sea levels, tropical forests and savannahs spread across the exposed Sunda shelf that stretched out almost endlessly before them punctuated by large lakes that lay in the present Gulf of Thailand and Java sea permeated by long and very large rivers. Frequent river crossings, paddling round large lakes and down rivers may have been where these people became familiar with and regularly used watercraft as well as learning the vagaries of large water bodies prior to longer sea voyaging.

For those continuing their journey into the icy wastes and Tundra of eastern Siberia, hunting large reindeer, saiga, mammoth and woolly rhinos of the northeast required a different strategy. Once killed the next fight was with competing bears, snow leopards, Siberian tigers and wolf packs, all animals so very common then but in danger of extinction today, who also wanted the fresh kill as well as a taste of the killers themselves. It could have been now that the first attempts at domesticating the wolf were made as a valuable hunting aid and ally, that was increasingly habituated to humans by it scavenging round their camps and occasionally being fed by humans, had its tentative beginning. It was the beginning of a special symbiosis that would bind dogs and humans together forever. It is surprising that moving north and over the Bearing Strait occurred when it seems to have been done. Moreover, far more adaptive strategies and biological adaptations were required by modern humans to move into those frigid regions particularly during the MIS 4 and 2 glacials than into the tropics.

The two groups were now big game hunters in the north, and small game and coastal, lacustrine and riverine hunter-fishers in the south that could arguably an easier way forward. In the north hunting was a premium together with the culture, weaponry and planning required to obtain success in those special conditions. Predatory animals were a problem for southern expansion groups with tigers, large pythons, angry rhinos and crocodiles and the occasional poisonous snakes, insects and vegetation as well as diseases like malaria. That could have killed some as well as conferring an adaptive immunity on others through natural antibodies and the development of balanced polymorphisms such as thalassaemia and sickle cell anaemia (see Chapter 7). Nevertheless, southern movements were comparatively easy, where at least the temperature would not kill you and you were not required to live in freezing darkness for many months with many hungry predators as adversaries.

Rainforest and tropical savannah demanded different strategies and tools. Surviving on Sunda with its widespread coastal, lake and river work brought about tool design to make and build watercraft. That required different material culture design and manufacture. In the light of these issues it is likely that while new implements may have evolved through the contrasting circumstances of the two groups, such as the use of bamboo in the south and bone and ivory in the north, it is possible that aspects of traditional stone tool making and type may have been lost or replaced with other materials and different weapons/tools. While stone was always used when available, its scarcity in some areas in both places, particularly in rainforests, made bamboo a good substitution because while stone would be found it may not have been ubiquitous or the right type but bamboo was, it was virtually everywhere. A sharpened end on a bamboo pole made a spear tip just as effective as any stone-tipped spear and was just as easily sharpened and resharpened without the need to search for, and likely not find, the required stone, put the effort into flaking it, with its share of discarded waste flakes, and make a fresh spear tip and haft it onto the spear. Split bamboo has razor sharp edges and made excellent knives. Other perishables were almost surely utilised for knives such as shell and people would use whatever came to hand. But after living in a tropical rainforest then moving across a large water gap to Australia, the useful wood and bamboo species were left behind and a fresh start with manufacturing material culture was required. In effect, crossing the equator was like moving through a *material culture sieve*: some skills were lost as people moved into Sunda's tropical rainforests and others were gained. On reaching Australia, there was another change with a greater reliance on stone artefacts with the loss of the useful bamboo and other vegetable materials. They began to use stone again for all basic knives, blades and scrapers, particularly when they moved inland, but with possibly with a reduced skill base caused by loss of those who how to make them and lack of practice generally, and a fall back to a more basic stone flake or sharp edge. The ground stone axe may have been one outcome of using new materials and using new skills the rainforests had taught them and the oldest evidence in the world for such weapons has been found at the M at Madjedbebe in Arhem Land (Clarkson et al., 2017).

The life experience of our two migratory groups determined what they needed and the technology and materials used each reflecting their independent journeys with an equally independent life experience. In this way two different cultural histories were founded and it was those differences that each took to their respective destinations. If that was so, we should not expect to see any cultural and artefactual reflection of African typologies arriving in Australia or California. So, what might we expect to see of the tool kit and culture of the people that arrived on Australia's shores? Not much, except for the obvious use of basic blades, knives and sundry sharpened flakes. We should not expect to find purposeful designs associated with modern human cultural activity elsewhere, in North Africa, or the Middle East, with their distinctive well-fashioned blades and knives in the form, for instance, of the Middle Palaeolithic Mousterian, or Middle or Upper Palaeolithic Aterian of North Africa, or the Late Palaeolithic Aurignacian of Europe or Sohan of India? There is some difficulty in fully translating or interpreting many stone artefacts found associated with the earliest sediments in the Willandra or Madjedbebe but they do not really look like any of those listed.

We are often limited to ethnohistoric or modern applications when translating the use of a flake or blade but it is a best guess as to what objects might mean or how they were used. There is a problem, however, with extrapolating modern tool use back tens of thousands of years. A spear tip is a spear tip; an arrow head is similar. Wooden spears are wooden spears but many basic flakes and other debitage are not so easily placed in a techno–cultural context particularly those with a basic generic appearance that could, and probably did, have many uses but in the mind of the maker did not have to have an appealing design.

It is also difficult to interpret what people meant or were thinking as they prepared for and buried their dead. Why did they treat or bury a body in a particular way? It is only too easy to put an archaeological interpretation in place of the ceremonial organiser living tens of thousands of years ago: what would *I* think and why would *I* bury or prepare a body in a certain way? Human palaeocerebral processes are difficult to know, try interpreting Palaeolithic art! Although we are often eager to extend burial practices into 'belief systems' and 'deliberate ritual' perhaps with the use of the concept of an 'afterlife', those interpretations may or may not be correct. The above outlines some of the problems faced in archaeological interpretation when considering the culture of modern humans as first Australians and the cultural baggage they may have brought with them. It also highlights who were the first Australians. What do we see here that illuminates the cultural side of those first arrivals coming from Africa, albeit that they may represent people living many generations after the first people that left the African continent and even many generations after the first arrivals in Australia?

I now want to add a postscript to the neat story of dispersals north to Arctic wastes and south to the tropics. Of particular concern now is the timing of northern human dispersal to the Americas, which is claimed to have taken place well before humans reached Australia instead of long after. After finishing this book and thinking that I had probably said a few things that seemed like 'science fiction', a bombshell of 'science fact' has dropped. Humans or possibly archaic hominins, are now believed to have reached California, USA between 120 and 140 ka (Holen et al., 2017). The evidence comes from the Cerutti Mastodon site where it has been claimed in situ hammer stones and stone anvils together with stone flakes and butchered and smashed mastodon (*Mammut americanum*) bones have been found. Collectively, these artefacts and bones have been claimed to represent a butchering site. Radiometric dating of Cerutti (130.7 + 9.4 ka) is believed to be reliable and that conclusion has been reached using multiple lines of corroborating stratigraphic and geochronological evidence. The controversy may rest on the interpretation of what constitutes hominin butchering of the mastodon bones. This is not something I want to go further with this sensational but controversial find, which if true would put people not only in North America but also far south in California. That would require a radical rethink of the abilities of humans, their adaptive development, who they were, where and how they were able to travel and when. But the greatest mystery would be: who were they? Cerutti's earliest date (120 ka) could reflect an onward movement of modern humans like those who reached Fuyan Cave (120 ka). Using the oldest date (140 ka) could also mean the arrival of archaic humans from northeast China, perhaps those descended from or associated with the descendants of Dali/Jinniushan, Penghu types or even Denisovans. The possibilities of who it is are only one of the puzzles of this find. Another is, whoever it was, how did they get there? Either they crossed the Bering Bridge by boat at a low stand during MIS 5 (130–120 ka or by crossing the bridge by walking during an Ice Age (MIS 6 140–135). If it was made at the end of MIS 6, that would mean it was more likely to be an archaic people who had biological and cultural high-latitude adaptive qualities enabling them to live in subpolar conditions and thrive. But if that was so, how far back before this date did they have did they develop such adaptive qualities? That is, with such an underestimation of modern human or late archaic hominin capabilities, it shows that moderns or archaics could have moved anywhere during MIS 5d or e and that includes Australia. The final statement for now is that if this site represents what is claimed then whoever it was, they went extinct and the Americas waited for a further human migration.

Whether the evidence from Cerutti has been correctly interpreted or not, what we know in archaeology is that there is always something to learn as analytical techniques constantly improve and the search continues. Research continues to progress. We now have plant DNA identified in cave sediments (Willerslev et al., 2003) and identification of human DNA in the same sediments where humans lived but left no other evidence of occupation (Slon et al., 2017). That has identified Neanderthal DNA at cave sites in Spain, Croatia, Belgium, France and Denisovan DNA in the Denisova Cave in the Altai. Not so many years ago these results would have called pure science fiction!

LOOKING AT THE TRIP

Today we live with speed: fast cars, trains, information networks, computers and air travel. Anybody that has travelled by air from Europe to Australia recognises the distance albeit at 900 km per hour in a comparative straight line and, besides falling out of the sky, without obstacles and ever-present danger. But even that prevents us recognising the real distance. If we try and visualise the trepidations and effort needed to undergo the same journey on foot it would put our usual complaints about jet lag and how long we sat in a cramped seat into perspective. Perhaps we cannot envisage such a journey by foot, we are too removed from such arduous travel. Even without encountering other people, fierce animals were present in greater numbers than today and don't forget the snakes and bugs. Imagine, crossing mountains or large rivers or walking around them and then crossing large water gaps that separated islands and you from Australia. Perhaps the best way to envisage the journey to Australia is to contemplate where and when we will die. For most people, this is a very difficult an unfathomable concept to contemplate and indeed while most might give it a fleeting thought occasionally they don't really think or worry about it at all, they just keep on living. I believe this is the closest we can come to understanding how Late Pleistocene super nomads would have behaved towards thinking about an end point of their journey or a destination: there is no end, you just keep going, it is a natural way of life and you don't think about the end.

The dates we have for the AE4 exits encompass *early*, *middle* and *late* phases spanning 140–60 ka, each comprising a series of pulses rather than a constant flow. Whatever the reason for leaving Africa it must have occurred spasmodically possibly prompted by erratic climatic subcycles and swings associated within MIS 5 to MIS 3 glacials and interglacials. The time taken for modern humans to complete their dispersal across Asia to Australia could have taken decades at their quickest and perhaps hundreds or even thousands of years at their slowest with different pulses moving at different paces and taking different routes. So, how do we assess those journeys and the effects on the travellers? The short answer is that we don't or can't; there is no point, although the treks occurred I am convinced they took some considerable time and I have outlined why elsewhere in this book. In short, the reason is that without a goal; taking random paths; journeys consisting of short hops and slow meanders; assuming only a few people participated at any one time; not knowing the world in front of them and failures and occasional group/band extinctions, makes it that way. The next question is: if the journey took thousands of years then what could have happened to the travellers in terms of culture, morphology and biological makeup?

What did the first people look like as they waded ashore on our northern coast? The scarcity of skeletal evidence for hominins between the Middle East and China prevents us knowing. But we can make some assumptions about what they looked like. Individuals like Liujiang, WLH 3 and perhaps WLH 50, are possibly the only evidence of them. Even the modern human teeth from Fuyan Cave, can't help us with the general morphology of their owners, we just know they were from *H. sapiens*. If modern humans had mixed with archic peoples we cannot know what such a founder population may have looked like when they eventuall arrived in Australia. The possible compinations and archaic types they could have mixed with make this so. The choice they could have encountered includes: Neanderthals, Denisovans, Maba types, the descendants of Dali and Jinniushan, *H. floresiensis*, Penghu people, Indian archaics and possibly others that inhabited East Asia that we have yet to discover. The fundamental developments and lineages that existed among the archaics themselves is also another unknown although what we probably can eliminate are Solo hominins, the cornerstone of the Multi-regional Hypothesis in Australia.

THE TIMES THEY ARE A-CHANGING

The 1980s and 90s brought change in palaeoanthropology, particularly in the study of the origins of modern humans. During that time, our understanding of modern human evolution was totally revised in rapid time. The changes particularly affected palaeoanthropology in Australia because of the 'robustic-ity conundrum' that loomed large here. I want to briefly outline that because it affects us to this day. The title of Bob Dylan's (1964) song, leads this section, and many lines in it can be applied to this story. But one line is worth quot-ing '... *don't criticise what you can't understand...*' Many here were guilty of doing just that and I was one. That line sums up the story of the fundamental changes taking place over the last 50 years including the genomics revolution. Around 40 years ago, it was dawning on some that Neanderthals were no longer the forbears of modern Europeans, an idea that had been a mainstay of human evolutionary thought for decades. That idea was Eurocentric as anything else together with the fact that Neanderthals were the only premodern candidate at the time. A darker possibility was an idea held by some that modern humans evolved from Europe and nowhere else because who else would be fit enough to produce the modern European. That idea was a throwback to the notion of European superiority and colonialism as, indeed, the Piltdown hoax intended to show, particularly for Britain. The evidence then pointed to Neanderthals evolving into modern humans like Cro-Magnon, who was represented by a middle-aged man dated between 45 and 21 ka discovered in France during road construction in 1868. Cro-Magnon showed Neanderthals changing into modern people and so at that time the former were regarded as our direct ancestors. But that evolutionary jump was always a puzzle because there were no intermediate fossils between Neanderthals and Cro-Magnon. It was assumed Cro-Magnon's appearance marked an extremely rapid morphological change occurring in less than 20,000 years which, at the time, was the time gap believed to exist between the two groups. It was also the best candidate of our ancestral lineage because Neanderthals were a natural precursor but we now know the two groups over-lapped and there were no other European fossils of that age that looked so mod-ern. The irony is, Cro-Magnon man does represent our direct ancestor but not by descending from Neanderthals. The news that Neanderthals were a separate but parallel species from us was indeed a minor revolution in palaeoanthropology in the 1970s. Instead of all ancestral roads leading to modern humans there were others that lead to similar species that began a story that continues today. Over tens of thousands of years, several types of non-modern people had evolved, lived their lives and gone extinct, and not contributed anything to the modern human line. However, we learnt that we lived side by side with Neanderthals for some time and that we learnt other species of human were also around that we lived beside.

For me the real revolution began in 1982 with the work of geneticists Rebecca Cann and Allan Wilson. Cann and Wilson's findings were focussed on the study of human mtDNA gathered from modern people from around the world. Their results were beginning to show all humans, wherever they lived, were related as a single group stemming from a common African ancestor 120,000 years ago, a date that has since been recalculated several times, is not agreed upon and been pushed 60–80,000 years further back. But importantly, genomic work was beginning to show modern humans did not descend from

H. erectus groups living in different parts of the world, an idea upon which the Multiregional Hypothesis and Regional Continuity arguments were based. Instead it was showing that there was only one ancestral erectine hominin that we all had all stemmed from and who lived in Africa. The rest is history, a convoluted history indeed, and many have written about it so I will not go over it here. I will say, however, that the work of Cann and Wilson began a revolution in palaeoanthropology that has since been carried on by hundreds of geneticists and archaeologists. Their work eventually made people think differently about the accepted model of human evolution based on fossil evidence alone. The mtDNA studies suggested a very different story from the one the fossils seemed to tell although the fossil record was still to be interpreted properly. New ideas emerging from genomics seemed to reduce the strength of interpretations and observations made on fragmented fossil crania alone. Fossils would always be the mainstay of human evolutionary study, but many believed DNA research was the thin end of the wedge and that it would end palaeoanthropology as it was known then by studying fossils. Comparing a few fragmentary crania often separated by tens of thousands of years, spread across thousands of kilometres, some on different continents, had always been naturally fraught with error. Now genetic study highlighted that error and made its conclusions seem irrefutable. Ironically, DNA results brought disappointment mixed with disbelief from those who studied fossils alone and who probably feared this marked the end of their style of palaeoanthropological research; even worse, their careers! However, people do not give up their ideas easily. It takes intestinal fortitude and a reliance on others to forgive to say you were wrong and had wrong ideas, but you have taken the cure and you are alright now. Those schooled in the older styles of physical anthropology were the most sceptical and that scepticism was underpinned by a limited understanding among non-geneticists of the complex and burgeoning world of genetic study and a sublime faith in fossils. Those who supported the regional continuity or the multiregional hypothesis were struck hardest when they realised their ideas were no longer correct. Some capitulated quicker than others. I was probably one of the slow ones, but mainly because of the unresolved puzzles of robusticity here in Australia that have not gone away even with the advent of the out of Africa idea and genetics. Some fossils in Australia showed they were surely not uniformly modern or directly descended from modern people only and that is still the case.

The mtDNA work was joined by the study of human nuclear DNA in the hope of mutually supporting mtDNA results but that did not always occur, particularly regarding genetic distancing which gave a temporal estimate of stages in human evolutionary separation. Nevertheless, information from the two types of DNA produced new ideas, even if geneticists did not always agree between themselves. That provided some insight into our past and pushed the boundaries of data extraction from fossil bone itself which, at times, showed how wrong fossil study alone could be. Nuclear DNA results could also be cross-checked with what was being learned from the Human Genome Program which took 13 years to complete. Direct comparison of fossil and modern DNA could now take place as extraction methods also improved forcing its exponential development as a useful technique in human evolutionary study. It was not long before DNA was extracted from a Neanderthal arm bone. Genetics would now make it very apparent how complex modern human evolution was as well as informing us we carried 2–4% of Neanderthal genes. A human ancestor soon emerged we had no idea existed, the Denisovan. In that case DNA was extracted from a small piece of finger bone and a single tooth, there was no other skeletal evidence of them. Even more astounding was the realisation that archaic humans had existed alongside us in the last 100,000 years, some we had virtually no meaningful fossil evidence for. Results showed we had all started in Africa and then had somehow replaced all other archaic human groups living outside. Genetic study showed how the fossil record was even more patchy, inconsistent and in some ways unreliable than was assumed previously. For some these changes were hard to accept. Many entrenched ideas were being overturned too quickly. Some always reject rapid change and so it was in palaeoanthropology, particularly as theories were collapsing and reputations were at stake. Initially, palaeoanthropologists divided into two camps: those that believed what the fossil record told them and those that accepted the genetic evidence. At the time, nobody suggested you could combine the two. However, the former believed the fossil story made more sense, it was logical and the evidence was undeniable whatever the genetics said, and that was the case in Australia.

The more we learned about our DNA and of everyday puzzles concerning disease and general inheritance the more genetics was gaining credence as an important and indeed vital tool in understanding our past, if suitable DNA could be extracted from fossil humans. Genetic work slowly took over from the old metrical and non-metrical skeletal studies of the past. No matter what way they were manipulated by statistics and modelling, some had always believed them to be poor methods for explaining human variation, evolution, ancestry and lineages with any certainty, but it was all we had. So, their interpretative strength is now waning against the new and constantly improving genetic methods that delve into ancient human biology itself rather than its morphological by-product. But fossils do not surprise us anymore they just present us with challenges to expect and accept the unexpected and to keep more of an open mind than we have had in the past. We humans have been a variable lot over hundreds of millennia, some faded away to extinction just like other animal species and we are beginning to learn about them through DNA analyses.

Nevertheless, genetics still has its limitations and one of them is time. While DNA extraction from bone now extends to almost half a million years ago, successful extraction and useful data is subject to bone history. Leaching, contamination as well as the fossilisation process itself can all give erroneous results or no results at all. Moreover, gene mutation rates, used for constructing genetic distances between species and phylogenetic relationships, are still being researched with variations occurring through variable mutation rates.

All this together with the common fragmentation met with in fossil material and bone degradation means it is less likely to contain DNA, but methods are constantly improving which makes one wonder what will be possible in the future. One future avenue is the possible extraction of DNA from China's fossil crania like Mapa, Dali, Jinniushan and other enigmatic archaics, important fossils that probably feature in Australasian human evolution more than we might think at present.

A question that arose from the out of Africa idea was how could African people change into all the different groups we see around the world in such a short time. It was felt that the vast differences in appearance and culture around the world could not have emerged in 60,000 years or less, and I too, was puzzled by this. We know that modern populations are the product of subtle and sometimes not so subtle mechanisms of adaptation and regional evolutionary processes and that includes a variable input by the *9EM*. Environmental, climatic and pathogenic adaptation and the subsequent lifestyles of people have been major contributors to how we look and how our genes have expressed themselves which is spatially and temporally different. Until recently the speed of anthropomorphic change, such as eye colour and shape, skin and hair colour, stature and metabolic differences all of which now differentiate people around the world, was believed to have been a slow process, perhaps over tens or even hundreds of thousands of years. Many doubted such changes could emerge comparatively rapidly, in what might be termed the blink of an eye in geological and evolutionary terms. But as I mentioned above, it is possible that some of those differences might have travelled or even been gained on long dispersal routes and, of course, through mixing with archaic groups. In short, how did so many biological variations and adaptive qualities enable people to live in such widely different environments and how did we manage to look so different in such a comparatively short time? Well, genetic research on ancient bone is beginning to tell us.

Thick, straight black hair only developed within the past 30,000 years due to the *EDAR* gene. A lighter skin colour seems due to mutations producing alleles called *SLC24A2* and *SLC24A5* (Mathieson et al., 2015). The *HERC2/OCA2* mutation has resulted in blue eyes and both these changes probably took place only 10,000 years ago, in people living near the Black Sea although the earliest Britain living 10,000 years ago found in Gough's Cave in Cheddar Gorge had dark to black skin and blue eyes. It seems that the early and middle Holocene seems to have been the crucible for many of the changes that demarcate our appearances. Later, during the Neolithic, with the migration of farming groups from west Asia, we developed the ability to break down milk sugar (lactose) with the emergence of the enzyme lactase, also brought into Europe by the Bell Beaker people after 4.5 ka within whom it had evolved. Selection in relation to variation in dietary or environmental sources of vitamin D produced the variants in alleles of *DHCR7* and *NADSYN1*. Other genes that appear in the Holocene provide an adaptation to various diseases such as leprosy, mycobacterial and coeliac disease as well as ulcerative colitis, while much earlier Neanderthals suffered from disabilities such as Crohn's disease and autoimmune disorders that we inherited when we mixed with them.

It is reasonable to think that if modern humans could change so much in such a short time, it was possible that hominins in Lower Pleistocene migrations could also undergo similar 'rapid' changes as they lived in new climatic and environmental regimes through which they travelled, but of course, they did not mix with others because there were no others, or were there? If such changes can take place so quickly we must ask: was it possible for early hominins or *H. erectus* to take on regional morphological, biological and metabolic differences as well as adapt quickly in other ways. If true, then there could also have been differences in these traits between different *erectus* populations spread across the world. While those early hominins had the world to themselves, and did not have other different populations to mix with so we might assume, remixing among *erectus* groups separated for tens of thousands of years was a possibility and their biological and adaptive histories would have contributed to variations among them if nothing else. We know those hominins present skeletal variations, at least in the cranium, so we can no longer firmly discount the idea that erectines could adapt to live in difficult environmental circumstances such as cold or high latitude climates. The Chinese erectines living around Zhoukoudian certainly experienced cool to very cold conditions at times. Moreover, the findings about our recent, rapid evolutionary changes must suggest that similar changes could easily have occurred more than once among earlier hominins over the last 2 million years as they moved around.

As I hinted earlier, Australia's first people must have been, to some extent, the product of their journey. It is easy to believe that journey would impose changes on their culture and, for some, their genetic composition, even without mixing with archaic hominins. The longer the journey took the more likely that would have occurred, possibly adding or subtracting to their original social, cultural and biological composition. Of course, one of the influences could have come from who they met on their journey and how many people they lost. Dispersal time must have been different for different groups and depended on the route taken and environments and terrain they encountered. Perhaps these effects altered the appearance of those that reached the end from those who had begun the journey. We might consider one change that took place following modern human populations leaving Africa. That was the meeting and mixing of modern humans with Neanderthal people. While our present Neanderthal DNA component is only between 2 and 4%, the first few generations of the original mixed groups would have had a much more equally balanced DNA profile between the two and that would have included forging a different cranial morphology than either of the original groups. Subsequent reduction of the Neanderthal component took place over many generations as we eliminated deleterious and otherwise unwanted Neanderthal genes from our genome. It also depended on no further mixing with Neanderthals only other modern humans among other factors such as incompatibility between the long-separated genomes and the possibility of first generation infertility particularly in males.

While many of those who believed in Regional Continuity have abandoned such 'heretical practises', fundamental questions, such as the rapid development of morphological variation among today's humans discussed above, remain about the out of Africa idea. Some might say that until they are answered other hypothesis cannot be properly laid to rest, but I will return to that issue later. It is worth considering that Regional Continuity might still be very much alive if it were not for the progress of genetics. However, that would still not make Regional Continuity the right theory, we just would not know any better. A change in our understanding came only because of palaeo-DNA work, that showed Regional Continuity to be wrong, at least in part. However, the spirit of Regional Continuity is not quite laid because the origin of Australia's robusticity is still not solved, although it might yet have an answer that looks a little like Regional Continuity. For example, DNA study has yet to examine the genetic identity of many regional fossil hominin crania. It may turn out that the story of modern humans in some places could be a blend of old and new ideas like Continuity by the mixing of archaic and modern people. Without the ability to extract ancient genes from fossil bone we would still have two strongly held and very opposed camps of thought, for and against the out of Africa hypothesis. It is a shame that at present we do not have DNA from emergent modern humans living between 250 and 200 ka and archaic individuals like Dali, Jinniushan, Maba and Penghu. I suspect that would add extremely valuable data to this story and might even bring forth some radical changes to it.

It seems people reached Australia long before they reached Europe, and that is unexpected with Europe's proximity to Africa. The oldest evidence of <65 ka was predicted by others but untill 2017 it was not confirmed empted by others (Webb, 2005; Malaspinas et al., 2016). The question, who were the first people to land in Australia, was once easy to answer, that is no longer the case and it might be about to get harder. Rightly or wrongly, many working in Australian archaeology see ~50 ka years as the earliest time of the first arrivals beyond which nobody worth their academic salt would venture. But with emerging dates from China and North America, as discussed above, a first entry into Australia could be much earlier. The date mentioned in Chapter 4 from modern human sites across the region (Jwalapurum, India, 80–75 ka; Liujiang 68 ka and Fuyan Caves, China 120–80 ka; Callao Cave, Philippines, 68 ka) and Madjedbebe show conclusively that modern humans were always closing in on Australia much earlier than 65 ka but those of >100 ka seems to underpin the possibility of an even early, perhaps much earlier landing in Australia. The dates from California, if true, suggests coastal travel even in very cold conditions may have been possible and generally broader seafaring skills and navigation were not far away from development, particularly in those working their way through South East Asia as part of the tool kit of the approaching people. The Fuyan Cave and Cerutti evidence could mean a 100 ka Australian entry much more likely if modern human migrations kept moving and were capable of open ocean travel. It certainly shows modern humans must have overlapped with archaic populations in the Far East which makes it more than likely that mixed archaic/modern people could have travelled elsewhere, even to Australia as mixed modern-archaic groups or even unmixed archaic people, nothing would surprise me. And it firmly indicates modern humans left Africa between 100 and 120 ka and made it all the way to East Asia making it more likely that Jebel Faya (128 ka) is a modern human site and a stepping off point for those modern humans on their way to South Asia and the Far East.

All those possibilities open questions about when Australia first received humans, who they were, what their migratory history was and whether they were of a mixed or unmixed origin. If there were African exits from 130 ka, reaching Jebel Faya at 128 ka and China 10,000 years later and with follow-up dispersals at 100 ka and 70 ka, a series of migrations could have made it this far, which could include mixed and unmixed people. We can only assume there were at least three chances for modern humans to arrive here before 70 ka plus the possibility of other sporadic African exits that we are unaware of who travelled in this direction. For the moment, however, a 65,000-year occupancy marks the earliest arrival until further evidence shows otherwise. That also means the oldest skeletal remains must represent the earliest Australians we know of for now. But while the Willandran individuals represent the oldest skeletal evidence for the first people, it is clear they are perhaps as much as 20–25,000 years after the first human entry.

People entering Australia brought with them Neanderthal genes, and others also brought Denisovan genes. That fact shows fecundity among people separated for hundreds of thousands of years even though it has been suggested that resulting generations may have suffered a certain amount of male infertility. Various artefact assemblages found along likely migration routes have a variety of designs and craftsmanship that seem to change retrogressively from the characteristic tool kit of contemporary modern humans in Africa and the Middle East. Their refinement and technique seems to be lacking from the earlier assemblages as modern humans crossed the world. Their tools also do not form a continuation of style and craftsmanship and that might suggest loss of tool making skills compared to modern human assemblages found in Europe, the Middle East, among the Aterian of the North African Maghreb, assemblages found in South African coastal sites and even in South Asia. The earliest Australian assemblages have none of the distinctive styles that are seen in any of these areas. One possibility is modern human travellers slowly lost skills or took up others as they moved away from Africa. Alternatively, those leaving Africa may not have had the same skills as those living over 7000 km away in South Africa or in the northwest, which seems doubtful and contradicted somewhat by assemblages found in Arabia. Certainly, artefacts found on the earliest Australian sites have no form or style that might align them with examples from any of those that occur in coastal and north east African and the Arabian Peninsula sites dated to 70–80 ka. It is possible that isolation, separation and parting from or the death of skilled tool makers without those skills being passed on may account for this. Perhaps functionality supplanted style as necessity while moving neded a 'quick fix' in terms of a cutting edge or knife. Whatever it was, the very different style of basic manufacture among early Willandran artefacts, as well as other early Australian sites, and those in Southern or Northwest Africa, that are over twice as old, suggest a loss of manufacturing skills over time and space. Another way of losing such skills was adopting implement design from those they encountered (or mixed with) either because they were easier to make or required less work to achieve the same result or do the same job. But, again, this brings us to who were the first people who entered Australia.

During the last 60 years, the date of Australia's first arrivals has changed dramatically from 6000 years in 1960 to 16 ka in 1962, 26 ka in 1969 and 32 ka in 1974. Then, redating of the Lake Mungo lunette sediments in 2002 pushed it back to 42 ka (Bowler et al., 2003). Dates from other archaeological sites around Australia have shown people did not just arrive 8,000 years earlier they had spread out across the continent by 50 ka. Now we have 65 ka at least for the first landings and there seems no reason it will go much farther back. The question of who those people were is demanding a rethink mainly because of evidence from many regions both close to and far from this continent. While a wide variety of evidence has emerged relating to the emergence of modern humans, it has become more complicated and that includes their dispersal patterns, the timing of those dispersals, who they met with and their capabilities. Moreover, we can never be sure what evidence will emerge day to day and in what direction it might take us. The Australian story depends entirely on these issues and their implications, now more than ever before. It is a particularly fascinating story but those issues are very complex and no doubt similar issues were in play elsewhere.

Now we should meet the oldest people found in Australia so far.

REFERENCES

Ambrose, S.H., 1998. Late Pleistocene human population bottlenecks, volcanic winter, and differentiation of modern humans. Journal of Human Evolution 34, 623–651.

Armitage, S.J., Jasim, S.A., Marks, A.E., Parker, A.G., Usik, V.I., Uerpmann, H.P., 2010. The Southern route "out of Africa", evidence for an early expansion of modern humans into Arabia. Science 331, 453–456. doi: 10.1126/science.1199113.

Augustin, L., Barbante, C., Barnes, P.R.F., Barnola, J.M., Bigler, M., Castellano, E., et al., 2004. Eight glacial cycles from an Antarctic ice core. Nature 429, 623–628.

Barham, L.S., 2002. Systematic pigment use in the Middle Pleistocene of south central Africa. Current Anthropology 31 (1), 181–190.

Beyin, A., 2011. Upper Pleistocene human dispersals out of Africa: A review of the current state of the debate. International Journal of Evolutionary Biology 2011, 17.

Blinkhorn, J., Achyutan, H., Petraglia, M., Ditchfield, P., 2013. Middle Paleolitich occupation in the Thar Desert during the Upper Pleistocene, the signature of a modern human exit out of Africa? Quaternary Science Reviews 77, 233–238.

Boivin, N., Fuller, D.Q., Dennell, R., Allaby, R., Petraglia, M., 2013. Human dispersal across diverse environments of Asia during the Upper Pleistocene. Quaternary International 300, 32–47.

Bowdler, S., 1977. The coastal colonisation of Australia. In: Allen, J., Golson, J., Jones, R. (Eds.), Sunda and Sahul: Prehistoric studies in Southeast Asia, Melanesia and Australia. Academic Press, New York, NY, pp. 205–246.

Bowler, J.M., Johnston, H., Olley, J.M., Prescott, J.R., Roberts, R.G., Shawcross, W., et al., 2003. New ages for human occupation and climatic change at Lake Mungo Australia. Nature 421, 837–840.

Breeze, P.S., Drake, N.A., Groucutt, H.S., Parton, A., Jennings, R.P., White, T.S., et al., 2015. Remote sensing and GIS techniques for reconstructing Arabian paleohydrology and identifying archaeological sites. Quaternary International 382, 98–119.

Clottes, J., 2016. In: Martin, O.Y., Martin, R.D. (Eds.), What is Paleolithic art? Cave paintings and the dawn of human creativity. The University of Chicago Press, Chicago, IL.

Clarkson, C., Jacobs, Z., Marwick, B., Fuallager, R., Wallis, L., Smith, M., et al., 2017. Human occupation of Northern Australia by 65,000 years ago. Nature 547, 306–310.

Crassard, R., Petraglia, M.D., Drake, N.A., Breeze, P., Alsharakh, A, Scheittecatte, J., 2013. Middle Paleolithic and Neolithic occupations around Mundafan Palaeolake Saudi Arabia: implications for climate change and human dispersals. PLoS One 8 (7), 1–22.

Dennell, R.W., 2010. The colonization of "Savannahstan: Issues of timing(s) and patterns of dispersal across Asia in the Late Pliocene and Early Pleistocene. In: Norton, C.J., Braun, D.R. (Eds.), Asian paleoanthropology from Africa to China and beyond. Springer Science & Business Media, Dordrecht, pp. 7–30.

Dennell, R., Roebroeks, W., 2005. An Asian perspective on early human dispersal from Africa. Nature 438, 1099–1104.

Dortch, C.E., Muir, B.G., 1980. Long range sightings of bush fires as a possible incentive for Pleistocene voyages to greater Australia. The Western Australian Naturalist 14 (7), 194–198.

EPICA Community. (see Augustin et al., 2004).

Faure, H., Walter, R.C., Grant, D.R., 2002. The coastal oasis, ice age springs on emerged continental shelves. Global and Planetary Change 33, 47–56.

Fernandes, C.A., Rohling, E.J., Siddal, M., 2006. Absence of post-Miocene Red Sea land bridges, biogeographic implications. Journal of Biogeography 33, 961–966.

Feuerriegel, E.M., Green, D.J., Walker, C.S., Schmid, P., Hawks, J., Berger, L.R., et al., 2017. The upper limb of *Homo naledi*. Journal of Human Evolution 104, 155–173.

Fisher, E.C., Bar-Matthews, M., Jerardino, A., Marean, C.W., 2010. Middle and Late Pleistocene paleoscape modelling along the southern coast of South Africa. Quaternary Science Reviews 29, 1382–1398.

Fleitman, D., Burns, S.J., Neff, U., Mangini, A., Matter, A., 2003. Changing moisture sources over the last 330,000 years in northern Oman from fluid-inclusion evidence in speleothem. Quaternary Research 60, 223–232.

Fleitman, D., Burns, S.J., Pekala, M., Mangini, A., Al-Subbary, A., Al-Aowah, M., et al., 2011. Holocene and Pleistocene pluvial periods in Yemen, southern Arabia. Quaternary Science Reviews 30, 783–787.

Fu, Q., Hajdinjak, M., Moldovan, O.T., Constantin, S., Mallick, S., Skoglund, P., et al., 2015. An early modern human from Romania with a recent Neanderthal ancestor. Nature 524, 216–219.

Fu, Q., Li, H., Moorjani, P., Jay, F., Slepchenko, S.M., Bondarev, A.A., et al., 2014. Genome sequence of a 45,000-year-old modern human from western Siberia. Nature 514, 445–449.

Garcia, E.A., 2012. Success and failures of human dispersals from North Africa. Quaternary International 270, 119–128.

Geyh, M.A., Heine, K., 2014. Several distinct wet periods since 420 ka in the Namib Desert inferred from U-series dates of speleothems. Quaternary Research 81 (2), 381–391.

Geyh, M.A., Theidig, F., 2008. The Middle Pleistocene Al Mahrúqah Formation in the Murzuq Basin, northern Sahara Lybia evidence for orbitally-forced humid episodes during the last 500,000 years. Palaeogeography, Palaeoclimatology, Palaeoecology 257, 1–21.

Groucutt, H.S., Scerri, E.M., Lewis, L., Clark-Balzan, L., Blinkhorn, J., Jennings, R.P., et al., 2015a. Stone tool assemblages and models for the dispersal of *Homo sapiens* out of Africa. Quaternary International 382, 8–30.

Groucutt, H.S., White, T.S., Clark-Balzan, L., Parton, A., Crassard, R., Shipton, C., et al., 2015b. Human occupation of the Arabian Empty Quarter during MIS 5 evidence from Mundafan Al-Buhayrah, Saudi Arabia. Quaternary Science Reviews 119, 116–135.

Haslam, M., Clarkson, C., Petraglia, M., Korisettar, R., Jones, S., Shipton, C., et al., 2010. The 74 ka Toba super-eruption and souther Indian hominins, archaeology, lithic technology and environments at Jwalapuram Locality 3. Journal of Archaeological Science 37, 3370–3384.

Haslam, M., Clarkson, C., Roberts, R.G., Bora, J., Korisetta, R., Ditchfield, P., et al., 2012. A southern Indian Middle Palaeolithic occupation surface sealed by the 74 ka Toba eruption: Further evidence from Jwalapuram locality 22. Quaternary International 258, 148–164.

Holen, S.R., Deméré, T.A., Fisher, D.C., Fullager, R., Paces, J.B., Jefferson, G.T., et al., 2017. A 130,000-year-old archaeological site in southern California USA. Nature 544, 479–483.

Hublin, J.-J., Ben-Ncer, A., Bailey, S.E., Freidline, S.E., Neubauer, S., Skinner, M.M., et al., 2017. New fossils from Jebel Irhoud Morocco and the pan-African origin of *Homo sapiens*. Nature 546, 289–292.

Jaubert, J., Verheyden, S., Genty, D., Soulier, M., Cheng, H., Santos, F., 2016. Early Neanderthal constructions deep in Bruniquel cave in southwestern France. Nature 534 (7605), 111–114.

Laird, M.F., Schroeder, L., Garvin, H.M., Scott, J.E., Dembo, M., Radovc̆ic̆, D., et al., 2017. The skull of *Homo naledi*. Journal of Human Evolution 104, 100–123.

Leslie, D.E., McBrearty, S., Hartman, G., 2016. A Middle Pleistocene intense monsoonal episode from the Kapthurin Formation, Kenya, Stable isotopic evidence from bovid teeth and pedogenic carbonates. Palaeogeography, Palaeoclimatology, Palaeoecology 449, 27–40.

Liu, W., Jin, C.-Z., Zhang, Y.-Q., Cai, Y.-J., Xing, S., Wu, X.-J., et al., 2010. Human remains from Zhirendong, South China, and modern human emergence in East Asia. PNAS 107 (45), 19201–19206.

Liu, W., Martinón-Torres, M., Cai, Y-j., Xing, S., Tong, H-w., Pei, S-w., et al., 2015. The earliest unequivocally modern humans in southern China. Nature 526, 349–352.

Lombard, M., 2012. Thinking through the Middle Stone Age of sub-Saharan Africa. Quaternary International 270, 140–155.

Malaspinas, A.S., Westaway, M.C., Muller, C., Sousa, V.C., Lao, O., Alves, I., et al., 2016. A genomic history of Aboriginal Australia. Nature 544, 207–214.

Marchi, D., Walker, C.S., Wei, P., Holliday, T.W., Churchill, S.E., Berger, L.R., et al., 2017. The thigh and leg of *Homo naledi*. Journal of Human Evolution 104, 174–204.

Marean, C.W., Bar-Matthews, Bernatchez, J., Fisher, E., Goldberg, P., Herries, A.I.R., et al., 2007. Early human use of marine resources and pigment in South Africa during the Middle Pleistocene. Nature 449, 905–908.

Mathieson, I., Lazaridis, I., Rohland, N., Mallick, S., Patterson, N., Roodenberg, S.A., et al., 2015. Genome-wide patterns of selection in 230 ancient Eurasians. Nature 528, 499–503.

McBrearty, S., Brooks, A.S., 2000. The revolution that wasn't: a new interpretation of the origin of modern human behaviour. *Journal of human Evolution* 39, 453–563.

Mishra, S., Chauhan, N., Singhvi, A.K., 2013. Continuity of microblade technology in the Indian subcontinent since 45 ka: Implications for the dispersal of modern humans. PLoS One 8 (7), e69280. doi: 10.1371/journal.pone.0069280.

Oppenheimer, S., 2012. The single southern exit of humans from Africa: Before or after Toba? Quaternary International 258, 88–99.

Petit-Maire, N., Carbonel, P., Reyss, J.-L., Yasin, S., 2010. A vast Eemian palaeolake in Southern Jordan (290N). Global and Planetary Change 72 (4), 363–373.

Petraglia, M.D., Ditchfield, P., Jones, S., Korisettar, R., Pal, J.N., 2012. The Toba volcanic super-eruption, environemental change, and hominin occupation history in India over the last 140,000 years. Quaternary International 258, 119–134. doi: 10.1016/j.quaint.2011.07.042.

Petraglia, M.D., Alsharekh, A., Breeze, P., Clarkson, C., Crassard, R., Drake, N.A., et al., 2013. Hominin dispersal into the Nefud Desert and Middle Palaeolithic settlement along the Jubbah Palaeolake, Northern Arabia. PLoS One 7 (11), e49840, doi.otg/10.1371/journal.pone.0049840.

Reyes-Centeno, H., 2016. Out of Africa and into Asia: Fossil evidence and genetic evidence on modern human origins and dispersals. Quaternary International 416, 249–262.

Reyes-Centeno, H., Hubbe, M., Hanihara, T., Stringer, C., Harvati, K., 2015. Testing modern human out-of-Africa dispersal models and implications for modern human origins. Journal of Human Evolution 87, 95–106.

Richter, D., Grun, R., Joannes-Boyau, R., Steele, T.E., Amani, F., Rué, M., et al., 2017. The age of the hominin fossils from Jebel Irhoud, Morocco, and the origins of the Middle Stone Age. Nature 546, 293–296.

Rito, T., Richards, M.B., Fernandes, V., Alshamali, F., Cerny, V., Pereira, L., et al., 2013. The first modern human dispersals across Africa. PLoS One 8 (11), e80031. doi: 10.1371/journal.pone.0080031.

Rose, J.I., Usik, V.I., Marks, A.E., Hilbert, Y.H., Galletti, C.S., Parton, A., et al., 2011. The Nubian complex of Dhofar, Oman: An African middle stone age industry in Southern Arabia. PLoS One 6 (11), e28239. doi: 10.1371/journal.pone.0028239.

Rosenberg, T.M., Preusser, F., Risberg, J., Plikk, A., Kadi, K.A., Matter, A., et al., 2013. Middle and Late Pleistocene humid periods recorded in palaeolake deposits of the Nafud Desert, Saudi Arabia. Quaternary Science Reviews 70, 109–123.

Slimak, L., Svendson, J.I., Mangerud, J., Plisson, H., Heggen, H., Brugère, A., et al., 2011. Late mousterian persistence near the Arctic circle. Science 13, 338–344.

Slon, V., Hopfe, C., Weis, C.L., Mafessoni, F., de la Rasilla, M., Laleuza-Fox, C., et al., 2017. Neanderthal and Denisovan DNA from Pleistocene sediments. Science 27, http://dx.doi.org/10.1126/science.aam9695.

Szabo, B.J., Haynes, C.V., Maxwell, T.T., 1995. Ages of Quaternary pluvial episodes determined by uranium-series and radiocarbon dating of lacustrine deposits in Eastern Sahara. Palaeogeography, Palaeoclimatology, Palaeoecology 113, 227–242.

Timmermann, A., Freidrich, T., 2016. Late Pleistocene climate drivers of early human migration. Nature 538, 92–95.

Vaks, A., Bar-Mathews, M., Mathews, A., Ayalon, A., Frumkin, A., 2010. Middle-Late Quaternary palaeoclimate of northern margins of the Saharan-Arabian Desert, reconstruction from speleothems of Negev Desert, Israel. Quaternary Science Reviews 29, 2647–2662.

Watts, I., 2009. Red ochre, body painting, and language: Interpreting the Blombos ochre. Botha, R., Knight, C. (Eds.), The cradle of language, 2, Oxford University Press, Oxford, pp. 93–129.

Webb, S.G., 2005. The first Boat People. Cambridge University press, UK.

Webb, S.G., 2013. Corridors to extinction and the Australian megafauna. Elsevier, New York, NY.

Willerslev, E., Hansen, A.J., Binladen, J., Brand, T.B., Thomas, M., Gilbert, P., et al., 2003. Diverse plant and animal genetic records from Holocene and Pleistocene sediments. Science 300, 791–795.

Williams, S.A., García-Martínez, D., Bastir, M., Meyer, M.R., Nalla, S., Hawks, L.R., et al., 2017. The vertebrae and ribs of *Homo naledi*. Journal of Human Evolution 104, 136–154.

Part II

People at the End of the World

Chapter 5

Dreaming Lakes: History and Geography of the Willandra System

HERITAGE AND WORLD HERITAGE

The criteria for the Willandra's World Heritage listing required the area to possess outstanding universal cultural and/or natural heritage values. That was satisfied and world heritage status was conferred in 1981. The words used to demonstrate its scientific and cultural value stated:

> *The Willandra Lakes system stands in the same relation to the global documentation of the culture of early* Homo sapiens *as the Olduvai Gorge relates to hominid origins.*

  Some might say that comparison was somewhat overstated, but the finishing line for the longest continuous journey of world exploration by modern humans was somewhat unique and those two places marked the beginning and end points of that journey over 2 million years. The Willandra region has the potential to show us much more about those earliest travellers and who they were as well as provide a greater understanding of our origins. Scientifically, the region is not only important because of what has been found there but also because of what we might find there in future particularly regarding recent archaeological finds elsewhere on the continent and developments in modern human evolution around the world. However, it appeals directly to the resumption of research that may uncover a human history taking us back to the last interglacial among the Gol Gol sediments formed during MIS 5. Modern humans reached here before they ever reached Europe or the Americas and the Willandra region remains in the forefront of Australia's scientific and cultural investigations. It remains a

Made in Africa. http://dx.doi.org/10.1016/B978-0-12-814798-6.00005-4

challenging monument marking the chance to know who we are, what our deep history is about and how old the human history of this continent is. In short, Australia is not a real outpost but an essential piece of the jigsaw of modern human origins as anywhere else. It is also a testament to our persistence and determination as a species as anywhere else. What has been discovered in the Willandra is part of the economic, technological, cultural, biological and social adaptation of people in migratory transition and after their arrival in Australia. My long-time good friend and colleague Scott Cane has called those people *Super Nomads* (Cane, 2013).

HISTORY AND SCIENTIFIC DISCOVERY OF THE WILLANDRA

Before scientific work began in 1969, oral reports and written documents indicate that local graziers, government land inspectors and Water Resource Commission staff had previously recognised the large shrub and grass covered plains in the Willandra region as once being lakebeds fed by water from Willandra Creek. The presence of ancient mussel shell and the outline of the Willandra Creek encouraged these early interpretations. There were also land inspection reports made on the old *Gol Gol* holding between 1875–1900 that referred to areas of open plains and adjacent dune ridges as being extinct lakebeds. The *Gol Gol* property then encompassed part of Lake Garnpung, most of Lake Leaghur and Lake Mungo. Reports contain references to a portion of a lease called *Gal Gal Hills*, which is drawn as a line of hills resembling a lunette and most likely refers to the Lake Mungo lunette. The Mungo lunette has also been called the 'Walls of China' because of its broken appearance and battlement-like residuals. Also, Chinese labourers employed on *Mungo* sheep station, whose lease included the lunette except a small section at the southern end owned by *Jounli station*. They built the large and impressive *Mungo station* shearing shed in the 1880s using logs of Murray pine (*Callitris glaucophylla*) that grew at the back or eastern side of the lunette. The wheel tracks from the large drays they used to fetch the wood were, until recently, still visible and were shown to me in 1982 by Alex Barnes owner of the adjoining *Joulnie station*.

Peter Clark interviewed Angus Waugh (1890–1987) an old grazier on *Clare station* located 50 km east of Lake Mulurulu just before Waugh passed away. According to the old man, Aboriginal people had a specific word that referred to Lake Garnpung. Unfortunately, he could not recall the name but stated that it referred to Garnpung as a lake. Angus Waugh's father new the name as did his contemporary, an old lessee of *Gol Gol* Len Carrol. It appears that all the Willandra lakes were generally referred to as lakes in the early days of European settlement. It is, therefore, likely that the last Aboriginal people who lived in the Willandra Lakes region also referred to the dry lakes by specific names as Angus Waugh remembered (Peter Clark, personal archives).

Further evidence that the Willandra was a recognised lake system prior to 1968 came from several retired members of the NSW Water Resources Commission who visited Lake Mungo in 1980. They stated that around 1939 a feasibility study was carried out by engineers from the commission to see if the system could be reactivated using Lachlan River floodwaters, replicating the late Pleistocene situation. A similar study was also completed at that time for the Menindee Lakes water storage scheme 120 km northwest of the Willandra Lakes. However, the Willandra study showed that there was not enough Lachlan discharge to maintain adequate high-quality water reserves for irrigation.

The Willandra did not become the subject of scientific research until 1968 when Jim Bowler, then a postgraduate scholar in the Department of Biogeography and Geomorphology at the Australian National University, began work there. As a PhD student of Dr. Joe Jennings, Bowler was encouraged to investigate a series of dry lakes spotted by Jennings from an aircraft while Jennings was returning to Canberra from Broken Hill. Bowler became specifically interested in how the hydrological information contained within the lake sediments could be used to interpret the nature of environmental changes that took place particularly during the last ice age over the whole of southeastern Australia (Bowler, Jones, Allen, & Thorne, 1970). Subsequently he surveyed the region naming each of the largest lakes in the Willandra system as well as recording information regarding environmental change, lake history and riverine hydrology in the last glacial/interglacial. This information became the primary subject of his PhD dissertation.

During his 1968 fieldwork, Bowler made a number of discoveries the significance of which went unrecognised until some months later. Most of his Willandra work was carried out on eroding lunettes along eastern lakeshores as he searched for clues to past climate and lake hydrology. A substantial part of the work involved the stratigraphic interpretation and dating Aeolian sediments contained within the lunettes as well as lacustrine deposits. Those sediments were identified as a depositional sequence related to the filling and drying of the lakes and thus a regional picture of late Pleistocene lake history was revealed. During his work Bowler also discovered well preserved archaeological evidence of camp sites and hearths, with the remains of meals eaten long ago including shell middens. Radiocarbon dating of some of these sites resulted in dates beyond 30 ka, doubling the age of the previous oldest evidence for humans in Australia that stood at 16 ka. Intense archaeological interest only arose, however, after he located an eroding block of carbonate at the southern end of the Lake Mungo lunette. The block contained burnt and highly mineralised bone fragments that he thought could be human (Fig 5.1). Even more important was the block's sedimentary context that suggested this discovery could be 20–30 ka. Jim Bowler reported the find to archaeological colleagues at the Australian National University and a small team was assembled consisting of himself, his PhD supervisor, geomorphologist Joe Jennings, prehistorians John Mulvaney, Rhys Jones and Harry Allen and biological anthropologist Alan Thorne. The team drove to Mungo where they identified the remains as human. John Mulvaney produced an old brown suitcase that in the name of science he duly emptied of his clothing. The carbonate block was put in that and it was taken back to the Australian National University where

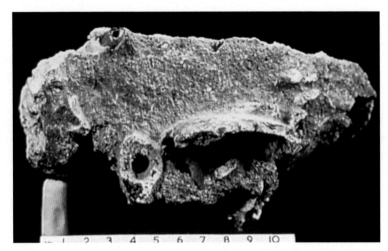

FIGURE 5.1   The burnt bone of WLH 1 embedded in a carbonate block. *(Source: Photo A. Thorne)*

Mulvaney was professor of prehistory. During the next few weeks, Thorne took over and slowly separated bone from its calcium carbonate matrix using a weak acetic acid solution. After close examination and slow reconstruction of the 200 plus fragments, the individual was assembled over 6 months. The result was found to be a young female initially called 'Mungo 1' now identified as WLH 1 and which is also called 'Mungo woman' or 'Mungo lady', the latter particularly used by the indigenous people of the region. A few fragments of a second individual were found as part of the burial and these are known as WLH 2. The discovery of WLH 1 literally not only turned Australian prehistory upside down first by its initial age of 25 ka but also because of a series of important skeletal features described in Chapter 6.

Lake Mungo's scientific importance was again underscored 5 years later in 1974 when another individual was found. That too was discovered by Jim Bowler, again on the southern end of the Lake Mungo lunette about 500 m north of WLH 1. A rain storm had washed sand from a shiny smooth cranial bone surface. This time the body was not in fragments as WLH 1 had been. Excavation revealed a near complete skeleton of a man in an extended burial position. The body had been laid slightly on its right side with its hands clasped in its lap and knees slightly raised. Significantly, this man had been buried after being covered with red ochre brought from 200 km away and he had been placed in the same late Pleistocene beach sediments as WLH 1. Given the number WLH 3 ('Mungo man'), examination of his features and the process of his burial revealed a number of profound things about him, would reveal the complexity and diversity of his culture and that of the earliest people but these and the WLH 1 details are described in Chapter 6.

THE SOUTH END OF THE LAKE MUNGO LUNETTE, WITH THE WLH 1 BURIAL PLACE MARKED BY NEAREST STAKE AND THE WLH 3 BURIAL MARKED BY AN ARROW. *(SOURCE: PHOTO S. WEBB)*

The 1970's marked the hay-day of research in the region with a variety of researchers involved. Harry Allen undertook PhD work in the Darling Basin during 1969–71. His Willandra work extended over the whole region but he examined only 12 sites spread across Lakes Mulurulu, Garnpung, Leaghur, Mungo and the Chibnalwood lakes details of which were described in his doctoral thesis *Where the Crow Flies backwards (1974)*. During 1974–76, Wilfred Shawcross excavated a large trench on the southern end of Lake Mungo lunette while Isabel McBryde worked on Lake Arumpo in 1975–76. Mike McIntyre began his PhD work in the region in 1976 but although he carried out an extensive site survey across the region his doctoral work was not completed and there are no published details of its results. The following are some of the early publications and reports that refer to research carried out in the Willandra: Allen (1972, 1974); Barbetti and Allen (1972); Bowler et al., (1970); Bowler et al. (1972); Bowler and Thorne (1976); Clark and Barbetti (1982); Dowling et al. (1985); Jones (1973); Mulvaney (1973); Thorne (1975); some of these references are unpublished reports and PhD dissertations, others take the form of very short papers.

Ironically, the 40,000 year plus record in the Willandra has been revealed almost totally through the agency of erosion rather than the archaeologists trowel, but interpretation of those finds is another matter and certainly one that requires revisiting. Nature has exposed the long past of the original peoples living in the region showing their ancient camps, artefacts, fire places and burials scattered along the old lake shores and other places. I was once told by an elder from one of the traditional tribal groups in the region that she believed the ancient ancestors had come back bringing their culture with them to show us who they were and how they lived. In her words they wanted to teach us. That is certainly one way of explaining the rich range of culture and knowledge that has emerged and continued to emerge in the region. There has been luck, however. That physical presence of the people living on the old beaches around the lakes required being buried beneath crescentic structures that fringe the eastern edge of the lakes now referred to as lunettes. Those lunettes became prominent repositories of the archaeological history of human occupation over thousands of years. They were built by the accumulation of windblown sediments forming time capsules holding the records of human occupation when lakes were full and fish, crustaceans and shell fish lived in them. The fresh water also attracted animals and birds to the lake edges and back swamps around the lakes so the larders of the inhabitants held great bounty and fresh drinking water was plentiful. But it was not always like that. In fact, there was a time when benign living conditions became harsh or almost impossible to live in as the climate changed, lakes dried and aridity took over the region. As climate cycles continued so the lakes filled and dried and in response the lunettes grew as a layer cake with different sediments being blown onto the beaches that corresponded to a wet or dry time. Sediments from the dry lake surface became sandwiched between those produced during wet time. This process slowly buried the living areas and burial places of generations of people who camped on along the beaches.

The vast majority of recorded survey work was carried out during 1979–85 by Peter Clark, resident archaeologist. Some of his data are used in this book and acknowledged as such. Since the mid-1980s archaeological work has been limited with years going by without any being undertaken. The major reason for that is described below. It was a disappointing time to see the largest archaeological area in Australia neglected and after the earlier discoveries and the valuable and unique insights that were gained about Australia's earliest people. There was little doubt that the region and lakes beyond Mungo formed a vitally important scientific and cultural area that could teach us a great deal more about Australia's earliest human history than we might contemplate.

THE WILLANDRA LAKES SYSTEM

The first people crossed the continent either by an inland route or by the scientifically favoured route of moving round the coasts and exploring up rivers they came across on the way. Rivers were small highways allowing internal exploration far inland in many instances. They often offered better provisions than the coast with its limited fresh drinking water and high-power surf beaches. Rivers provided drinking water, fish, shellfish, water birds, birds' eggs, reptiles and marsupials of various sizes, all forming a great attraction for exploration. One place they could have reached in this way was the Willandra Lakes and large bodies of fresh water in the lakes would have been an added bonus, making them ideal places to camp beside.

When full the lake system comprised at least 13 interconnected lakes ranging in size from six to 350 km² covering 3600 km² in southwestern New South Wales (Fig 5.2). The region lies within the greater 1.1 Mkm² Murray–Darling basin that contains four of Australia's largest river systems. They carry water from the east and northeast, west into the Murray River that flows into South Australia, exiting at Lake Alexandrina and Encounter Bay. The largest and longest of Australia's main rivers is the Murray (2508 km) and Darling (1545 km) systems, the other two include the Murrumbidgee (1485 km) and Lachlan (1339 km) rivers. It is not surprising that in such a dry continent this region should have attracted the early people to move inland. The fossil Willandra Creek no longer runs but at 45 ka it branched north and west from its Lachlan River feeder just east of the present town of Hillston. During glacial periods the Lachlan received melt water from a periglacial catchment in the southeastern Australian highlands and Snowy Mountains. At those times glacial runoff swelled the Lachlan enough to activate Willandra Creek that flowed west in a big arc. Willandra Creek's course took it to the Willandra Lakes through which it flowed filling the lake cyclically at different times. Those fluctuations reflected wet–dry cycles that lasted from >60–14 ka as ice-melt run off in the Lachlan watershed fluctuated. Similar fluctuations probably occurred back to the last interglacial (125 ka) and even earlier but that possibility has yet to be studied fully.

Six of the largest lakes are strung along the lower half of the fossil Willandra Creek. From north to south they include Lakes Mulurulu, Garnpung, Leaghur, Mungo and Arumpo. Small lakes between include the Chibnalwood Lakes, that formed within Lake Arumpo; the Prungle Lakes lie south of Arumpo and the small twin lakes of Pan Ban and Baymore are situated between Mulurulu and Garnpung (Fig 5.2). Lake Garnpung was the *largest lake (500 km²) that held around 5 million cubic metres of water when full*. When water flowed in the Willandra Creek the lakes were transformed from dry basins into a closely positioned string of freshwater oases that at times became bigger than present day shorelines might suggest. The lakes dried from south to north as the Willandra Creek retreated. That left Lake Mulurulu, the most northerly of the large lakes the last to dry out when glacial melt-water ceased to flow at the end of the last ice age (~14 ka). A major tectonic movement in the Holocene diverted the Lachlan River near Hillston permanently preventing water from ever again flowing into Willandra Creek.

LAKE FUNCTION AND HISTORY

During the late Pleistocene the Willandra lakes were set within a changing semiarid/arid landscape at times little different from today. The region became much more arid during glacials particularly in the region surrounding the lakes contrasting with the high lake levels and their fringing vegetation. This was encouraged by the fresh water which had the effect of making the lakes and their immediate surroundings 'islands' of habitation. Recent research has shown

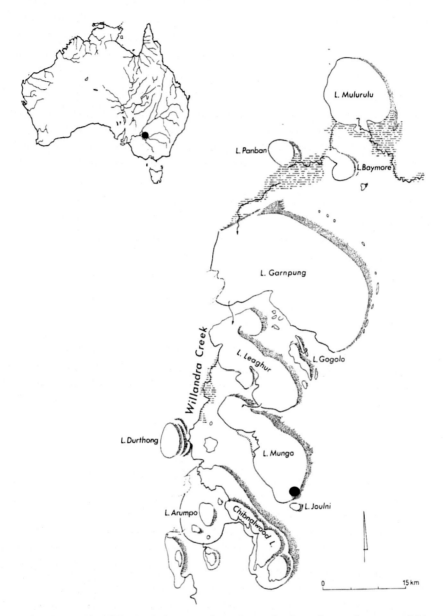

FIGURE 5.2 **The Willandra Lake system, large dot marks photo place on front page of this chapter.** *(Source: After Peter Clark.)*

that Lake Mungo and other lakes went through a 'megalake' stage around 24 ka (Fitzsimmons et al., 2014). Exaggerated pulsing of downstream flood water from the Lachlan River and along the Willandra Creek could have expanded the lakes beyond their normal shoreline boundaries for some time. During normal creek flows the region formed a tight knit, highly productive area. It was a place of rich biological productivity as well as refuges along the Willandra Creek corridor. Back swamps around the lakes harboured a wide variety of terrestrial animal species such as Quolls (*Dasyurus sp.*), bandicoots (*Isoodon obesulus, Perameles, sp., Macrotis lagotis*), bettongs (*Bettongia pencillata, B. lesueur*), possums (*Trichosurus vulpecula*), wombats (*Lasiorhinus krefftii*), wallabies (*Onchyogalea fraenata, Lagostrophus fasciatus, Lagorchestes leporides*), kangaroos (*Macropus gigateus, M. robustus, M. rufus, M. altus, M. cooperi*), and birds Table 5.1. During megalake times increased wetland areas increased all food resources substantially as back swamp areas expanded and new regions became flooded, enhancing the region as a focus for human occupation and activity. Food and freshwater were an obvious draw card for humans, making the area hard to move away from. Geomorphological work at Lake Mungo shows that 64–40 ka was largely a lake-full period, although lake levels fluctuated with three short periods of reduced water level during that time.

Lake Mungo was the smallest of the five main lakes and technically it was a backwater as it did not fill directly from the Willandra Creek as other lakes did. Water entered through a small interconnecting channel from the lower end of the next lake to the north, Lake Leaghur. We can only assume Mungo filled when Leaghur had enough water to overflow through the Mungo connecting channel, but those levels needed to be maintained to keep water in Mungo. The level of the Leaghur–Mungo connecting channel is above the Willandra Creek, so flow into Mungo could only occur during times of high creek flows which also signal good flows necessary in the Lachlan River. Therefore, Lake Mungo may have been dry when water was available in Willandra Creek and in other lakes. Considering these issues we can see how much water was required to form Lake Mungo 'megalake'!

Intermittent drying of Mungo must have seen people move to other Willandra lakes as well as more distant lakes in the region. Archaeological sites around the lakes have often been interpreted as people just choosing to camp on one lake rather than another or spreading out. Instead, the pattern observed might reflect people moving from one lake to another, particularly from Mungo to Leaghur or Garnpung, when Mungo was dry. After 40 ka Lake Mungo continued to receive water but entered a period of low lake level with regular drying, a lake full phase around 30 ka and a megalake stage at 24 ka. Lunette formation has a long history and is intimately tied to enviroclimatic conditions. Sporadic wet/dry cycling allowed prevailing winds from the southwest to blow sediments onto the eastern shoreline. Washed quartz beach sands accumulated during wet times and peletal clays during lake dry phases, blown up from deflating lake floors and deposited on eastern shores. Over time a layer cake formation accumulated as the lunette grew, curling around the lake's eastern shore. Over thousands of years, various coloured sediments accumulated forming a late Pleistocene environmental timeline of lake history. The coloured sedimentary units have been given names corresponding to local sheep stations.

TABLE 5.1 Archaeological Faunal Listing for the Five Major Willandra Lakes

Marsupials	Common Name	MUL	GA	LE	MUN	AR	N
Dasyurus maculatus	Spotted-tailed Quoll	1	2		1		4
Dasyurus geoffroyi	Western Quoll			3	2		5
Dasyurus sp.	Native cat	1	3		2		6
Dasycercus cristicauda	Mulgara	1	1	1	1		4
Isoodon obesulus	Southern Brown Bandicoot	1					1
Perameles bougainville	Western Barred Bandicoot	2	1		1		4
Perameles sp.	Bandicoot	2	3		1		6
Macrotis lagotis	Rabbit-eared bandicoot (bilby)		12	4	1		17
Lasiorhinus krefftii	Hairy-nosed wombat	1		5	12	3	21
Trichosurus vulpecula	Brush-tailed possum	1					1
Bettongia penicillate	Brush-tailed Bettong	2	4		1		7
Bettongia lesueur	Burrowing Bettong	2	9	2	5		18
Bettongia sp.	Bettong			7	7	3	17
Caloprymnus campestris	Desert Rat-kangaroo			1			1
Lagorchestes leporides	Eastern Hare-wallaby		3		1		4
Lagorchestes sp.	Hare Wallaby		1	1	8	1	11
Lagostrophus fasciatus	Banded Hare-wallaby	2	4				6
Onychogalea fraenata	Bridled Nail Tail Wallaby				1	1	2

Genus/Species	Common Name	MUL	GA	LE	MUN	AR	N
Macropus gigateus	Eastern Grey Kangaroo	1	1		1		4
M. robustsus	Common Wallaroo	1	2				3
M. rufus	Red Kangaroo	1	1	1	1		4
M. altus	Extinct[a]	1	1				1
M. cooperi	Extinct[a]	1					1

ANIMALS EXTINCT BEFORE 1788[1]

Genus/Species	Common Name	MUL	GA	LE	MUN	AR	N
Thylacinus cinocephalus	Tasmanian Tiger	1			1		2
Procoptodon goliah	Giant Short-faced kangaroo		7	3	5		15
Protemnodon sp.[a]	Giant grey kangaroo		1				1
Zygomaturus trilobus	Diprotodontid (32 ka)			1			1
Sthenurus	Giant Short-faced kangaroo				1		1
Sacophilus harrisii	Tasmanian Devil	1	5	1	4		11
Genyornis newtonii	Giant Rattite flightless bird		1	1	3		5
Macropus ferragus	Wallaby				1		1

Birds	Common Name	MUL	GA	LE	MUN	AR	N
Dromaius novaehol-landie	Emu	2	6	4	7	2	21
Podoceps sp.	Grebe	1					1
Ardeidae sp.	Heron	1					1
Columbidae sp.	Dove	1					1
Anatidae sp.	Duck			1			1

(Continued)

TABLE 5.1 Archaeological Faunal Listing for the Five Major Willandra Lakes (cont.)

Fish, Shellfish, Crayfish	Common Name	MUL	GA	LE	MUN	AR	N
Maccullochelia peeli	Murray Cod	1	1		1		3
Macaquaria ambigua	Golden Perch	2	13	5	5	1	26
Vellesunio ambiguus	Fresh water mussel middens	2	23	16	7	2	50
Alathyria jacksoni	River mussel		2	1	1		4
Corbiculinid sp.	Snail		4	1	3	1	9
Limnionid sp.	Snail			22	7	2	31
Gastropod sp.	Snail			1		2	3
Cherax destructor	Yabbie	1	9		2		12

Source: After P. Clark.
[a]Uncertain (unknown) timing of the extinction of these species.

Lunette erosion has exposed the coloured layers to varying degrees by breaking down internal structures by wind and rain, forming deep gullies that have cut into the lunette structure. Gullying accelerated during the last 140 years through grazing domestic stock. In a few places the original lunette construction and size can be seen where little or no erosion has taken place. Erosion has also led to formation of spectacular 'residuals' or islands of the original lunette. They consist of harder sediments left behind and some tower-like ancient wrecked castles casting a colourful but eerie and lonely feel across the landscape at sunset. Looking out across the dry lake bed in the quiet of evening from a high spot on the lunette the scene can give the viewer an uncanny feeling of time, deep time (Fig 5.3).

Besides forming a climatic record and the story of lake history, the lunettes have become unique time capsules preserving the human history of the region. As sediments gradually accumulated into lunettes so humans made their camps along the lake shores and the remains of those camps as well as burials were gradually stored within them. The resulting archaeological debris have provided a series of proxy indicators of the life of the people. Archaeological remains include fires, activity areas, artefact and flake scatters, tool manufacturing places, dietary debris consisting of marsupial and fish bone, freshwater mussel shell middens, burial places and a nearby stone quarry. Hunting and gathering strategies and food preferences can be constructed from these remains which continue to erode from lunettes on all the lakes forming an expansive display of late Pleistocene occupation.

The remains of small mammals and, birds, emu eggs, freshwater mussels (*Velesunio ambiguus*), lizards and golden perch (*Plectroplites* spp.) in hearths and shell middens dated to before 34,000 BP attest to the varied and protein-rich diet enjoyed by the Willandra inhabitants in the Pleistocene (Allen, 1972).

FIGURE 5.3   Evening on the Lake Mungo lunette showing Late Pleistocene living surfaces and residuals. *(Source: Photo S. Webb).*

Large animals and megafaunal remains are, however, conspicuously absent from the food debris. There are some interesting contrasts in the distribution of food debris, which currently defy explanation. Clark (1987) reported an extensive hearth at the northern end of the Lake Mungo lunette, dated to 34,000 BP, which contains nothing but golden perch (*Plectroplites ambiguus*). Similar hearths at the southern end have a predominance of small mammals and a total lack of fish bone. In respect of the food quest, the pattern of dental attrition in WLH 3 may indicate the manufacture of nets (see Chapter 6).

Although hard times must have occurred on occasion, starvation and nutritional inadequacy were probably rare. Thus, attainment of natural body growth and development and the maintenance of good health and well-being must have been the norm. Indeed, this is confirmed by the evidence in the skeletal collection of quick and accurate healing of fractures, while the strong structural composition of long bones attests to well-built limbs, and prominent areas of muscle attachment to ample muscle development. The range of stature is wide (151–178 cm), but from what has been said already, the small size of some individuals may have less to do with nutritional inadequacy than with natural factors of bodily proportion. There are pathologies, such as fractures and osteoarthritis similar to those of hunter-gatherer populations elsewhere.

Archaeological material scattered across the ancient living areas has often raised more questions than answers, particularly concerning the origin and appearance of the first inhabitants and what their material culture might mean. At the same time people have left behind fascinating clues about their culture, belief systems, ritual and ceremonies and the complex language they must have used to convey their culture to each other. That must have concerned how they should dispose of the dead and why certain types of burial practice must be carried out and maintained. The same material has also left behind unequivocal evidence of the morphological contrast among the inhabitants, a topic that has generated much more heat than light so far. However, that puzzle, I believe, is close to being solved.

The sedimentary history of the region is not quite as simple as the above description might convey. Detailed geomorphological interpretation of lunette composition has largely taken place only on Lake Mungo, particularly at its southern end. As explained before, Lake Mungo has a different hydrological history from the other lakes and in some respects is a backwater lake that might mean its sedimentary history is not necessarily synchronised with other lakes. Therefore, we cannot be certain whether that history is reflected in other lakes although it is suggested it might be in a gross fashion. Logically, however, a lake with a different hydrological profile and filling mechanism, from other lakes may well have a different sedimentological profile from others. A much wider geomorphological study of all lakes is required to compare, correlate and contrast their independent and synchronous histories. Lake Mungo stratigraphy is often extrapolated to other lakes but usually with caution. Understanding the sedimentary history of this enormous region has been pushed along by the dogged persistence of Jim

Bowler who researched that history through examining stratigraphic and lake-fill sequences, particularly at Lake Mungo, during the early days of investigations in the area. But even he would concede the work is a long way from complete.

Much of Lake Mungo's 30 km lunette is eroded. In some areas it has almost disappeared, eroding down to the 50–40 ka level where WHL 1 and 3 were found, and in some places even exposing the ancient Gol Gol layer at the base. The lower Mungo level shows a confusing picture of a deflated surface with stone artefacts, mammal, fish and human bone, campfires and other ancient debris. Confusingly, that is mixed with the odd fence post and rusted wire from long disused boundary fences. The resulting pancaked history only adds to the confusion of interpretation which bedevils clarifying artefact sequencing and temporal and stratigraphic positioning of skeletal remains. Work has focussed mostly on the southern end of the lunette because of the WLH 1 and 3 discoveries for which dates and contexts were eagerly sought. Since then Lake Mungo has taken centre stage in regional archaeological investigation. The value of the lunette's southern end arose again in 1987 when another burial emerged between WLH 1 and WLH 3. It was a young adult cranium buried on the same stratigraphic level as WLH1 and 3 and was given number WLH 135. It is described as far as possible in Chapter 11.

THE AGE OF THE COLLECTION

The age of the Willandra collection is crucial to its fullest interpretation and placement in a palaeoanthropological context and a chronology for the stratigraphic units within the Lake Mungo lunette has been constructed over the years to aid this. Originally, the basis for archaeological dating was from Bowler's geomorphological work. (Bowler, 1971, 1973, 1976a, 1976b, 1980, 1983, 1998; Bowler & Thorne 1976; Bowler & Wasson, 1983; Bowler et al., 1970, Bowler, Thorne, & Polach, 1972). He was the first person to establish a chronology for the stratigraphic sequences that were apparent in the blow outs and eroded gullies of the Lake Mungo's lunette. Much of his evidence for dating his geomorphological sequence was provided by dates from shell middens and hearths. Since 1974 over 250 mainly radiocarbon dates have been determined for sites around the lakes that have come from those as well as fish remains and the comparatively few excavations. Most dates are now thought to be unreliable or inaccurate for various reasons and a new dating program is required throughout the region using more recent and much more accurate geochronological techniques.

Dating the collection initially relied on radiocarbon dating particularly of shell and fire places. Skeletal remains were always difficult to place in temporal context because only three individuals were found in situ, all at the southern end of the Mungo lunette. They were associated with the sedimentary equivalent of Bowler's Mungo lacustral phase that ended about 36 ka. Attempts to date five other individuals from the series (WLH 9, 23, 24, 44 and 122) have proved equivocal. They were published over 25 years ago (Webb, 1989) but are no longer thought valid. Nevertheless, almost all individuals in the collection are heavily mineralised and were found eroding from late Pleistocene sediments predating the final lunette formation and lake drying that took place around 17 ka. This conclusion is based on considerations relating to bone collection. They include the position of specimen recovery, bone condition, fragmentation patterns and bone surface erosion, degree of mineralisation, mineral staining and the colour and form of carbonate encrustation. The highly mineralised, red stained cranial and postcranial fragments of WLH 52 match the deep red sediments of the oldest sedimentary level, Gol Gol. If that is the case then this individual may go back beyond 60 ka, but further research is needed.

GEOMORPHOLOGICAL FACTORS

Recent reassessment of Lake Mungo's stratigraphy against a background of earlier research shows that the timing of major sedimentary sequences required updating. The basic Willandra stratigraphic sequence now goes back >120 ky with the dark red lowest Gol Gol levels. Major lunette sedimentary sequences from youngest to oldest are: Mulurulu (13–16 ky), Zanci (18–25 ky), Arumpo (25–32 ky), upper Mungo (32–40 ky?), lower Mungo (40–65 ky?) and Gol Gol (MIS 5). Many dates supporting the age of this stratigraphic sequence have been obtained from shell middens and hearths, pointing to a firm occupation of the region over almost 50,000 years, although there is evidence that it is far older than that (see below). These dates not only testify to a well-established human occupation but also to the well adapted population to the semiarid conditions that may indicate how long they lived there which is a factor how much earlier people arrived on this continent. The broad variety of archaeological evidence shows the skeletal remains testify to a late Pleistocene population throughout the lake system living there on an almost continuous basis, whether lakes were full or dry, perhaps till 17–16 ka when permanent drying of the lakes required they shifted towards the Murray and other large continuously flowing rivers.

The length of time bone has remained in the ground and the conditions that have affected it during burial are important factors for its survival. The longer a bone has been buried in the Willandra the more likely that it is mineralised. Percolation of ground and rainwater over thousands of years through soils containing carbonates and silicates promotes silicification throughout the whole bone, carbonate deposition on external surfaces and its organic components are altered with apatite taking up fluorine to produce inorganic fluorapatite. The strength of the processes in the Willandra region was exemplified when 56 samples of unburnt bone were analysed to determine their nitrogen content, as an indicator of organic content and thus its suitability for radiocarbon dating. Only two of the samples, WLH 15 and 55 had more than the minimum 0.2% nitrogen required, so radiocarbon dating could not be pursued.

Almost all bone in the collection is covered with calcium carbonate to some extent. It ranges from a thin wash covering the whole bone in a layer of hard material several millimetres thick. Its colour varies from light grey to yellow/ orange, red and dark brown and is associated with lunette sediments. Permeable sandy substrate and plentiful primary carbonate associated with enhanced rainfall set thick secondary carbonate horizons. These were ideal conditions for bone preservation. As lakes dried and salinities increased, earlier lunette sediments were covered by a thick sequence of wind-blown sandy clays, culminating in the deposition of the Zanci unit. With the absence of lake biota, the supply of carbonate was greatly reduced. Additionally, the carbonate which did occur existed as extremely fine grains locked up within clay pellets. The combination of reduction in primary carbonate, decreased permeability due to higher clay content, and lower effective precipitation resulted in thinner, less well-developed, secondary carbonate horizons forming in soils laid down since lunette deposition ceased. In places, the lunettes have become severely deflated and that includes some of the Mungo lunette. That means metres of calcareous soil overlying the Zanci unit have been removed as well as upper Mungo soils. The origin of carbonate-incrusted bone found on the surface of the Mungo unit suggests it is from that unit rather than anything younger that has deflated from the younger Zanci unit.

Calcareous soils continue to form especially on sandier substrates. Away from the large lunettes, in areas of more complex sedimentary formation, such as the Garnpung/Leaghur inter-lake area, there is close proximity of sediments and calcareous soils which differ in both age and provenance. In such areas the possibility exists that bone interred in soils formed during the Holocene could obtain a thin coating of secondary carbonate, be reworked and left as a lag deposit on lower, older sediments leading to stratigraphic misinterpretation. Dark blue–black patchy manganese staining or spotting is present on some bone often beneath overlying carbonate. Manganese staining is common on late Pleistocene skeletal material from elsewhere, is a ubiquitous mineral, commonly found in soils having a high calcium carbonate component.

Erosion exposes camp fires still containing crustaceans and the skeletal remains of water birds, large golden perch (*Macquaria ambigua*) and Murray cod (*Maccullochella peeli*). Measurements of the diameter of vertebra from these fish suggest they would have exceeded 20 kg. Freshwater shell middens are still found where people formed them along ancient beaches over perhaps 50 millennia ago. The bones and teeth of a variety of small macropods and wombats also protrude from ancient hearths all testifying to the nutritional variety and abundance that was available. Hundreds of fossil footprints of 20 ka ice age occupants going about their daily activities have also been found. Some form long meandering track ways made by men, women and children indicating a variety of activities of people of all ages. Some show men running fast presumably hunting in a group chasing game. They present a unique social gathering and bring to life the people then while making a unique human connection across millennia with those who visit the region today (Webb, Cupper, & Robbins, 2006; Webb, 2007).

Although much of the Willandra collection is fragmentary, and its 20,000 years younger than Australia's oldest archaeological date, it still represents a cultural and biological shadow of those first 'super nomads', who roamed the world, crossing continents and sailing the ocean. It makes a connection with other cultural and environmental data that are also visible across the system. Their bones represent the success of that journey and provide a tantalising glimpse of the cognitive and spiritual life of some of the earliest people living on this continent. The burials have told us much about the people themselves: their morphological characteristics, activities, health, society, beliefs, rituals and ceremony, craft skills and the undeniable use of complex language. The baggage they brought with them is starkly evident of cultural imperatives that were part of their culture at the time. We can see their ceremonies that gave them a believe system that linked them to a world beyond life. That evidence gives strong substance to the social fabric of people that lived there not long after the first arrivals landed on our northern coasts. They also brought with them the genes of others who they had encountered on the way and which gave them a unique diversity in morphology and possibly culture.

The Willandra Lakes Collection in World Context

The Willandra Lakes collection has a broad sweep of values that bring together a unique record presenting cultural, anatomical and environmental pictures of Australian history. But primarily it represents not only the oldest remains yet found it also represents the skeletal remains of all our ancestors. Those were the ancestors that populated the world and laid the foundations of society every-where. We don't know what they really looked like but one thing we can be sure off they did not look like anyone alive and the planet today.

That is a fascinating and very special thing to think about. The collection was seen originally as a valuable resource for unlocking the origins of the first arrivals, but with the passing of time the value of the collection has broadened from that narrow view of its worth. Now we know those remains represent not just the first people but show a broader story of the trans-world movement of people. When the first Willandra humans were discovered we did not know about the story of modern humans as it has unfolded over the last 30 years. That emerged essentially after the Willandra collection had been placed under a research moratorium enforced by state and federal legislation and it was lost to further research. That has resulted in the Australian skeletal evidence being eliminated from a world study of all modern humans and in a broader context that involved recent developments in the story of modern humans. The ability to delve into and extract palaeobiological information from a collection like the Willandra collection has grown exponentially in recent years. So, what does the cultural and biological information from the collection mean for a group that was spatially and temporally far from its origin with many generations separating them from those who first left Africa.

Undertaking such a spectacular journey could impose socially adaptive traits as well as cultural and behavioural changes along the way. That may have included morphological changes all of which would mean those who finished the journey were not the same as those who started it. While some individuals in the Willandra collection might represent the oldest fossil humans yet discovered here, they do not represent the first people to reach Australia. The WLH 1 and 3 skeletons were buried at Lake Mungo at least 28,000 years after the first arrivals (Bowler & Magee, 2000; Bowler et al., 2003; Clarkson et al., 2017). Just because we know much more about modern humans and their origins than we did 25 years ago, it does not exclude the possibility of other people who lived in the region arriving here even earlier, at the same time or subsequent to modern people. Because of our awareness of the existence of several archaic hominins in the region, perhaps more than ever we cannot exclude the possibility of those people or their genes, in one guise or another, entering Australia. Moreover, they could have done that some time before the arrival of modern humans. Indeed, what we have learned in recent years about such people existing at the same time as modern human moved through South East Asia has increased the likelihood of that possibility. But even if modern humans were the first arrivals, they did not resemble anything like modern Aboriginal people, particularly if the journey here took place in a short time frame.

Individuals referred to as 'robusts' were generally tall, had a rugged build, long, muscle-crafted crania with thick cranial walls, pronounced brow ridges, broad faces, large teeth and palates and rugged areas of muscle insertion and tori around the skull. In complete contrast the gracile people had modern human cranial architecture, were delicately built, with thin cranial walls, no brow ridges, high oval, rounded crania, smooth areas of muscle insertion, smaller dentitions and were generally shorter than robusts. It is now apparent they were modern humans, descendants of those who left Africa, crossed Asia, sailed the Timor Sea and made other water crossings to Australia. Their culture included: complex burials (indicating a belief system), ritual and ceremony in which they used ochre, they had a complex language and advanced weaponry including spear throwers and ground stone axes and they did not hunt megafauna.

Changing Course

The Willandra collection became almost entirely responsible for how we saw the earliest arrivals and, for some, they represented how modern humans had evolved around the world. Thus, the Regional Continuity theory was stretched to other regions by those who sought explanations for the emergence of modern humans elsewhere. Subsequent research during the last couple of decades has proved that theory wrong. It had explained, for example, the contrasting morphologies found in the Willandra but the story was far more complex than it was believed to be at the time. The robust appearance of some Australian fossils remains a puzzle, however, and is patently not explained by evoking a wide range of variation among modern humans. Perhaps I should qualify that. From our present knowledge it may also be likely that the robust appearance of some Australia fossils could have an explanation in the mixing of modern humans with archaic populations that are now known to have inhabited various parts of Asia and elsewhere that might very well produce a population with a wide range of variation.

One activity fraught with danger in archaeology is piecing together a puzzle without all the pieces, that was the problem with Regional Continuity and holding faith in that theory was just a stage in knowing the actual story. Thirty years ago, the modern human genome was unknown or at least in its infancy. We did not know about genetic distancing and the close genetic relationship between all the world's people and the breakthroughs in genetic sequencing and other research had not yet arrived. The discovery that genetic material could be retrieved from human bone, let alone ancient bone, was deemed science fiction by some. I remember receiving a comment from a prominent archaeologist of: *'pure science fiction'* when I told him of the recent identification of blood residue on stone tools, first published in *Science* by Tom Loy in 1982. What would the same person have said at that time about retrieving genes from Neanderthal bones? *Tempus fugit*; time has flown and naturally the physical evidence of the story of humanity has grown as well as the technical ability to retrieve and interpret more evidence from the material we gather. With those advances we have seen the human story change dramatically and in turn it has changed the possibilities for establishing the origins and story of us all including the First Australians.

The world of archaeology has changed exponentially over the last 40 years in a way it never did or could before. It did not change because it faced what now seems significant barriers to the scientific methods it employed, interpretations made on the meagre evidence it dealt with and the limited capabilities of its methods but things have changed. To go further, we also needed to wait for certain breakthroughs in our ability to read the past. Accurate chronological dating was one of the barriers, but better archaeological knowledge particularly in the realm of human evolution relied on obtaining more and different information from the meagre evidence that was discovered. The technological age has affected the dusty art of archaeology providing it with new cutting-edge tools to see into the past in a way once only dreamt of, even in 1980. Breakthroughs came by riding on the back of technological advances particularly in genetics, chemistry and physics as well as many other scientific tools such as ground penetrating radar, the GPS, improved geochronology and above all computer technology. All of these have contributed profoundly to our understanding of the past so that we can see deeper and more accurately into it and read more from the hidden record of preserved objects and people as never before. A good friend of mine (another archaeologist) once said to me with a smile, we will eventually be able to hear the tune whistled by the ancient potter as they formed their pot on a whirring stone wheel in Nubia. *Pure science fiction*, or is it…? It certainly was at the time my friend said it, I am not so sure now, perhaps anything is possible in archaeology.

The above sounds like digression you might say: not really. Every week we hear of another archaeological find or breakthrough that has changed old interpretations of what we thought the human story was about and it makes writing a book like this particularly hard. I have had to make some large changes to this book since I first sent it to the publisher in the middle of last year. Sometimes changes take place so fast it is difficult to keep up with emerging ideas and the recombination of new and old data that follow. As an immediate example, time has certainly moved on since the first archaeological discoveries were made at Lake Mungo 46 years ago. But the scientific assessment has not. The greater meaning of the human skeletal collection, the recovery of more skeletal evidence and the palaeoenvironmental pageant assembled from the human and geological evidence of the region has stopped. One reason for that has been a moratorium on scientific study enforced by the Aboriginal community for many years that stagnated further biological anthropological research in a broad sense and prevented us releasing the ancient knowledge that could help us further understand the first people who arrived here. Nevertheless, the archaeological significance of the Willandra Lakes region remains the same as it did when it was first discovered, the only thing is, we cannot unlock more of it at present and that has been the position for the last quarter of a century.

Australia and eventually South America were end points of the movement of modern humans out of Africa. Migration through island South East Asia to reach Australia was somewhat different from the land-based movement they had undergone previously and those in the north continued to some extent into the Americas. They probably skimmed the coast tracking very close to the mainland to reach North America. For those continuing south moving great distances across land was now replaced with island hopping where the use of water craft was the only way forward and that was different again in the repertoire of modern human transport. Somehow, I think they were ready for an ocean crossing perhaps having cut their teeth on smaller journeys among the islands of the Philippines or among other islands of the region not very different from the strategy of their kinfolk roving close to the Arctic. They had accomplished the required skills and marine knowledge to use that form of transport by at least 67,000 years ago as the Callao Cave evidence seems to suggest.

Those changes are dramatically reflected in the layer cake composition of sediments that make up the lunette construction around all lakes. The coloured coded quartz and clay bands visible in the eroded and etched lunette face starkly reflect severe climatic changes as lakes periodically dried and filled rhythmically moving to the drum of fluctuating world temperatures and climate. Over 100 ice ages occurred during the last 2½ million years. During each one, environments and their animal and plant populations responded in various ways in different combinations in different places across the planet. They promoted both speciation and extinctions. Landmasses changed their ecology and environments as ice shifted south from high northern latitudes covering large parts of Siberia and Europe and blotting out Canada completely. Lowering sea levels fastened continents together, merging islands adding Japan to China, Britain to Europe, Indonesia to Malaysia and Papua New Guinea to Australia to mention only a few examples. As all this took place a group of people, living at the height of the last glacial, camped close to one of the lakes, carrying out their daily activities of hunting and gathering and leaving their footprints for all to see 20,000 years later. As those people lived their daily lives, in central Australia deserts changed to savannahs straddled by massive lakes fed by equally large river systems and just as quickly they became deserts again bigger than before with vast empty lake beds and dry river channels.

The arrival of anatomically modern humans in Australia terminated one of humanity's longest and toughest journeys humans of any kind had ever made. Their arrival in Australia marked the end of that journey which began in Africa. I began this book in Kakadu National Park and the artistic record of the past. Australia's engravings and paintings span an extraordinary long time and undoubtedly some of it is of great antiquity. We might ask how much of it arrived as cognitive and cultural baggage with the first arrivals? The art of Arnhem Land could be contemporary with European cave paintings, we just don't know because we cannot date it, nevertheless it is often reported as the oldest painted art in the world but that cannot be proven just yet. The idea art arose when populations reached a critical number has been proposed. However, we are seeing in the bi-hemispheric juxtaposition of artistic capabilities that in both Europe and Australia at the same time. If it did not, then its appearance is not connected to population growth because it cannot be expected that there were many people in Australia for thousands of years after the first landings. Much is made of European cave art and rightly so. The beautiful depictions of animals and the variety of styles and decoration is a stark tribute the European Magdalenians. Art in the Kimberley and Arnhem Land is no doubt the first pictures of Australia. My reading of recent commentary suggest that such art began with the first European *Homo sapiens*, marking Europe as a centre of innovation and cognitive inspiration spawned by people of that time, that is if Neanderthals did not beat them to it. The argument is reminiscent of cultural diffusionist arguments of the late 19th and early 20th century that claimed cultural innovation always began in Europe and moved from there to other parts of the world. Australia might be a modern human outpost but it was not a backwater. It is clear we are part of an artistic movement among modern humans but rather than comparatively few caves in an area the size of Victoria, Australia has thousands of such art sites scattered across a continent the size of Europe and many must belong in the Late Pleistocene. The people that arrived here had nothing to do with the culture of Europe or even Africa. They were a product of their journey, particularly its last phases, which took them across large oceanic gaps imbuing them with even greater abilities than they started out with. Their accumulated experience over generations as they moved, survived and wandered across vast tracts of country between here and Africa built their character and the foundations of a particular culture. One that had not left Africa but formed on the way as they accumulated an ability to cope, survive and transmit culture to future generations. By the time they reached Australia this ability was in full swing. These were the *super nomads* a term which could not be applied to them when they began their journey but it fits the bill when they finally landed on our shores.

The Earliest Arrivals, So Far

The Willandra Lakes mark one of the finishing posts of modern human movement out of Africa. Except for rats and bats, humans were the only mammals to reach Australia, until somebody brought their pet dog here on their raft around 5000 years ago. The mere fact that people reached here using rafts indicates what an undaunted and determined group they were, moving across the world till they could go no further than Australia's south, west and east coasts a few hundred kilometres further on from Lake Mungo. Landing somewhere on Australia's north coast was only the beginning of the end for the new arrivals. Depending on where they landed they then began a 3–4000 km walk to reach the south of the continent, the equivalent of walking from the Persian Gulf to the Burmese border, the real journey must have been longer. It was as if these explorers were driven and needed to continue, at least while an empty landscape lay before them. People eventually reached the Willandra Lakes. Australia's first 40,000-year date for human occupation came from the Willandra, but importantly it was where the earliest physical evidence of the people themselves was found and that situation continues. The date of 40 ka from Lake Mungo became a widely accepted measurement for the length of time had been here and it has retained that status for the last 40 years, becoming a chantable corner stone for Aboriginal land rights, marches and social campaigning. In that time, few expected Australia's human occupation would reach much beyond 40 or even 50,000 years (O'Connell & Allen, 2015). Recent genetic studies suggested that a *successful* modern human movement out of Africa did not occur before 60–70 ka although we know some ventured out at least 180,000 years ago, and they placed the first arrival around 50 ka (Tobler et al., 2017). However, any successful exit, ~100 ka, should have placed an arrival towards 80-70 ka or even earlier and we know now that is the case.

Recent evidence has put the Willandra Lakes into perspective. We now know they are part of a settlement pattern around Australia probably completed by 50 ka. Sites have included Devil's Lair in the far southwest of Western Australia and, unsurprisingly, sites in the Kimberley and Pilbara in northwestern Australia and Warratyi, in the Flinders Ranges of South Australia (Turney et al., 2001; Hiscock, O'Connor, Balme, & Maloney, 2016; Hamm et al., 2016). But the recent re-excavation of Madjedbebe in Arnhem Land has shown this once controversial site that was called Malakananja is 65 ka (Clarkson et al., 2015, 2017). Well-made artefacts such as spear points, ochre pieces, the use of mica and a variety of ground stone axes, have also been found there. Ironically, 30 years ago, this site (then known as Malakananja) was original believed to be that old but the 60 ka date obtained then was rejected by many who did not accept people could have reached Australia by that time. They argued there was something wrong either with the dating or the sedimentological taphonomy or both. This now well-dated site, independently dated by two laboratories, is now the oldest evidence for humans who crossed the Wallace Line taking long ocean boat or raft trips; constructing adequate craft for such a crossing and continuing a worldwide dispersal to its limits. Nevertheless, even with older sites than those so far found in the Willlandra Lakes region, each is a single site, but the Willandra consists of clusters of sites over a comparatively small area that is without parallel in Australia, and there are undoubtedly more to be discovered. But what can we make of all this evidence?

People reached Australia by 65 ka. That was something totally rejected by most archaeologists till now, although the recent discovery of several 50 ka dates here softened that attitude somewhat. How much further did they penetrate the continent and is it possible they landed even earlier and elsewhere on the continent? I would suggest their capabilities and skills at 65 ka were probably no different from those 100-120,000 years before, some they possibly developed when they lived and moved through South East Asia and Sunda, so making an earlier journey here was always possible. I want to reiterate that with modern humans in southern China by at least 120 ka, that possibility seems very likely. That also raises the question regarding mixing of modern humans with relict archaic populations in South East Asia, China and elsewhere and the bringing of those genes onto this continent. However, we should consider survival. How many landings might there have been before or at 65 ka is unknown as well as the number of people involved. The smaller the numbers of both watercraft and people the bigger the chances of group extinction in the isolation of a vast continent. Talking of extinction, the emergence of this date shows a very long overlap of at least 35,000 years between humans and megafauna here, and that means it is now even less likely that humans were the direct cause of Australia's megafaunal extinctions (Webb, 2013).

REFERENCES

Allen, H. R. (1972). *Where the crow flies backwards (Unpublished PhD thesis)*. Canberra: Australian National University.

Bowler, J.M., 1971. Pleistocene salinities and climatic change, evidence from lakes and lunettes in southeastern Australia. In: Mulvaney, D.J., Golson, J. (Eds.), Aboriginal man and environment in Australia. Australian National University Press, Canberra, pp. 47–65.

Bowler, J.M., 1973. Clay dunes, their occurrence, formation and environmental significance. Earth-Science Reviews 9, 315–338.

Bowler, J.M., 1976a. Recent developments in reconstructing late Quaternary environments in Australia. In: Kirk, R.L., Thorne, A.G. (Eds.), The Origin of the Australians. Australian Institute of Aboriginal Studies, Canberra, pp. 55–77.

Bowler, J.M., 1976b. Aridity in Australia, age, origins and expression in Aeolian landforms and sediments. Earth-Science Reviews 12, 279–310.

Bowler, J.M., 1980. Quaternary chronology and palaeohydrology in the evolution of Mallee landscapes. In: Storrier, R.R., Stannard, M.E. (Eds.), Aeolian landscapes in the semi-arid zone of south eastern Australia. Australian Society of Soil Science, Riverina Branch, Wagga Wagga (NSW), pp. 17–36, Proceedings of a conference held at Mildura, Victoria, Australia on October 17–18, 1979.

Bowler, J. M. (1983). Lunettes as indices of hydrologic change, a review of Australian evidence. *Proceedings of the Royal Society of Victoria,* 95(3), 147–168.

Bowler, J. M. (1998). Willandra Lakes revisited, environmental framework for human occupation. *Archaeology in Oceania,* 33(3), 120–155.

Bowler, J. M. & Magee, J. W. (2000). Redating Australia's oldest human remains, a sceptic's view. *Journal of Human Evolution,* 38(5), 719–726.

Bowler, J. M. & Thorne, A. G. (1976). Human remains from Lake Mungo, discovery and excavation of Lake Mungo III. In: Kirk, R.L., Thorne, A.G. (Eds.), The origin of the Australians. Australian Institute of Aboriginal Studies, Canberra, pp. 127–138.

Bowler, J., Wasson, R.J., 1983. Glacial age environments of inland Australia. In: Vogel, J.C. (Ed.), Late Cainozoic palaeoclimates of the southern hemisphere. A.A. Balkema, Rotterdam, pp. 183–208, SASQUA international symposium, Swaziland, August 29–September 2, 1983.

Bowler, J.M., Jones, R., Allen, H., Thorne, A.G., 1970. Pleistocene human remains from Australia, a living site and human cremation from Lake Mungo, Western New South Wales. World Archaeology 2 (1), 39–60.

Bowler, J.M., Thorne, A.G., Polach, H.A., 1972. Pleistocene man in Australia, age and significance of the Mungo skeleton. Nature 240, 48–50.

Bowler, J.M., Johnston, H., Olley, J.M., Prescott, J.R., Roberts, R.G., Shawcross, W., et al., 2003. New ages for human occupation and climatic change at Lake Mungo, Australia. Nature 421, 837–840.

Cane, S., 2013. First footprints. Allen & Unwin, Melbourne, Australia.

Clark, P., 1987. Report on a plan of management for the Willandra Lakes region, western New South Wales. New South Wales National Parks and Wildlife Service, Sydney, (Unpublished report).

Clarkson, C., Smith, M., Marwick, B., Fullager, R., Wallis, L.A., Faulkner, P., et al., 2015. The archaeology, chronology and stratigraphy of Madjedbebe (Malakunanja II): A site in northern Australia with early occupation. Journal of Human Evolution 83, 46–64.

Clarkson, C., Jacobs, Z., Marwick, B., Fullager, R., Wallis, L., Smith, M., et al., 2017. Human occupation of Northern Australia by 65,000 years ago. Nature 547, 306–310.

Fitzsimmons, K., Stern, N., Murray-Wallace, C.V., 2014. Depositional history and archaeology of the central Mungo lunette, Willandra Lakes, southeast Australia. Journal of Archaeological Science 41, 349–364.

Hamm, G., Mitchell, P., Arnold, L.J., Prideaux, G.J., Questiax, D., Spooner, N.A., et al., 2016. Cultural innovation and megafauna interaction in the early settlement of arid Australia. Nature 539, 280–283.

Hiscock, P., O'Connor, S., Balme, J., Maloney, M., 2016. World's earliest ground-edge axe production coincides with human colonisation of Australia. Australian Archaeology 82 (1), 2–11.

O'Connell, J. F., & Allen, J. (2015). The process, biotic impact, and global implications of the human colonization of Sahul about 47,000 years ago. *Journal of Archaeological Science,* 56, 73–84.

Tobler, R., Rohrlach, A., Soubrier, J., Bover, P., Llamas, B., Tuke, J., et al., 2017. Aboriginal mitogenomes reveal 50,000 years of regionalism in Australia. Nature 5434 (7644), 180–184.

Turney, C.S.M., Bird, M.I., Fifield, L.K., Roberts, R.G., Smith, M.A., Dortch, C.E., et al., 2001. Early human occupation at Devil's Lair, Southwestern Australia 50,000 years ago. Quaternary Research 55, 3–13.

Webb, S.G., 1989. The Willandra Lakes hominids. Panther Press, Canberra.

Webb, S.G., 2007. Further research of the Willandra Lakes fossil footprint site, southeastern Australia. Journal of Human Evolution 52, 711–715.

Webb, S.G., 2013. Corridors to Extinction and the Australian Megafauna. Elsevier, New York, NY.

Webb, S.G., Cupper, M., Robbins, R., 2006. Pleistocene human footprints from the Willandra Lakes, southeastern Australia. Journal of Human Evolution 50, 405–413.

Chapter 6

The Osteology of WLH 1, 2 and 3

Research on the Willandra skeleton collection has been undertaken for the last 30 years because of the moratorium that was placed on it. Nevertheless, when it was available for study, gathering useful data from it and assessing and interpreting what it meant was hampered by its poor condition. The use of such fragmentary and eroded material, that generally lacks most skeletal elements, is limited. Moreover, very few individuals have been or can be accurately dated and most of those have only an approximate date from assumed stratigraphic associations. The fossilised bone as well as its sedimentary origin from within or beneath lunette structures certainly places almost all remains within the Late Pleistocene, specifically between 50–20 ka. However, its antiquity and value as Australia's oldest human skeletal evidence remains, is unchallenged, if for no other reason than there is nothing else. What there was has been reburied under the wishes of the Aboriginal community. That makes the collection our most valuable human remains and cornerstone of the earliest human story here. It also places it as a valuable testament to the dispersal of modern humans and their settlement patterns. Unfortunately, the collection is going to effectively disappear as thoroughly as the Peking man collection did at the beginning of World War II. The Willandra collection will now join that collection as one of only two fossil collections of world importance to be lost to us. It will be

Made in Africa. http://dx.doi.org/10.1016/B978-0-12-814798-6.00006-6

reburied or put in a keeping place without access for scientific study. Moreover, future study of fossil skeletal biology from the region is yet to be guaranteed. Therefore, the collection is now more important now than ever before in order to record and publish all data possible from these important remains that are the heritage of all humanity.

All individuals have most of their skeleton missing and consequently have lost important diagnostic features due to their long interment, and erosion, fragmentation and bone scattering following exposure. Many individuals are represented only by small or single pieces of bone others consist of larger pieces. Occasionally, some fragments yield worthwhile measurements about the individual's general morphology, albeit limited, that can be compared with others. I first want to outline the methods I have employed to analyse these remains in the face of their incomplete and fragmented condition. All measurements were taken using digital callipers.

CRANIAL REMAINS

The Malar

The compact nature of the malar and supra-orbital preserves them over time better than other skeletal parts and there are nine malars in the collection. Malar structure can reflect general cranial robusticity and can be an allometric indicator of cranial morphology and structure. Therefore, two indices were developed to assess malar morphology: *Robusticity* and *Size/Length* (S/L) modules that maximise malar metrical information using four and three measurements, respectively (Fig. 6.1).

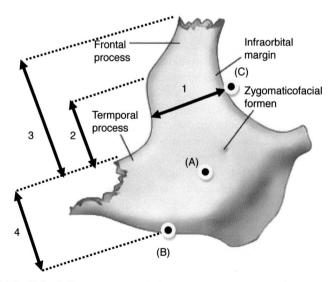

FIGURE 6.1   Malar indices measurement places.

S/L Module

1. The width of the malar body from temporal border to closest point on infra-orbital margin.
2. Height from the superior process of the temporal border to the marginal process.
3. Maximum malar height from midway on the inferior temporal border, vertically to the fronto-zygomatic suture, and
4. Height of the temporal process measured from the corner of the temporal process and malar body to a point directly below on the inferior or masseteric border.

Robusticity Module (RM)

1. Thickness through the malar tuberosity from its anterior facies to its temporal surface.
2. Thickness of the inferior border just superior to the margin.
3. Thickness of malar body measured obliquely through the malar body from the temporal surface to the infraorbital margin. (Same point as in *S/L* 1.)

Malar Morphology

S/L Module

The S/L module was constructed to determine overall malar size (Fig. 6.1). While the malars of WLH 1 and 2 have a similar form, the latter is more robust with a thicker inferior border and deep muscle insertion points at the insertion sites of *zygomaticus minor* and *major* and the masseter which are all particularly prominent. There is also a deep groove at the position of *levator labii superioris* below the inferior orbital margin. The superior edge of this groove continues laterally, reaching the supero-medial corner of the zygomatic tuberosity. It then curves superiorly and parallel with the orbital rim. The effect lends a rugged appearance to the anterior bone surface which contrasts with the flatter anterior surface of the WLH 1 malar. There is a more rounded appearance to the infero-lateral border of the orbit of WLH 2 and a prominent thickening of the marginal process of the malar body not present in WLH 1. These features on WLH 2 show a slight robusticity and slightly larger malar suggesting a gracile male.

Comparing eight individuals, WLH 50 does not stand out from the others in the same way it does for cranial thickness and supraorbital development. It is slightly larger, but not significantly so (Fig. 6.2). Most of the group clusters around the 19–20 value (Table 6.1). While WLH 1 (female) and WLH 2 (? male) stand out as small. WLH 3 groups with WLH 50 and other robust individuals instead of its expected grouping with WLH 1 and 2. That is probably because of body proportions rather than gracility. WLH 3 was much taller (173–175 cm) than WLH 1 (151 cm) and it seems the S/L module reflects that as well as the body proportions that go with stature which might explain the small S/L

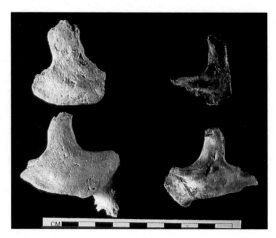

FIGURE 6.2   Malars from WLH 50 (*top left*), WLH1 (*top right*), WLH 3 (*bottom left*) and WLH 2 (*bottom right*).

TABLE 6.1 Malar Size/Length (S/L) Module and Robusticity Module (RM)

WLH No	Measurement codes							S/L module	RM
	1	2	3	4	A	B	C		
1	10.0	12.0	–	20.3	6.0	4.8	6.0	14.1	5.6
2	12.5	13.0	17.0	18.2	8.8	4.3	8.0	15.2	7.0
3	15.6	13.0	24.0	25.0	9.1	5.4	7.0	19.4	7.2
11	8.9	–	–	26.0	7.5	–	6.0	–	6.8
27	15.5	14.2	21.1	–	10. 8	–	9.0	19.2	9.9
28	15.5	11.0	20.1	24.9	7.4	–	11.0	17.9	9.2
50	17.9	16.0	22.0	25.8	14.0	7.9	13.0	20.4	12.5
67	15.2	16.2	28.1	–	9.3	–	8.5	19.8	8.9
102	14.1	15.1	20.7	27.5	11.5	5.4	7.0	19.4	8.0

module values of WLH 1 and 2. There is one additional member of the short-stature group, WLH 11. Unfortunately, only two of the four S/L measurements are available and their stature may have been similar to WLH 3indicated from comparing the two available measurements with those of WLH 1, 2 and 3.

Robusticity Module (R/M)

The R/M reveals a wide range of variation in the series with WLH 50 the most robust and WLH 1–3 the least. These results reflect results from the supraorbital module. WLH 1 and 2 are separated in this comparison that might emphasise gender difference with WLH 2 positioned towards maleness. The rest are scattered both sides of the line with graciles below and robusts above. The position of WLH 50 strongly emphasises that its body size (S/L module) is similar to others in the middle grouping although it does stand out somewhat. The gracility and short stature of WLH 11 suggested in Chapter 10 are supported by this analysis and is similar in its general degree of gracility to WLH 1. There is a consistency in the separation between robust and gracile individuals. Not all graciles are female; correspondingly, the data reveal a vast difference in male morphology within the Willandra Lakes series suggesting that the males, WLH 2 and 3, cannot represent the same population as the WLH 50 male. Further, the wide separation of the two groups makes it highly unlikely that they represent a wide range of variation as some have suggested.

Supraorbital Development

Besides describing certain features of WLH 1, 2, 3 others are made mention of as a comparison. The supraorbital region was assessed using medial, middle and lateral thickness defined as follows:

- Medial is from medial of the supraorbital notch, close to the nasal bones and vertically aligned with the medial wall of the orbit (Fig. 6.3);
- Middle thickness is vertical and lateral to the supraorbital notch halfway along the superior orbital margin;
- Lateral is a superolateral measurement from the orbital plate of the frontal, inside the free border, and through the zygomatic trigone.

A supraorbital module (SOM) multiplies medial, middle and lateral measurements, divides them by three and then 100 (Fig. 6.3). There were 11 individuals for which these three measurements were taken and another five with two dimensions. Four individuals are included where a third dimension could be estimated (Table 6.2).

There are two basic types of browridge development. One is a medial development around the superciliary ridges (WLH 19) and the other occurs where the glabella is also enlarged boosting the image of a torus. Although a full toral arch is not seen, the lateral brow can be more prominent than the middle section (WLH 18, 69 and 100). Lateral thickness is usually restricted to around the

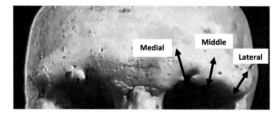

FIGURE 6.3   **Brow ridge with the position of measurements used in the SOM.**

TABLE 6.2 Brow Ridge Thickness of Some Individuals in the WLH Collection

WLH no.	Medial	Middle	Lateral	SO module
1	5	3	5	0.3
3	8	6	8	1.3
11		7	7	1.2
18	17	9	13	6.6
19	16	16	8	6.8
45	–	6	7	–
50	22	22	12	19.4
51	9	4	6	0.7
67	9	7	9	1.9
69	19	10	13	8.2
72	11	6	6	1.3
73	14	8	7	2.6
100	14	8	11	4.1
101	14	9	9	3.8
124	13	9	10	3.9
134	14	5	6	1.4
Range	9–22	4–22	6–13	–
X	12.8	8.4	8.7	–
S.D.	4.7	4.7	2.6	–

zygomatic trigone forming a bulbous angular process above the fronto-zygomatic suture and accentuating postorbital constriction. Reduced supraorbital thickness between the middle and lateral sections further allows a smooth frontal slope to the central area. The Willandra brow pattern occurs in the Kow Swamp and Coobool Creek populations but their SOMs are far smaller than that of WLH 50.

The range of supraorbital development in the series is considerable (Table 6.2). At one end, the supraorbital region of WLH 68 is less than the 0.3 of WLH 1 so it is not included in the analysis. WLH 1 has the smallest measured module which contrasts strongly with those of WLH 18, 19 and 69, although they are far less than the 19.4 of WLH 50. Large brows usually indicate an overall robust-icity among fossil hominins. The suite of cranial traits supporting that among Willandrans include a thick cranial vault, marked cranial buttressing and rugose areas of muscle insertion around the cranium and on the basi-occipital surface. Gracile individuals contrast strongly from the robust form with thin, rounded cranial vaults; an absence of rugged muscle markings; a smooth nuchal area and no superciliary brow development. The lack of supraorbital development is uniform, with no section larger than the other two, adding to a very oval cranial shape in WLH 1, 3, 11, 51 and 68. One exception is WLH 134, whose right superciliary ridge is pronounced compared to its overall gracile appearance. Some individuals lacking brow development are female, but the WLH 3 male is an exception.

There is high positive correlation ($r = 0.85$) between cranial thickness and supraorbital development among 16 individuals that is probably related to allometric factors. An interesting difference between WLH 3 and 134 is the underdeveloped brow ridge in WLH 3 but a large maximum cranial thickness (11 mm) for an otherwise gracile skeleton. WLH 134, on the other hand, presents the reverse trend. Another example is WLH 69 which has a well-developed supraorbital region, although its cranial vault is not as thick as might be expected. On the other hand, WLH 19 is also generally robust, but excels in cranial thickness compared with WLH 69.

Vault Thickness

Only WLH 1, 3 and 50 have large portions of their crania remaining which prevents a proper comparison of their cranial thickness with most other individuals in the series. I deal with WLH 50 in a separate chapter (see Chapter 7) so it's not included except for comparative purposes. Two methods of recording were used: 'primary' and 'secondary' measurement points. Primary measurements are taken at standard anatomical points, such as bregma or the frontal bosses (numbered 2–8, Fig. 6.4). While measurements at asterion (point 7) and at inion (point 9) have been recorded where possible, in many crania these areas have undergone disproportionate thickening. Therefore, they have been excluded from determinations of maximum cranial thickness and the range of vault thickness for the series. Secondary points of measurement were taken at the thickest place on any part of the vault and that is most often quoted in discussion of maximum vault thickness. This concept was introduced to allow many individuals lacking primary points to be included at some level in vault thickness

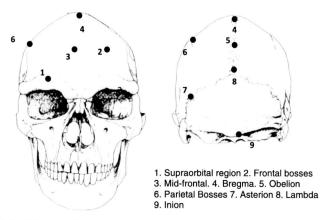

1. Supraorbital region 2. Frontal bosses
3. Mid-frontal. 4. Bregma. 5. Obelion
6. Parietal Bosses 7. Asterion 8. Lambda
9. Inion

FIGURE 6.4 Primary points of vault thickness measurement.

analysis. However, overall vault thickness of some individuals should be treated with caution because of the random nature of secondary thickness points. Maximum secondary thickness measurement can still be used as a general indicator of minimum vault thickness and, thus, development in an individual.

Vault Composition

The collection has an extremely wide range of cranial thickness (Table 6.3). The maximum thickness range using secondary measurements is 5–15 mm which falls within the thickness range using g seven primary anatomical points

TABLE 6.3 Cranial Thickness and Vault Composition in the WLH Collection

WLH no.	ICT	OCT	Diploe (mm)	Vault % diploe	TVT
1	1.5	2.5	2.5	50.0	8
2	3.5	4.0	4.0	25.0	8
3	1.0	1.5	1.5	72.7	11
9	1.0	1.0	1.0	–	5
10	<1.0	2.0	2.0	40.0	5
11	<1.0	1.0	1.0	86.7	7
12	<1.0	2.5	2.5	60.0	10
13	1.0	2.0	2.0	42.9	7
15	<1.0	1.0	1.0	85.7	7
16	2.0	2.0	2.0	40.0	10
17	1.5	2.0	2.0	53.3	7
18	1.0	2.5	2.5	74.1	13
19	2.0	2.5	2.5	50.0	15
20	2.0	3.0	3.0	60.9	11
21	1.0	2.0	2.0	46.7	7
22	2.0	3.0	3.0	66.6	12
23	1.0	1.0	1.0	45.0	10
24	<1.0	1.5	1.5	60.0	10
26	<1.0	1.5	1.5	77.8	9
27	1.5	2.0	2.0	71.4	14
28	2.0	4 .0	4.0	50.0	12
29	<1.0	2.0	2.0	62.5	8
42	1.0	2.0	2.0	70.0	10
43	<1.0	<1.0	<1.0	72.2	9
44	<1.0	<1.0	<1.0	71.4	7

(Continued)

TABLE 6.3 Cranial Thickness and Vault Composition in the WLH Collection (*cont.*)

WLH no.	ICT	OCT	Diploe (mm)	Vault % diploe	TVT
45	1.5	2.5	2.5	64.3	14
46	1.0	3.0	3.0	70.0	10
47	1.0	1.0	1.0	57.1	7
48	1.5	1.0	1.0	80.0	5
49	<1.0	2.0	2.0	75.0	8
50	1.0	1.0	10.0	73.7	19
51	<1.0	1.0	1.0	60.0	5
52	2.0	2.5	2.5	44.4	9
53	1.0	2.5	2.5	62.5	8
55	1.0	1.0	1.0	64.3	7
56	2.0	4.0	4.0	40.0	12
58	1.5	2.0	2.0	60.0	10
63	2.5	7.5	7.5	48.0	12
64	<1.0	1.0	1.0	66.6	9
67	<1.0	1.5	1.5	61.1	9
68	1.5	1.5	1.5	35.7	7
69	1.0	3.0	3.0	80.0	10
72	–	2.0	2.0	80.0	10
73	1.5	1.5	1.5	50.0	8
75	1.0	2.0	2.0	77.8	9
98	2.0	2.0	2.0	55.6	9
99	1.5	3.0	3.0	55.6	9
100	2.0	5.0	5.0	46.2	13
101	2.0	4.5	4.5	44.4	13
102	1.5	4.0	4.0	58.3	12
120	1.0	1.0	1.0	71.4	7
122	<1.0	2.0	2.0	58.3	6
123	1.0	1.0	1.0	42.9	7
124	2.0	4.0	4 .0	52.2	11
125	1.5	2.5	2.5	66.6	9
127	1.0	2.0	2.0	65.0	10
128	1.0	2.0	2.0	50.0	6
129	1.0	2.0	2.0	52.2	11
130	1.5	1.5	1.5	55.6	9
134	<1.0	2.0	2.0	66.6	6

ICT; Inner cranial table, *OCT*; outer cranial table, *TVT*; total vault thickness.

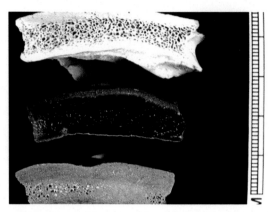

FIGURE 6.5 Cranial vault composition and thickness in (top to bottom) WLH 22, 28 and 63.

(4–19 mm). WLH 50 thickness is exceptional but some individuals in the Willandra series almost match its thickness at one or other anatomical point. The next thickest measurement is 14 mm on the parietal bosses and asterion of WLH 27 and 100, respectively. Another five individuals (WLH 19, 22, 28, 45 and 101) have vaults 12 mm thick at one or other of their primary points, but using a secondary measurement, four others (WLH 18, 56, 63 and 102) are 12 mm and over. There are 16 individuals with a maximum vault thickness of <8 mm. Using primary points, WLH 1, 67 and 68 have a mean thickness of <7 mm and have uniformly thin crania, with ranges of 4–8, 5–9 and 4–5 mm, respectively.

Generally, both thick and thin crania have uniform vault thickness. There is wide variation in thickness structure in the series with 70% having a 1–2 mm inner table and 25% an inner table of <1 mm. The thickest inner table is 3.5 mm but the outer table is nearly always thicker (<1–7.5 mm) with WLH 63 the thickest (Fig. 6.5, *bottom*). Diploic bone ranges from 2 mm (WLH 2, 10 and 68) to 14 mm in WLH 50. Diploe comprises 40%–80% of cranial thickness except for WLH 1 and 68. While vault composition is similar for most individuals, differences in the ratio of compact to cancellous bone do occur in others. Usually, individuals with thinner cranial vaults have less diploic or spongy bone. There are others with very thick inner and outer bone tables with less than half their cranial thickness composed of spongy bone (Table 6.3), (Figs. 6.6–6.12).

Material Culture, Dental Attrition and Infection

WLH 3 has an unusual toothwear pattern. It occurs especially on the left molars in the form of an acute buccally directed slope on M_1 and M_2 with associated alveolar gum disease. The wear slopes almost to the alveolar rim of the socket (Figs. 6.16 and 6.17). A similar pattern has been observed on mid- and late-Holocene dentitions from the Murray River. There, processing of Typha bull rush (*Typha latifolia*) produced the attrition pattern. In WLH 3 the wear is generally very uneven between molars with M3 having a slightly different pattern,

FIGURE 6.6   WLH 1, right lateral view.

FIGURE 6.7   WLH 3, left lateral view.

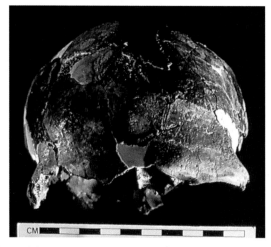

FIGURE 6.8   WLH 1, frontal view.

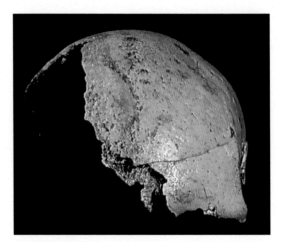

FIGURE 6.9   WLH 3, frontal view.

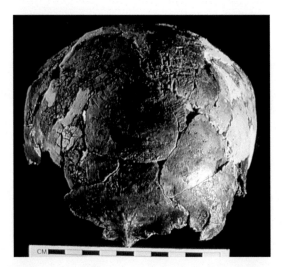

FIGURE 6.10   WLH 1, posterior view.

although matching that on the M_3 of the opposite side. A x10 magnification of the polished occlusal surfaces of M_1 and M_2 shows striations cut into the surface enamel that run parallel with the slope. The grooves and the unusual pattern of sloping attrition suggests that vegetable or plant fibre was repeatedly drawn across and down the teeth, as it was later in the Murray, and that hard quartz grains (plant phytoliths) cut into the enamel surface producing the grooves. The M_1 and M_2 on the right side of the mandible are missing but because the pattern on both M_3's is similar it seems likely that the right M_1 and M_2 would have had the same pattern (Tables 6.4 and 6.5).

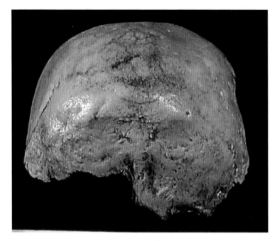

FIGURE 6.11   WLH 3, infero-posterior view.

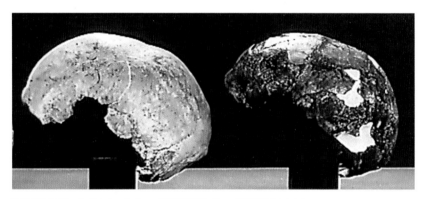

FIGURE 6.12   Left lateral view of WLH 3 (*left*) and WLH 1 (*right*).

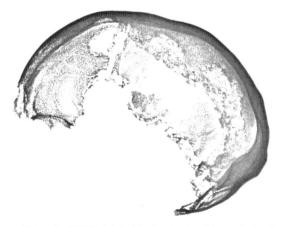

FIGURE 6.13   Radiograph of WLH 3, left side showing the thin vault structure.

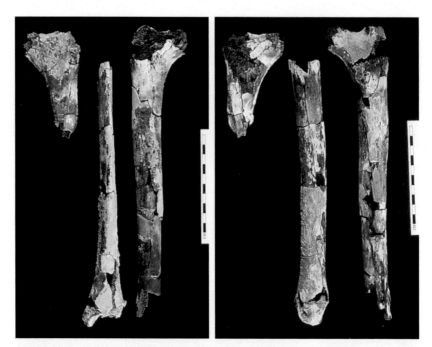

FIGURE 6.14   WLH 1 lower limb bones, anterior view (*left*) and posterior view (*right*). Note smashed and cremated condition associated with ritual breaking of all skeletal elements during the burial ceremony.

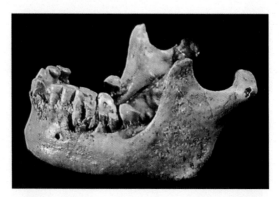

FIGURE 6.15   Lateral view of WLH 3 mandible showing periostitis and recession of the M_1 and M_2 sockets.

The strong use-stress implied by this wear suggests mechanical forces on the jaws that in the long term would alter the contiguity of the tooth in the socket, something also observed by me among Murray River people. Here we see the typical resorption of the alveolar rims of M1 and M2 that must have undergone similar stressors that induced subsequent periostitis around the alveolar neck and

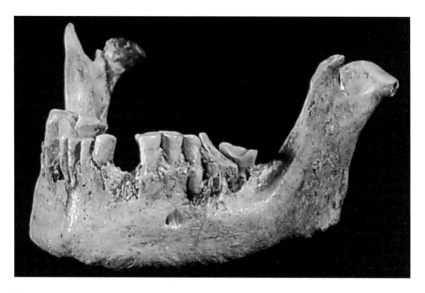

FIGURE 6.16   WLH 3 mandible showing oblique molar wear on M₁ and M₂ and resorption of alveolar bone around the left canine socket.

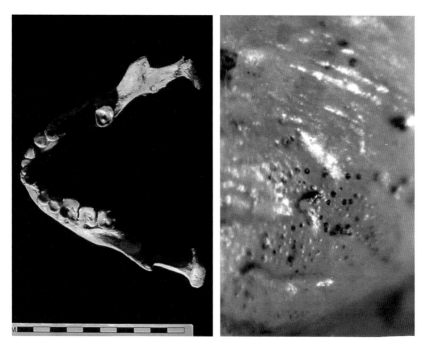

FIGURE 6.17   Superior view of the WLH 3 mandible showing the M₁ and M₂ buccally sloping attrition (*below left*). At right x20 magnification of the occlusal surface of WLH 3, M₂ showing incised grooves running parallel to slope wear.

TABLE 6.4 Craniometric Data for WLH 1 and 50

Cranial measurement (mm)	WLH 1	WLH 50
Galbella-Opisthocranion	181	216
Glabella-Inion	170	198
Glabella-Bregma	–	127
Glabella-Lambda	173	211
Lambda-Bregma	106	133
Lambda-Inion	66	72
Lambda-Asterion	–	90
Lambda-Opisthion	95	–
Lambda-Opisthocranion	47	57
Basion-Opisthion	36	–
Basion-Sphenobasion	18	–
Bi-Asterion	103	130
Bi-Coronale	110	–
Bi-Frontotemporale	90	–
Bi-Infratemporal Crests	84	–
Bi-Sphenion (estimated)	–	125
Bi-Stephanion	103	98
Post-orbital constriction	17	113
Bi-Euryon	–	152
Inion-Opisthion	40	–
Opisthion-Opisthicranion	64	–
Basi-occipital Breadth	32	–
Maximum bi-parietal breadth	130	145
Maximum bi-temporal lines	–	112
Maximum post-orbital breadth	–	111
Maximum posterior frontal breadth	–	123
Maximum supra-orbital breadth	107	131
Maximum breadth Bi-temporal lines	–	112

tooth roots. This type of dental condition is characteristic of laterally directed tooth 'wobble' caused by persistent and increasing bucco-lingual rocking of the tooth in response to vegetable matter processing by drawing it in an inferolateral manner across the occlusal surfaces of the molars. Pressures generated during this behaviour loosen the teeth in their sockets and they begin to rock laterally.

Table 6.5 Metrical Data for WLH 3 Mandible

WLH 3 mandible	mm
Length	112
Symphyseal height	33
Symphyseal thickness	15
Coronoid projected height	62
Corpus height	33
Corpus thickness	15
Corpus projected height	85
Bi-condylar breadth	129
Bi-gonial breadth	109
Gonial angle	33°
Ramus projected height	59
Ramus breadth (maximum)	39
Ramus breadth (minimum)	32
Condyle length	105
Condyle breadth	23
Condyle projected height	56
Sigmoid notch length	13
P_1–P_2 length	30
M_1–M_2 length	11

Continued processing loosens the tooth and allows food and vegetable matter to lodge in the socket between the gum and tooth. The downward movement of the material towards the tip of the root encourages an infection of the alveolar bone and sets up an infection at the root tip (apical periostitis), all of which adds to the general stressing of the socket bone. The personal estimated age of WLH 3 suggests that fibre stripping probably occurred over many years and the pathological processes were similarly long term.

The wear pattern then suggests this man was shredding vegetable fibre using his back teeth. He was teasing out fibres probably for him or others to weave into nets, bags or baskets for carrying objects or trapping birds or fish in and around the Willandra Lakes. It is interesting that this type of manufacturing was not necessarily left up to women as observed ethnohistorically. One possibility is that the severe osteoarthritic destruction of his right elbow (his throwing arm) made it impossible for him to usefully hunt anymore but he could still contribute by providing shredded vegetable fiber which he processed using his left arm and left dentition. (See below). People may have used vegetable fibre bags when collecting shellfish (*Unio* spp.) that they brought back to their camps, the remnants of which now appear in middens along eroding lunettes and other places around the fossil lakes. Processing of plant fibre by those living close to large bodies of water would not be unusual. But to find evidence of this activity as a skeletal marker on a 50–60-year-old man, buried beside an inland lake 42,000 years ago forms a fascinating insight into the life of modern humans not long after they are supposed to have moved into Australia and 25,000 years before the lake finally dried.

Osteoarthritis as a Reflection of Material Culture

WLH 3 has a series of osteological changes indicating severe osteoarthritis. The worst involves joint surfaces of the entire right elbow joint. They include ankylosing (or fusing of the joint) of adjacent bones, eburnation (polishing) and pitting of joint surfaces (severe wear) with some bone loss. The proximal ends of the radius and ulna and distal end of the humerus display such advanced changes (Figs. 6.18–6.21). Erosion of articulated surfaces prevented proper anatomical association of the elbow and wrist preventing full joint movement. Other changes include marginal osteophytic growth in the form of new bone proliferation on the medial and lateral sides of the distal humerus and along the margins of the olecranon, the trochlear and radial notch (Figs. 6.20 and 6.21). Such bone growth was a product of inflammatory processes from aggravated tissues of the joint capsule from the ongoing degenerative process and joint use. The articular circumference of the radial head shows eburnation indicating the loss of the articular cartilage buffer in the radial notch producing bone-on-bone contact and the subsequent polishing of adjacent surfaces.

FIGURE 6.18   Osteoarthritic pitting on the distal right radius of WLH (*left*).

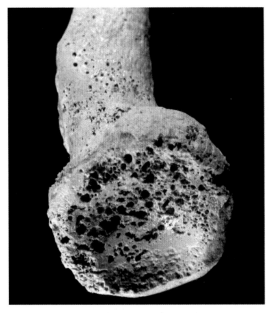

FIGURE 6.19   Pitting of radial head foveal surface and marginal breakdown.

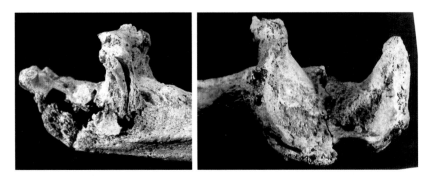

FIGURE 6.20   Medial (*left*) and lateral (*right*) views of trochlear notch from WLH 3 right elbow showing scrolling of proliferative bone along articular and marginal edges.

Normally, a secondary process then occurs when thin eburnated bone begins to collapse. That process slowly leads to a general breakdown of bone surfaces as pits become contiguous.

The suite of arthritic changes in WLH 3 indicates a chronic condition occurring over many years during which time the loss of articular cartilage between joints has eroded joint surfaces and allowed the head of the radius to leave a ground circular impression on the humeral capitulum as well as scour the radial fovea surface (Figs. 6.20 (*right*), Figs. 6.21 and 6.22). Scrolled bone proliferation has formed on the lateral margin of the capitulum, filling the coronoid fossa

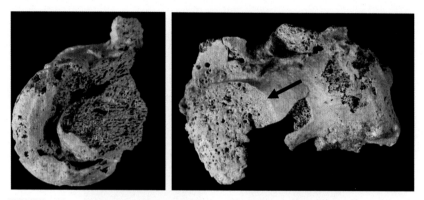

FIGURE 6.21   **Medial view of distal humerus (*left*) with epicondyle surrounded by proliferative bone growth.** Anterior view (*right*) of the distal articular humeral surface showing eroded circular scar (*arrow*) made by the head of the radius after loss of articular cartilage that was followed by erosive pitting.

FIGURE 6.22   **The elbow joint of WLH 3 in articulation displaying complete osteoarthritic degeneration and remnant ankylosing with erosion of all articular surfaces.**

and following the trochlear margin (Fig. 6.21 *left*). That process ankylosed the radial head to the radial notch which would have immobilised the lower arm preventing rotation and supination movement. Consequently, proper arm rotation was prevented through this severe and chronic condition.

Such bone degeneration at the elbow and wrist indicates heavy stressing and repetitive use of the right arm. That activity included rapid elbow extension and pronation of the lower arm leading to stressing at elbow and wrist. Long-term use of the arm in this way can result in almost total removal of articular cartilage across all bones of those joints with the radius particularly undergoing heavy rotational stress. The set of traits observed on WLH3 have been termed 'atlatl elbow' by others in observations on the elbows of North American native people who used the atlatl spear thrower (Angel, 1966; Ortner, 1968). Although WLH3's arthritic condition may have been exacerbated by an additional infection of the humerus, possibly though continued use of the arm, the main cause of bone degeneration and alteration was the particular pattern of stress imposed through behaviour which is consistent with using a spear thrower. Another identical example of this type of severe osteoarthritis, with joint destruction and ankylosing of the right elbow, occurs on another male person WLH 152 (see Chapter 10).

The severe osteoarthritis in the right arm of WLH 3 is as fascinating as his molar tooth wear pattern and tooth evulsion. They all provide valuable evidence for behaviour, material culture use and ceremony and culture among these earliest arrivals in Australia. A detailed description of this pathology has been published previously (Webb, 1995) but I mention it here to put it into a more holistic presentation of culture in Australia at 45 ka and because of the evidence of complex weaponry among Australia's earliest modern humans which was thought to be 12,000 years earlier at the time of the first publication. Another reason is that it reflects the advent of the spear thrower possibly 8000 years before it is recorded in Europe 35,000 years ago. Our earliest evidence for the spear thrower in Australia is presently dated to 19 ky. Besides being the oldest evidence for the use of a spear thrower in Australia and possibly the world, the evidence from the Mungo Man opens some interesting possibilities about the early weaponry used in Australia.

The pattern of osteoarthritis in WLH 3 may not be solely a product of using a spear thrower. The condition is unusual in its features even for *extreme* use of the limb. In my experience, this extreme form of osteoarthritis is rare although not unknown in Aboriginal skeletal remains from anywhere in Australia, which is unusual in view of the common use of the spear thrower in many parts of the continent. I, therefore, want to consider another possible cause, direct or contributory, for this condition: *infectious osteoarthritis*. It is based on several features occurring in and around the mid-shaft of the right humerus. The proximal half of the shaft has a large oval hole on the medial side exposing the medullary cavity, possibly caused by interment processes and there is no obvious sign of pathology around the hole, but some small patches of fine, cancellous bone on the deltoid tuberosity might be periostitic in origin. Whether this opening was a cloaca to release infectious fluids or became enlarged when thin, weak surrounding periostitic bone was removed during interment, cannot be determined. However, the cortical bone around this feature is unusually thin with trabecular bone comprising most of its thickness.

Inferiorly, the shaft has much thicker walls narrowing the medullary cavity. Cortical bone thickness here ranges from 4.3 mm to 8.5 mm, whereas higher up on the shaft it is less than 2 mm. Part of the thickening comprises trabecular bone, which takes the form of an *involucrum* normally associated with a local-ised infection. Besides these features there is nothing about the general appear-ance of the bone to suggest that an infection did occur, nor is there any scar tissue, trauma or other discontinuity that might have caused an infection. Peri-ostitis is present around the distal end of the left humerus, particularly within the olecranon fossa and the posterior surface of the lateral epicondyle. This would suggest a more generalised infection of the skeleton although minimal in extent. There is no evidence of acute or chronic suppurative osteomyelitis in the right arm, alternatively, the case might be one of non-suppurative osteomyelitis which lacks pus formation and consequently the associated scarring and obvi-ous alteration of bone architecture. It affects the long-bone shaft mainly and is characterised by new bone formation, increasing cortical width and narrowing of the medullary cavity.

Chronic, non-suppurative osteomyelitis, however, is rare. It may have been even rarer, or non-existent, 42,000 years ago. There is enough evidence, however, to tentatively suggest that there may have been an infection of the right upper arm of WLH 3 that altered cortical and trabecular structures and in the form of periostitis on other long bones. Infection in the right arm, however, seems to have progressed to a non-specific osteitis following a definition used previously on more recent Aboriginal skeletal remains (Ibid, 1995). Osteoarthritis can also occur through infection of the synovial pint. Upon entering the joint space, it causes pus production and destruction of the articular cartilage where it is rapid. Destruction of the articular surfaces, the process becomes chronic, and arthritic bone changes occur and bony ankylosis is frequent. The lack of evidence for pus production at the site is at odds with this diagnosis, however. Nevertheless, it is possible that a non-specific infection may have aggravated mechanical changes taking place in the right elbow joint from the wear and tear of using the spear thrower. So, continued use of the arm, for throwing a spear or using a spear thrower, exacerbated the condition, hence the physical pattern of degeneration.

Besides an extremely arthritic right elbow, the skeleton also shows extensive lipping on the thoracic and lumbar vertebrae. It appears as a series of osteo-phytes on both the superior and inferior borders of the centra of L2, L3 and T9. The condition is not unusual in a mature active male and is regarded as merely the effects of an active lifestyle. Almost complete ecto- and endocranial suture closure indicates that WLH 3 was an elder adult when he died, probably around 50 years of age which may have been a long life at those times.

RITUAL AND CEREMONY AMONG THE EARLIEST AUSTRALIANS

Dental Modification

The use of ochre is not the only indicator of ceremony and ritual among Wil-landrans. Pre-mortem loss of both mandibular canine teeth in WLH 3 has been described elsewhere (Ibid, 1995), but it is worth briefly mentioning it again in this discussion of Willandran cultural complexity. Tooth evulsion was practiced by many groups around the world in the past, including in the Magreb of North Africa where it goes back at least 21,000 years (De Groote & Humphrey, 2016). Reasons for the practise are not well understood but the practise among recent peoples was carried out when a chief dies, it was also meant to intimidate enemies, change language pronunciation, was carried out during mourning and as a tribal distinction from others in the same way head binding was carried out among Canadian Indians to distinguish one group from another. Ethnographic and ethnohistorical accounts of traditional tooth avulsion among recent male and female Indigenous Australians was usually part of an initiation or rites of passage among juvenile males and females. It is a ceremony that recognises their transition into the adult world, but it usually involves the removal of one or both upper and/or lower incisors (Campbell, 1981; Durband, Littleton, & Walshe, 2014). In Australia, the type and number of teeth extracted varies with both maxillary and mandibular teeth removed which is the practice favoured in Central Australia. During adult life resorption or remodelling of alveolar bone takes place where the tooth has been removed leaving a marked depression often topped with a ridge of bone between the remaining teeth either side.

WLH 3 shows all these characteristic features between the lateral incisors and the first premolar on both sides of the mandible although they are less distinct on the right side because of post-mortem bone loss. Alveolar shrinkage and remodelling of the canine socket has reduced alveolar height and mesio-distal thickness resulting in a ridge of bone between the lateral incisor and first premolar on both sides of the mandible. The incisors and premolars either side of the empty sockets have succumbed to mesial drift resulting in them leaning inwards as though over time they have tried to fill the resulting gap as though compensating for the missing canines. This feature together with alveolar bone resorption suggests this man's canines were lost in early life (Fig. 6.23). Even though canine teeth have the longest roots of any human tooth and are not usually lost

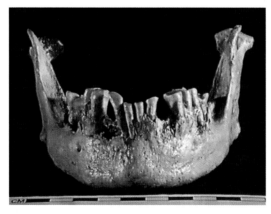

FIGURE 6.23   Anterior view of WLH 3 mandible showing missing canine teeth and incisor attrition.

under normal circumstances, an injury or blow to the face severe enough to remove a canine would also damage or remove teeth on either side and this has not occurred. There are also few, if any, reasons to lose a canine because of its strength and solid root, let alone lose both. So, bilateral loss of canines is unlikely, particularly with an identical pattern of socket resorption indicate a contemporary timing of their loss. Moreover, canine loss is not normally observed among skeletal populations with the odd exception of edentulous individuals and those with extreme forms of occlusal attrition, neither of these is the case here. The removal of canines is unknown in Australia although it has been recorded in prehistoric Cambodia (Domett et al., 2013). It is difficult to positively identify why WLH 3 is missing these teeth except by their deliberate removal and speculate that this practise may have been brought into Australia by the early arrivals, however, an indigenous origin may also be just as likely. It does suggest, however, that the WLH 3 example is not only unique in the type of tooth chosen but also is by far the oldest example yet found anywhere in the world and the first indication that modern humans might have developed it during their dispersals or soon after they landed.

The lower, central incisors of WLH 22 were also lost *pre-mortem* (Fig. 6.24). Complete resorption of both sockets has occurred, in a similar manner to that of WLH 3, suggesting that they were lost many years before death probably in late childhood or early adulthood. Nothing remains of the sockets themselves, which are filled with new spongy bone. Advanced resorption indicates a considerable time lapse since their loss/removal and that has led to a narrowing of mandibular width. The removal of the lower incisors alone was practised in central Australia, so their loss in WLH 22 suggests that lower incisor evulsion during initiation may go back a long way. Ritual and ceremony are strongly implicated in burial procedures of the Willandrans so it is not unlikely that other ceremonies such as initiation were also part of their cultural practises. Whatever the reasons were for tooth evulsion in WLH 3 and 22 it seems highly likely that it was part of some form of ceremony that took place over 40,000 years ago which makes it the earliest evidence for such a practise in the world at present. Two other individuals (WLH 11 and 20) also have *pre-mortem* tooth loss but this is confined to the molars.

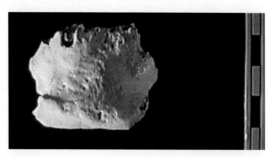

FIGURE 6.24 Mental eminence of WLH 22 showing pre-mortem loss of incisors with remodelling of the socket area.

Cremation

The following description of the WLH 1 disposal was written in the original report (Bowler, Jones, Allen, & Thorne, 1970:57).

> *The individual was cremated as a complete and fully fleshed cadaver, though the pyre was insufficient to achieve full incineration. There was a total and thorough smashing of the burnt skeleton, particularly the face and cranial vault. The ash and smashed bones were gathered together and deposited in a conical hole either beneath the pyre or immediately adjacent to it.*

The difference in colour between adjacent cranial bone fragments showed that the pieces lay in different parts of the fire and subjected to various temperatures. The WLH 1 bone colour ranges from dark black/deep blue to light grey, including white powdery calcination indicating burning hot enough to turn the bone to powder. Some long bone also displays distinctive transverse cracking, and show thorough carbonising right through the bone with charring of broken edges. The various blues and greys are indicative of two phases of burning at different temperatures and calcination is indicative of exposure to very high temperatures, usually greater than 800°C probably in a large, constantly maintained fire. The bone was covered with sand which later became impregnated with calcium carbonate leached from surrounding beach sands during rain.

The distribution of skeletal remains shows burials were normally carried out along the lake shores. The majority have been found along the southwestern part of Lake Garnpung, the Lake Gigolo area and the southern end of the Lake Mungo lunette. Cremated individuals are found in these areas but the association (if any) between area and cremation is not clear. This complex burial process implies ceremony and ritual and is probably the oldest evidence for the practise in the world. It is significant that when WLH 1 died the people believed she required not only burial but also it should be carried out in a certain way and accompanied with ritual. That indicates her people and their culture demanded more than just a hole in the ground or a cursory surface disposal, the body left for scavengers. This complex burial procedure also required a complex language for explanation of the reasons for and method of disposal which had to be carried out in a special way. This was not something just invented but was part of the culture of the group and probably of the wider population of which the group was a part. That further suggests that such burial ritual and ceremony would need to be transmitted to the younger generation to carry on the cultural practises of that group which also required complex language.

The WLH 1 burial is reflected in other individuals who also have the same bone colour changes, calcination and transverse cracking of long bones. They include WLH 2, 9, 10, 68, 115, 121, 122, 123 and 132. That might suggest cremation or deliberate burning of the dead was common practice, particularly among gracile people of which all these individuals are examples. There have been other remains with similar features that were not collected, one was of a similar gracile nature to WLH 1 that lay less than 100 m from where WLH 1 was found. WLH 68 consists of very gracile cranial vault fragments (4–5 mm thick) replicating those of WLH 1. Seven others (WLH 6, 22, 24, 63, 93, 120 and 126) have burn marks on some part of their surfaces but these are regarded as 'burnt' rather than 'cremated' because the scorching is not as thorough. They may have been partly cremated but because they do not show the characteristic colour variation described above they have not been regarded as cremations. It seems, however, cremation was practised by robust people also. Two individuals (WLH 28 and 63) display the characteristic features of cremation with typical colour changes of grey and blue/black.

Other Burial Practices

Besides cremation several other burial practices are worth noting. These include the extended burial of WLH 3 that was discovered only 500 m from WLH 1. These two different interment procedures highlight a contrast in funerary rites among these early modern humans. Whether these depended on gender is not clear because WLH 2, which is believed to be male, was found in association with the WLH 1 cremation. The similar chronological ages of WLH 1, 2 and 3 suggest several methods of internment were being used by these people at that time (42 ka) as flexible cultural practices possibly dependent on gender or status. Another aspect of their mortuary customs was bone smashing and this is obviously the reason for the lack of whole bones in the burials around the Willandra Lakes.

Observation of eroding burials over the last 30 years or so has shown that upright burial was also practiced as is suspected in the case of WLH 135 (see Chapter 10). In that case the position of the cranium suggested the body was placed in a grave in a sitting or crouched position so when the grave eroded the forward tilted top of the cranium was the first part of the body exposed. It would be nice to know when these various burial practices were introduced or changed and by whom. Did gracile or robust groups learn of cremation from the other or did they develop them separately? All that can be said is that the earliest Willandrans used several complex burial procedures that included extended and crouch burials and cremation with secondary smashing of bone on some occasions.

Bone Smashing Without Cremation

While bone fragmentation is mostly the product of natural agencies, it is my impression that deliberate smashing was carried out on some individuals in the series. Although fresh breaks occur on some remains (e.g. WLH 13, 15 and 17), they are not common. Most fragmentation occurs on major long bones, many of them with tough, thick cortices. Such bones cannot be fractured without striking a severe blow and are not easily broken. There is, moreover, evidence to suggest that much of the bone was broken before its exposure. Often, broken edges have been completely bevelled by erosion which has been covered by carbonate incrustation and/or carbonate or manganese staining. Such features are indicative of a broken bone at the time of burial and before the carbonate horizons were being formed within the sedimentary horizons, and long-term lunette stability is unlikely to have caused breakage. Evidence of such stability comes from the preservation of the very fragile WLH 3, which was still in good condition after 42,000 years of interment. The bone itself was very friable and subsequent loss of parts of the skeleton was unavoidable after excavation due to its tenuous condition with very thin cortices. That was unlike the many individuals with thick to very thick cortices, whose bones have broken into quite small pieces, for example WLH 16, 18, 19, 69, 106, 107 and 110.

The neat pattern of fracturing that produces rather straight, even edges on many bone fragments, suggests also that whenever the bone was broken, it was not fresh. If deliberate smashing did take place, it must have occurred sometime after the death of the individual perhaps as a funerary ritual taking place 6 months or more after death when the bone had dried. Such treatment should have left scars on the bone surface, although only one blow on dry bone might have been needed to produce complete fracturing throughout, so many blows were not required although a combination of weathering and calcium carbonate may have eroded or camouflaged such marks. Perhaps the most parsimonious conclusion is that while some fragmentation is due to natural causes before and after exposure of skeletal remains, this may not be the case for all fragmented bones. The suggestion is that some individuals were deliberately smashed before interment, but not cremated as others were. Bone smashing, whether accompanied by cremation or not, would render the integrity of the original skeleton prone to further fragmentation.

Ochre Use for Burial

A large amount of ochre was used to cover the body of WLH 3 after he was placed in the ground. It signals a prominent additional ritual practise for burials among these people. The nearest source of ochre to Lake Mungo is over 200 km northeast in the Barrier Ranges. The ochre used for the WLH 3 burial must have either been traded from those ranges into the Willandra for this and other purposes or they sought supplies and made the 400 km round trip to fetch it. The existence of trading 42,000 years ago could suggest another aspect of cultural complexity in place and supported by others living in other areas. It suggests a process of goods moving from one group to another in an established network and a tradition of reciprocation. That also might imply substantial group distributions with some degree of semi-permanent occupation across large tracts of country at this early period of continental settlement that could then infer a much earlier settlement of the region than even the Willandran people have led us to believe possible. Ethnological examples of trading and exchange networks usually involved a trading centre with reciprocal and widely accepted exchange mechanisms and procedures, suggesting a certain degree of 'settled' and structured society. Again, this implies a well-bedded culture of some duration that in itself indicates an even older time for first settlement in the region.

Ochre could also have been used for body decoration as part of ceremonial activities or for unknown secular purposes. Rock art sites are not known locally mainly because of the lack of suitable rock walls and rock shelters except in the Barrier Ranges. While art has not been found in association with the Willandrans, undated petroglyphs as well as painted motifs exist in those Ranges, particularly at Mootwingee. Dating this art is difficult if not impossible at present preventing us knowing whether the Willandrans, their ancestors or their contemporaries were instrumental in its production when collecting ochre, although they obviously knew where to find it. They may have left inscriptions as a calling card to perhaps a spirit being who they believed produced the ochre and who may have played an important role in the ceremonial and ritual of their burial practices. What we can say is that ceremony, ritual and complex social behaviour were definite aspects of life around the Willandra Lakes a long time before the last glacial maximum.

Willandran Cultural Complexity: A Holistic View

Cultural indicators among the Willandrans show a complex society with a rich ceremonial life. There are at least two different burial practices (perhaps three) with layered ceremony, and tooth avulsion as part of a rights-of-passage ceremony, all part of the ceremonial life of these people and in place by at least 42 ka. Ochre was not only being used in burial ceremony but also it is likely it was also used for body decoration. The rituals recorded here suggest the use of symbols and complex ritual activity pointing also to possession of complex language in order to transmit tradition from generation to generation, communicate ideas and explain ritual and ceremony as well as make sure ceremonial processes were carried out properly. But these aspects of the culture of the first Australians were likely to have been brought onto the continent. We cannot forget that modern humans in South Africa were carrying out very similar rituals using and incising ochre at 100ka, but that is not verylong before we now realise people first entered Australia 70 perhaps 80,000 years ago. (see Modern humans within Africa, Chapter 4).

The dental attrition of WLH 3 points to the use of the molar tooth complex in shredding plant fibre probably to make string or cord for weaving into bags, nets and bindings, with some inevitably used for resource gathering in the lakes and streams of the region, particularly the Willandra Creek itself. Hearth size and contents patterns not only confirm this but also suggest a family structure. Combined, these sources of evidence there is little doubt that they possessed social structures and networks as well as family groups that may have been organised into bands. Trade may have also been part of an extended regional social network that had emerged with long-term settlement some considerable time before 42 ka.

The severe osteoarthritis in the right elbow of WLH 3 points to the use of sophisticated weaponry, namely, the spear thrower. Whether they also had shields, hafted axes and stone points mounted on spears cannot be confirmed but the undoubted skill and social complexity of these people could easily suggests this was the case. Willandran material culture, ritual, ceremony and burial practices strongly point to Australia as being anything but a Late Pleistocene socio-cultural outpost. The Willandrans seem fully adapted to their surroundings that consisted of lacustrine and riverine ecosystems, marshes and semiarid backcountry. They developed an adaptive social system that allowed them to live in many different environmental settings in a strengthening ice age as well as fluctuating lake and river levels. Their adaptive lifestyle was underpinned by a very lively and spiritual existence focussed on half dozen fresh water lakes. At times, the lakes were permanent in terms of a human lifespan and even over many generations. The lacustran lifestyle among these Late Pleistocene Australians was very different from the dry inland way of life known by the later Holocene populations.

The Willandra Lake system, therefore, consisted of a unique set of environmental circumstances for people to adapt to and live in, almost as an oasis in the middle of a semiarid environment. After 17 ka the lakes began to dry permanently. Perhaps a similar area and lifestyle took place in another region of Australia; however we have yet to find it. Nobody then had a 'coastal economy' unless they lived on the coast. But the original coastal economy, if they had one as such, was easily transferable to river and lake systems where similar species and resources were present in fresh rather than salt water. Most people must have been generalist hunter-gatherers adapted and locked in to wherever it was they lived. The social and cultural wherewithal of these people was their main armament, if they had their culture they had their ticket to move and survive. However, the Willandra must have provided some with a very special place to live.

REFERENCES

Angel, J. L., (1966). Early skeletons from tranquillity, California. *Smithsonian Contributions to Anthropology.* 2, 1–15.

Bowler, J. M., Jones, R., Allen, H., & Thorne, A. G., (1970). Pleistocene human remains from Australia: A living site and human cremation from Lake Mungo, Western New South Wales. *World Archaeology.* 2(1), 39–60.

Campbell, A. H., (1981). Tooth avulsion in Victorian aboriginal skulls. *Archaeology Oceania.* 16, 116–118.

De Groote, I., & Humphrey, L. T., (2016). Characterizing evulsion in the later stone age Maghreb: Age, sex and effects on mastication. *Quaternary International.* 413, 50–61.

Domett, K. M., Newton, J., O'Reilly, D. J. W., Tayles, N., Shewan, L., & Beavan, N., (2013). Cultural modification of the dentition in prehistoric Cambodia. *International Journal of Osteoarchaeology.* 23, 274–286.

Durband, A. C., Littleton, J., & Walshe, K., (2014). Patterns in ritual tooth avulsion at Roonka. *American Journal of Physical Anthropology.* 154, 479–485.

Ortner, D. J., (1968). Description and classification of degenerative bone changes in the distal joint surfaces of the humerus. *American Journal of Physical Anthropology.* 28, 139–156.

Webb, S. G., (1995). Palaeopathology of Australian aborigines. Cambridge, UK: Cambridge University Press.

Chapter 7

One of a Kind? WLH 50

In my view, the greatest mystery to emerge in the Australian human fossil record is the individual WLH 50. Perhaps it deserves a better identification, possibly Willandra Man, but that might be confused for Mungo Man (WLH 3). For me, the mystery of WLH 50 is why I have given it its own chapter. The mystery is its extraordinary robust character that includes a thick cranial vault, prominent brow ridges, its large size and robust cranial morphology. Collectively they make it standout in the Willandra collection even if no other label is attached. The label of robust and an outstanding fossil are much deserved because it also stands out from all other Late Pleistocene fossil crania from this continent and South East Asia labelled modern human. There are, however, others in the Willandra collection that have similar characteristics, but they are more fragmentary even though they are in themselves worthy of an archaic label. But WLH 50 is just that bit closer to being an archaic individual.

While this chapter concentrates on WLH 50 other morphologically similar individuals will be referred to not only those from the Willandra collection but also from other parts of Australia and internationally. The reason is that in general terms WLH 50, while in many respects having a morphologically unique appearance, is the standard-bearer of an ancestral origin from which an enigmatic robusticity among other Australian fossil humans probably stems. That same robusticity has also been commented on as being somewhat unique among modern humans and late Holocene populations living elsewhere in the world. Although an element of robustness occurs occasionally among Holocene people, the degree and presence of WLH 50's features is a reminder that it is possible, if not likely, not all early arrivals were unmixed modern humans genetically untainted by their journey through regions where archaic humans still existed. Moreover, some of those early humans represented populations supporting a substantial archaic genetic component and that cannot be denied in WLH 50's case. It should be asked, however, whether its appearance is derived directly from archaic people or is a product of genetic mixing between modern and archaic groups at some point along the dispersal track by WLH 50's forebears as they passed through East Asia and crossed Sunda.

THE BEGINNING OF AN ENIGMA

In 1980 Rolf Hogan working on *Garnpung Station* close to Lake Garnpung 40 km north of Lake Mungo, found what he believed at first glance to be cow or horse bones. He drew up on his motorbike and looked down at them. There was something different about these bones and he had seen fossil bone before. He collected the bone and took it back to *Garnpung Station* at lunchtime and reported the find. The bones stayed at the house in a cupboard for some weeks under the stewardship of Ted Richardson, owner of *Garnpung Station*. Australian National University (ANU) Research Fellow, Jeannette Hope, who had been working in the region, and Mike Macintyre who was gathering data for his PhD in the archaeology of the region and also based at the ANU, eventually collected the fossil. I was working in Alan Thorne's lab in the Research School of Pacific Studies at the ANU, when one morning Macintyre, arrived holding a cardboard box. It contained the pieces of bone Rolf Hogan had found. Mike asked me if Alan was around but he was not. He showed me what he had brought. I peered into the box and looked at the grey stone-like appearance of the lumps of bone some covered in calcium carbonate. I was immediately astonished at the bone thickness and size. The pieces consisted of several large and some smaller pieces of cranium and other fragments including what looked like a large portion of elbow joint partly encased in the grey calcium carbonate. Other postcranial fragments were present but there was no sign of facial bone, mandible or teeth. When Alan arrived, he spoke to Mike in his office. Mike left and Alan literally locked himself away for the rest of the day: he was reconstructing the find. The box held WLH 50 (Figs. 7.1–7.4). Even in large fragments, it was something

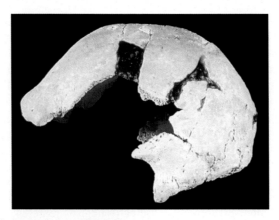

FIGURE 7.1   Left lateral view of WLH 50.

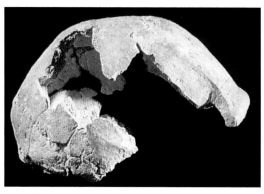

FIGURE 7.2   Right lateral view of WLH 50.

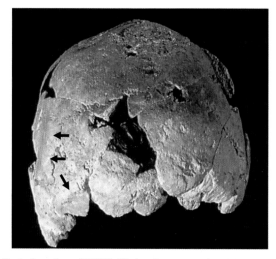

FIGURE 7.3   Posterior view of WLH 50 showing a prominent transverse nuchal torus (*diagonal arrow*) **and angular torus** (*two horizontal arrows*). Note also, rugged areas of muscular attachment on the fragmented nuchal surface, vault keeling and the widest cranial breadth is measured low down on the vault. Silica sheen is reflected in studio light either side and posterior to the sagittal peak.

which eclipsed anything I had seen in my work as Alan's research assistant and I had worked on the Kow Swamp people and handled the Coobool Crossing remains as well as many similar robust individuals. I had also watched Milford Walpoff reconstructing the Sangiran 17 *Homo erectus* cranium in the same lab using the 'art of the satay stick' as reinforcing scaffolding to hold bone in place during gluing (Fig. 7.5). But this new individual from the Willandra both Alan and I instantly recognised as something very different for an Australian fossil human.

FIGURE 7.4   **Superior view of WLH 50 showing marked postorbital constriction, partly fused coronal suture and no trace of the sagittal suture.**

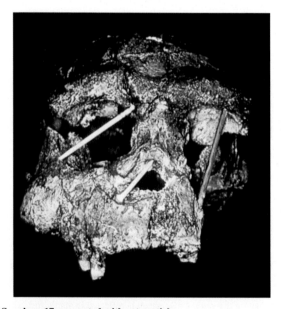

FIGURE 7.5   **Sangiran 17 supported with satay sticks.**

Unfortunately, WLH 50 consists of only a large calvarium, with the face, base, mandible and teeth missing. The fragments included pieces of long bone, metacarpals and the section of right elbow joint consisting of the distal end of the right humerus and proximal ends of the ulna and radius. This discovery, 15 years after Kow Swamp, reinforced Thorne's previously held conclusion, that people with a very robust morphology had arrived in Australia sometime in the Late Pleistocene. As I have said, other individuals discovered in the Willandra also display similar robust traits. They are, WLH 22, 28 and 63 and although fragmentary they display robust cranial features, particularly a prominent brow and a thick cranial vault but none equal WLH 50 for robustness (see individuals described in Chapter 10). But how could such robusticity be explained?

Only Regional Continuity could provide an explanation at that time and as an avid supporter WLH 50 was vindication for his support of this hypothesis. It was natural to believe that Java was the home of WLH 50's robust genes and they had somehow made their way to Australia, perhaps with descendants of *H. soloensis* but it certainly was not a *H. soloensis*. It was a logical idea that made sense at that time and for many years after.

The Morphology of WLH 50

WLH 50 is essentially a calvarium. The face and basi-cranium are missing, except for a few pieces of nuchal plate and left mastoid as is most of the temporals and sections of the parietals leaving only the frontal and occipital as complete cranial bones (Figs. 7.1–7.4). The mandible is missing, except for a posterior section of the left body that contains the only teeth, M_1–M_3 that was found several months later at the discovery site (Fig. 7.6). The vault surface is generally eroded with pitting of various diameters and endocranially there arachnoid pitting, and occipital erosion has removed marginal, broken edges. WLH 50 was a surface find that rendered it without any contextual setting and original interment features. Moreover, the lack of major parts of the skeleton, even sections or fragments of long bone, suggests that the cranium and elbow section were possibly transported to the place where they were found and did not originate there.

The cranium is large and long measuring 216 mm, 35 mm longer than WLH 1, and has a maximum bi-parietal breadth of 145 mm, 15 mm wider than WLH 1. The outer vault bone has patches of silica wash causing surface sheen in places

FIGURE 7.6   An in situ section of the left mandibular body of WLH 50 at M2 and M3 found after returning to the site several months after initial discovery of cranial pieces. The molar teeth are heavily worn from food containing abrasives. *(Source: S. Webb)*

(Figs. 7.3 and 7.4). The length of time silica is believed to precipitate on bone in this way has led to suggestions that WLH 50 is placed well within the Late Pleistocene (Bowler pers. comm.). Attempts at dating have not been successful so far (Caddie et al. 1987). WLH 50 is a fully mature male person probably >40 years. The coronal and lambdoid sutures are fully closed but can be traced, the sagittal suture is fully closed and difficult to discern. A rounded sagittal crest is seen in anterior view although it seems accentuated when viewed posteriorly (Fig. 7.3). Both parietals slope forming a sagittal keel but greatest cranial width is located low down on the temporal bones. The frontal bone rises smoothly and almost straight from the supraorbital region to bregma (Fig. 7.1). The outline then follows a smooth curve to bregma where it turns posteriorly into a long, shallow curve at obelion passing infero-posteriorly to Lambda. From there the line drops vertically to inion where a prominent but damaged transverse nuchal torus crosses the occipital something noted previously (Hawks et al., 2000). The surface of the remaining basi-occipital bone is roughened indicating large, broad and strong areas of nuchal muscle attachment. Cranial capacity is broadly estimated at between 1400–1500° cc.

The frontal sinuses of WLH 50 are large by any standard and contrast strongly with all others in the series (Fig. 7.7). Their size and shape correspond closely to that of *H. soloensis* with ranges of 14–18 mm-deep, 19–25 mm-wide and up to 17 mm-high compared with Chinese erectine sinuses at 4–15 mm-deep, 8–25 mm-wide and centrally located.

(Weidenreich, 1943:167; Weidenreich, 1951:252). But the thin, solid brow ridge and postorbital sulcus of some Chinese erectines may have restricted frontal sinus development. Weidenreich (1943:165) confirmed this pattern '...*It seems to me as if the tendency in pneumatisation were less pronounced in* Sinanthropus [Chinese erectines] *than in* Pithecanthropus [Javan erectines]'. Therefore, there are two differences in the frontal sinuses of Zhoukoudian and Ngandong people: in the former they are small and centrally located in the inter-orbital region while the rest of the supraorbital region is solid. The mastoid sinuses of WLH50 extend posteriorly to the occipito-mastoid border, superiorly

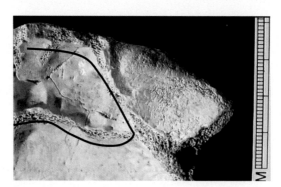

FIGURE 7.7   The large frontal sinuses of WLH 50 (*outlined*).

to the parietal notch and anteriorly above the supramastoid crest. The air cells are uniformly large with the largest 14 mm in diameter. Because of missing bone medial expansion of air cells cannot be assessed and it is often difficult to determine how far they rise superiorly within the temporal wall. Their subtle change in shape and size becomes smaller with increasing distance from the mastoid enabling them to blend with normal spongy bone. Nevertheless, they probably reach the parietal notch in WLH 50.

WLH 50 Archaic Features

Apart from its overall rugged appearance and large size WLH 50 displays a series of characteristics that are widely accepted as traits associated with archaic hominins. These include thick cranial bones; a prominent supraorbital torus; a sloping frontal bone; postorbital constriction; prominent bony areas of muscular attachment; a nuchal torus, an angular torus; a supramastoid torus as well as other prominent bony features around the glenoid fossa, basi-cranium and on the maxilla and mandible. Unfortunately, because the latter three missing from WLH 50, they cannot be examined (Figs. 7.8 and 7.9).

Therefore, only some of the above listed features will be presented and discussed. They may highlight their significance in the relationship of this individual to possible ancestral populations.

SUPRAORBITAL TORUS

An outstanding feature of WLH 50 is its well-developed supraorbital region that strongly accentuates the overall robusticity of the cranium. The brows are pronounced but not uniformly developed. The pattern presents massive thickening in the middle and medially, with the lateral one-third tending to a graduated thinning. At the same time, the lateral region is pronounced around the zygomatic trigone accentuating that area and emphasising postorbital constriction when viewed superiorly. That constriction is supported by a maximum postorbital breadth of 111 mm and a maximum supraorbital breadth of 131 mm. Brow ridges on archaic hominins are generally rounded on their anterior surfaces, but on WLH 50 they are somewhat flattened. The brow is massively thickened into a

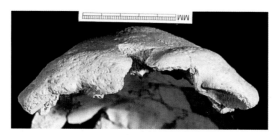

FIGURE 7.8 **Robust brow ridge development of WLH 50 emphasising medial and middle brow development and reduced lateral brow margins.** *(Source: Photo S. Webb)*

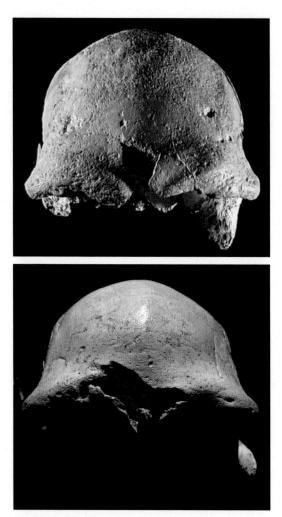

FIGURE 7.9   **An anterior view of WLH 50 (*top*) showing the silica sheen on its frontal, a large brow ridge and a gentle but prominent sagittal keeling.** Below is an anterior view of Atapeurca 4 (550 ka) for comparison of general morphology and archaic features of the two. *(Source: Photo S. Webb)*

torus along the medial two-thirds of its length and appears unlike any observed among other fossil or more recent Australians. However, there is a marked resemblance between WLH 50 and Atapeurca 4 as well as in cranial vault shape when viewed anteriorly. There is no suggestion of a direct link between these two individuals, however, they are compared from the standpoint of their resemblance and the general archaic morphology of WLH 50 as opposed to that recognised among modern humans (Fig. 7.9). Although bone at glabella is missing in both individuals, medial brow shape, including its infero-medial thickness at the broken edges, suggests it must have continued, swelling the glabella region and possibly forming a torus-like structure joining the brows (Table 7.1).

To assess the size of the WLH 50 brow, a module has been used consisting of multiplying medial, middle and lateral measurements and multiplying the

TABLE 7.1 Comparative Cranial Dimensions of WLH 1, WLH 50 and Coobool Crossing People

Cranial measurement (mm)	WLH 1	WLH 50	CC range[a]
Galbella-Opisthocranion	181	216	185–207
Glabella-Inion	170	198	–
Glabella-Bregma	–	127	110–128
Glabella-Lambda	173	211	181–207
Lambda-Bregma	106	133	107–129
Lambda-Inion	66	72	60–91
Lambda-Asterion	–	90	79–100
Lambda-Opisthion	95	–	95–120
Lambda-Opisthocranion	47	57	–
Basion-Opisthion	36	–	–
Basion-Sphenobasion	18	–	–
Bi-Asterion	103	130	103–120
Bi-Coronale	110	–	–
Bi-Frontotemporale	90	–	–
Bi-Infratemporal Crests	84	–	–
Bi-Sphenion (estimated)	–	125	–
Bi-Stephanion	103	98	87-119
Bi-Euryon	–	152	
Inion-Opisthion	40	–	39-55
Opisthion-Opisthocranion	64	–	–
Basi-occipital Breadth	32	–	–
Maximum Bi-parietal Breadth	130	145	–
Minimum Bi-temporal lines	–	112	73-102
Maximum Post-orbital breadth	–	111	85-107
Maximum Posterior frontal breadth	–	123	–
Maximum Supra-orbital breadth	107	131	101–116

[a]Males only.

sum by 100 and divided by three (see Supraorbital Index, Chapter 6). The WLH 50 brow module is 19.4 higher than any calculated for *H. soloensis* (Ngandong) (17.5), *H. erectus* (Choukoutien) (16.7) and Neanderthals (14.4), and greater than individual scores for Sangiran 17 (14.8) and Ngandong XI (14.1), the latter having the largest brows in the Ngandong series (Table 7.2). The closest individual to WLH 50 is Dali with 18.6. A more even brow development is

TABLE 7.2 WLH 1, 3, 50 Supra-orbital Measurements (mm) Compared with Other Hominins

Hominins	Medial	Middle	Lateral
WLH 50	22	22	12
WLH Series	5–22	3–22	5–12
Dali	21.0	19.0	14.0
Jinniushan	14.5	11.4	13.8
Jebel Irhoud	14.1	11.6	13.7
Leitoli 1	22.0	14.0	14.0
Tabun	17.3	12.3	12.2
Skhul	15.1	9.2	10.2
Qafzeh 9	17.0	5.0	7.8
Qafzeh 6	19.0	16.5	13.8
Shanidar	18.0	15.7	10.2
Amud	14.7	13.0	9.8
Sangiran 2	–	8	12
Sangiran 10	–	–	19
KNM-ER 3733	–	8	9
KNM-ER 3883	–	7	11
OH 12	–	–	10
Lake Ndutu	–	10.5	–
Bodo	18	14	13
Kabwe	22	17	16
Atapeurca SH 4	–	12	12
Atapuerca SH 5	15	14	15
Petralona	21	16	12
Arago 21	15	10	11
Steinheim	17	10	10
H. soloensis	14–17	11–14	16–22
H. erectus	12–20	5–17	11–15
Neanderthals	9–18	9–14	10–13

present for both the Chinese, Javan *erectus* samples and Atapeurca 5. That Javan sample has greater lateral brow development and smaller medial development which is opposite that of the WLH 50 pattern. Morphologically, WLH 50 has a similar pattern to the Mongolian Salkhit cranium (Coppens et al., 2008; Lee 2015; Tseveendorj et al., 2016) and Atapeurca 4. Salkhit displays prominent medial development but somewhat less robust than that of WLH 50 and its brow thickness is reduced more evenly medio-laterally than the WLH 50 brow and that seems to reflect that of Atapeurca 4.

The WLH 50 module is far bigger than any individual in either of the Late Pleistocene Australian populations of Kow Swamp and Coobool Creek that are well known for their general robusticity. It is also bigger than the Indonesian archaics and Neanderthals but it has a significant association with the later Chinese archaics such as Dali. WLH 50's supraorbital morphology and size is unique in Australia and is unlike any other Late Pleistocene hominin example from the region. The nearest hominin with a similar brow morphology (that is emphasised medially) and has three similar measurements (medial, middle and lateral) is the Dali male archaic cranium from China (Table 7.2). The brow morphology of WLH 50 alone makes it hard to ignore as an archaic marker and equally difficult to accept as a modern human characteristic, but it is also difficult to pinpoint its origin.

Angular and Supramastoid Tori

An archaic feature of the parietal bone is the angular torus....

   (Rightmire, 1990:226)

WLH 50 displays a prominent angular torus on its left parietal (Figs. 7.10 and 7.11). This feature appears as J-shaped and presents as a vertical, raised and thickened ridge. Its placement marks the posterior margin of the temporalis

FIGURE 7.10   **A prominent J-shaped angular torus (*arrowed*) on the left side of the WLH 50 cranium.**

FIGURE 7.11 **A supramastoid crest on the right side of WLH 50 (*right arrow*), the bone at this point is missing on the left side.** The left arrow points to the lateral extension of the transverse nuchal torus. *(Source: S. Webb).*

muscle attachment that curves antero-inferiorly ending at asterion. It extends posteriorly almost to the nuchal torus marking the edge of the distinctive fan-shaped fibres of the temporalis muscle angled downward and forward. A similar torus is seen on Choukoutien erectines, the Javan Ngandong series, Sangiran hominins, Arago and Petralona from Europe and several African hominins including Kabwe (Weidenreich, 1951; Rightmire, 1990).

Table 7.2 highlights the difference between WLH 50 and a range of archaic hominin groups including hominins from China, Africa and Europe. The comparison is made to show the archaic nature of brow development in WLH 50 and not to align it with any compared group or individual. Table 7.3 compares WLH 50 with others from the Willandra with measurable brows. Three others have large brows although smaller than WLH 50, while several have prominent but smaller medial development (WLH 18, 19 and 69), and four have much smaller brow thicknesses although still somewhat prominent. It is worth noting the great difference between these three groups, particularly the top four individuals with the most prominent brows and those with exceedingly small brows which are gracile individuals.

Three characteristics identified as archaic in the Kow Swamp population included transversely flat and sloping frontal bones, a pre-bregmatic eminence and a post-bregmatic saddle. A numerically larger Late Pleistocene population, the Coobool Creek group, found in1949–1950 by George Murray Black in the Kow Swamp area, also shares this frontal morphology. The shape likely stems from deliberate infant head deformation either by binding or manual pressure on the frontal and occipital bones of the head (Brown, 1981). That conclusion countered accepted wisdom that said it was *the* feature of the … *mark of ancient Java.* The Cohuna cranium, found long before Kow Swamp in 1925 on the north-western edge of Kow Swamp, has the same features and is undoubtedly part of the Kow Swamp group (Fig. 7.12).

TABLE 7.3 Brow Ridge Thickness among Individuals in the WLH Collection

WLH No	Medial	Middle	Lateral	SO Module
1	5	3	5	0.3
3	8	6	8	1.3
11	–	7	7	1.2
18	17	9	13	6.6
19	16	16	8	6.8
45	–	6	7	–
50	22	22	12	19.4
51	9	4	6	0.7
67	9	7	9	1.9
69	19	10	13	8.2
72	11	6	6	1.3
73	14	8	7	2.6
100	14	8	11	4.1
101	14	9	9	3.8
124	13	9	10	3.9
134	14	5	6	1.4
Range	9–22	4–22	6–13	–
X	12.8	8.4	8.7	–
s.d.	4.7	4.7	2.6	–

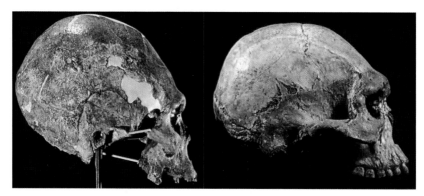

FIGURE 7.12 **Cohuna (*right*) showing sloping forehead, prognathism and prominent brow.** The same features can be seen on a cranium from Kow Swamp (*left*). (*Source: S. Webb*)

Cohuna has all the cranial characteristics of Kow Swamp people. Head binding was firmly rejected by Javan-link supporters, (Multi-Regionalists) but they missed the point. Head binding provided by-far the most likely explanation for the type of frontal flattening seen among Australians living in the last few thousand years of the Late Pleistocene. It looked different but more than that, it was always a puzzle why frontal flattening would have persisted from Early and Middle Pleistocene *H. erectus* in which a flattened frontal had more to do with smaller brain capacity and could not be expected to exist in bigger brained modern people without some form of artificial deformation being involved. The flattening was always too flat! What was interesting was the firm possibility of head binding among people that have never practised it in the Holocene. It showed a Late Pleistocene practice that seemed to be later abandoned. It came and went which expands our knowledge of palaeoculture reminding us of what other cultural practises, beliefs and ceremonies were lost over time because, unlike head binding, they were not preserved in some way. Robusticity occurred in other individuals from Baratta Station, Poon Boon and Bourkes Bridge, all in general close proximity to Kow Swamp, Coobool Creek as well as at Euston farther downstream in the Murray River valley.

But flat frontal bones were only one trait. Head binding did not explain other features such as a thickened cranial vault, prominent brow ridges, large teeth, broad, rugged faces and muscle markings all reminders of possible erectine ancestors or at least an archaic ancestor that had entered the lineage somewhere along the line. These other gross and discrete cranial features on early fossil crania have also been used in a looser fashion in the past to support arguments linking them with Indonesia's archaic hominins which includes thick cranial vaults on some individuals. Thick vaults occur in WLH 22, 28, 50 and 69 as well as among Kow Swamp and Coobool Creek individuals (Fig. 7.13). If it is an archaic trait passed down from earlier hominin

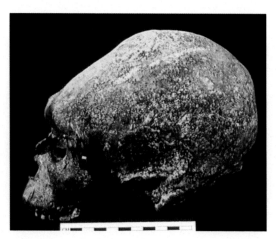

FIGURE 7.13 This Coobool individual has a flat frontal, pre-bregmatic eminence, and a shallow post-bregmatic saddle. There is also a general robustness around the brow and face and deep infra-glabella depression.

populations, structural differences such as thin inner and outer cranial tables enclosing a very thick diploe (up to 85%) in WLH 50 should be noted. That ratio is, however, different from Javan and Chinese hominin groups, and indeed among *H. erectus* from other parts of the world. Moreover, the range of cranial thickness among early Australians is far wider than any other population, archaic or modern. A robust supraorbital region among early Australians was also used in Regional Continuity arguments. While prominent brow development occurs in many Willandra hominins, particularly WLH 19, 50 and 69, it is different from standard regional archaic populations with a medial rather than a lateral thickening as seen in *H. soloensis*, nor does it match the even development pattern in Chinese *erectus*.

The combination of traits in Australian Late Pleistocene fossils suggests separate migration histories among the earliest people who carried genes for a robust cranial skeleton possibly favoured to some extent by selection within the new continent. The complexity of the process brought together and maintained a series of characteristics that, because of the comparatively few remains available to us, tend to continually confuse and mystify, rather than reveal patterns of precise external geographical affiliation. Of course, as the threshold of continental-wide occupation now pushes beyond 70 ka, it is becoming increasingly difficult to accept Kow Swamp, Coobool Creek and Cohuna as representing examples of the earliest morphology except as distant recipients of a robusticity original brought in probably before 65 ka. The intervening period between them and the original arrivals, therefore, represent over 3500 generations. In that time a substantial amount of internal genetic change, exchange and selection as well as genetic extinction could occur together with the mixing of any populations that also arrived during that time. In that case the continent would begin to impose its own adaptive themes on them. The possibility of subsequent migrations during those thousands of years bringing with them new genes in different combinations would add to confusion over origins and that seems to have taken place. While certain gross features among later people may reflect the remnant ingredients of an ancestral morphology, direct comparisons are apt to ignore the intervening environmental, biological and genetic history of movements within the continent. That included the evolutionary mechanisms (*9EM*), operating on small or semi-isolated populations. While not totally isolated, as Tasmanians undoubtedly were, it could be argued that Australia's inhabitants were not on the main highway of the gene movement increasingly happening elsewhere, instead they were at the end of an alleyway. Within Australia, we assume we had only a few people spread across or around the continent with contrasting environmental shifts taking place as the severest of Ice Ages (MIS 2) continued with cool conditions and aridity sweeping the continent lasting for thousands of years. That required good adaptive strengths among the people and with lowest sea levels between 30–20 ka, it was a time of optimised crossings from island Southeast Asia.

HOW OLD IS WLH 50?

The clear majority of Willandra fossils were found as surface finds and cannot be dated absolutely, and WLH 50 is one of them. It cannot be dated by any technique presently available because it is completely fossilised. It lacks any datable organic material although the silica skin covering its surface undoubtedly places it in the Late Pleistocene. Attempts at dating it have produced 14 ka, using the gamma spectrometer *U*-series method, and another of 29 ka was produced using electron spin resonance (Caddie et al., 1987). Some believe the latter to be probably a minimum date others feel even that is far too young. The area where WLH 50 was found was revisited a number of times and efforts were made to rediscover its stratigraphic origin as well as find other skeletal elements belonging to it. Some bone fragments were found including a piece of jaw (Fig. 7.6). All efforts to discover its exact sedimentary placement were inconclusive because it was a surface find. Another aspect to the mystery of the age of WLH 50 was raised when it was suggested that it had not come from the exact place it was found. It was suggested that it had been brought to the site from elsewhere as a manuport, that it had been found elsewhere and perhaps carried for some time possibly as a curiosity or for some other reason. Moreover, if that occurred it could have taken place in the late Upper Pleistocene and discarded where it was found in 1980. That is the problems with finding an indirect date for this individual and it requires a direct date on the cranium itself the only possibility.

CRANIAL THICKENING: PATHOLOGY OR MYSTERY?

Finally, I turn to WLH 50's cranial thickening and one worth examining at length. The greatest mystery about our greatest mystery is its extraordinary thick cranial bones. Other individuals in the series with thick vaults include: WLH 18, 19, 27 and 69. But WLH 50 has a uniform vault thickness and because many individuals in the series consist of only a few cranial fragments it prevents assessment of their uniformity of thickness. However, even cursory comparison of WLH 50 with others in the series and archaic hominins from around the world highlights its unusual size and construction. The human vault can thicken with age but the extent of thickening in WLH 50 is exceptional and cannot be accepted as a normal development at any age. Its thickness can be appreciated from the variation through primary anatomical points that shows it to be largely uniform (Table 7.4). Uniform thickening occurs in other Late Pleistocene Australians particularly among the similarly robust Kow Swamp and Coobool crania. However, they were not studied in detail or described in this regard when they were available and traditional owners have since reburied both populations. However, this Late Pleistocene morphology in some Australian fossils seems to have no comparison among Asian fossil humans from the same period or at any time. It also seems that among modern humans exaggerated cranial thickness has no real equivalent in non-pathological crania and even pathologically thickened individuals are not uniformly thick like WLH 50.

The only other example of a similar vault thickening is the early modern Singa skull from Sudan dated to around 135 ka (Stringer, 1979; Stringer et al., 1985). Its vault thickness measures 14 mm on the parietal and 15 mm at asterion compared to 16 and 17 mm on WLH 50, respectively. Singa also has thin bone tables (<1 mm) like those of WLH 50. Its diploeic thickness is also similar, around 80%, (WLH 50's 87.5% in some places) (Figs. 7.14 and 7.15). The major difference between Singa and WLH 50 is that Singa's vault is not uniformly thick (9 mm on the parietals).

TABLE 7.4 Cranial Vault Thickness (mm) of Willandra Individualsat Primary Anatomical points Only on Those that Three Measurements and Above can be Taken

WLH	A	B	C	D	E	F	G	H	Range
1	5	6	6	4	6	6	8	7	4–8
3	10	11	8	8	7	6	9	15	6–11
19	13	12	10	10	11	12	10	19	10–13
22	–	–	–	9	11	8	12	14	8–12
24	–	–	9	–	9^2	8	10	–	–
28	–	–	–	–	12	9	–	21	–
29	–	–	–	7	7	7	7	–	–
45	–	–	9	10	13	–	10	–	9–13
50	19	17	17	15	16	17	15	18	15–19
67	5	4	8	9	5	7	5	14	5–9
68	–	5	4	5	4	–	–	–	4–5
72	10	9	10	–	–	11	–	–	9–11
73	7	–	8	8	7	8	6	–	6–8
100	13	–	–	9	11	14	10	–	9–14
130	–	–	9	7	9	9	9	17	7–9

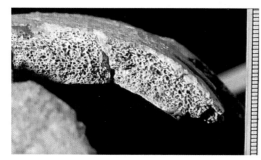

FIGURE 7.14 **Cranial thickness in WLH 50 consists almost entirely of woven (diploeic) bone with thin inner and outer cranial tables, the former is very thin.** WLH 50 vault composition (*right*).

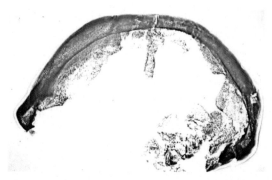

FIGURE 7.15   Radiograph of WLH 50 showing a general uniform thickness and pre-bregmatic thickening.

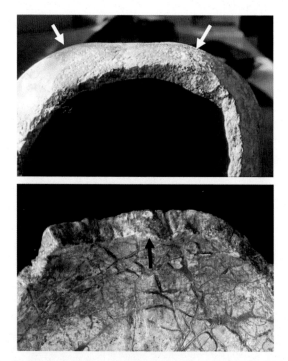

FIGURE 7.16   The vault of the Singa cranium (*posterior view*) showing thickened posterior border of the right parietal and thickened parietal bosses (*arrows*) resulting in a slight sagittal depression. *(Source: Photo S. Webb)*

Personal inspection of Singa in the Natural History Museum in London, allowed me to see its prominent bilateral thickening around the parietal bosses causing a surface bulge that accentuates a sagittal depression and increases cranial breadth at those points (Fig. 7.16). It is similar to recent crania with symmetrical osteoporosis and thickening of the parietal bosses with associated pitting that can also occur on the frontal bosses. There is, however, no such pitting on Singa's cranial surface although it is eroded which may have removed or blocked pitting. WLH 50's does not have any pitting of that type either. Singa also displays vault expansion of the parietal bosses. The comment was made in the published account saying *...with the exception of the diploeic thickening of the parietals, the Singa skull did not exhibit any of the other radiographic criteria associated with bone changes in anaemia* (Ibid 1985:354). While concluding that the thickening was unusual and the skull needed further study, it was suggested that *...certain cranial and endocranial characteristics may be due to a pathological cause...* but no specific pathology was suggested and no further study has been published (Ibid, 1985:357). My observations basically support those findings but radiographs of Singa taken some years ago are of poor quality and lack enough detail to show intracranial morphology. Further scans with modern equipment might help with a firmer diagnosis. Perhaps possible DNA extraction might also help with this mystery pathology, but until then the hyperostotic features do suggest Singa suffered some form of haemoglobinopathy (blood dyscrasia) or severe, chronic anaemia, that cause similar features among recent skeletal and extant modern populations.

Hyperostotic cranial vault structures can result from a number of pathologies, including Paget's disease, *hyperostosis frontalis interna* and *leontiasis ossea*. They can also result from a haemoglobinopathy, including the genetically determined balanced polymorphisms of sickle cell anaemia and other haemoglobin variants including thalassemia variants that can vary in severity. Changes to the cancellous diploeic bone and vault structure among those suffering Paget's disease include areas of rarefaction and new bone formation. *Hyperostosis frontalis interna* is largely localised particularly to anterior internal sections of the vault, and *leontiasis ossea* is very rare and was only occasionally encountered in the past. Both conditions are also characterised by substantial alteration of external and internal bone surfaces, they are not uniform and often grotesquely alter the appearance of the whole skull. Both the rarity and other distinctive features of the first three conditions above almost certainly preclude their implication in these cases.

Neither WLH 50 nor Singa have any sign of cribra orbitalia, another widely accepted sign of chronic anaemia, that can accompany symmetrical osteoporosis and other signs of haematological disease. What both fossils do show is an exaggerated thickening of the vault and an overwhelming predominance of cancellous over cortical tissue; I suggest similarities more than coincidental. Archaic hominin cranial thickening is almost always composed of thick cortical bone tables and minimal diploeic or trabecular bone. Thickened cranial vaults often seen on *H. erectus*, for example, are listed as a standard archaic feature, but vault composition among these and other hominins with thickened vaults is different. Firstly, cranial thickening is greater in WLH 50, and its cranial tables are much thinner, than the vaults of *erectus* or other hominins in whom the diploe usually makes up only one-third of the thickness or less with a little variation (Fig. 7.17). Therefore, the thickening of WLH 50 is not necessarily an archaic trait although its other robust traits might be considered as such. Is it possible, therefore, considering the combination of traits in these two individuals, their vault structure can be considered as evidence of a possible haemoglobinopathy?

It has been known for some time that genetically determined anaemias cause skeletal changes in the cranium (Hrdlicka, 1914; Cooley and Lee, 1925; Williams, 1929; Angel, 1964, 1967). Among those changes are thickening of the cranial diploe and symmetrical osteoporosis, a bi-parietal thickening and pitting, accompanied by cribra orbitalia, a pitting of the orbital plate of the frontal bone that is commonly associated with any form of anaemia. Bone alteration is due to haemapoetic hyperostosis triggered by inadequate or malformed red blood cells, typical in thalassemia and sickle cell anaemia. It results in a compensatory enlargement of marrow tissue responding to a requirement to produce normal blood cells that will replace the damaged cells produced by these hereditary diseases. The resulting enlargements of the diploeic space causes pressure atrophy, particularly of the outer table, resulting in resorption, thinning and, occasionally, complete destruction of compact bone, particularly of the thin orbital roof of the eye socket in the case of cribra orbitalia. The condition can affect all major cranial bones individually, but usually not the whole vault. WLH 50 does not appear to reflect such changes.

In their heterozygous form, the balanced polymorphisms are an adaptive reaction as a defence against malaria, normally endemic to areas between latitudes 45 N and 20 S. Radiologically, affected internal diploeic structures present a 'brush' or 'hair-on-end' appearance. The 'brush shape' is caused by formation of bone spicules within the diploe at right angles to the table, resulting from osteoclastic and osteoblastic bone cell activity during the proliferation process. Of all the possible osteological symptoms caused by the condition, only the

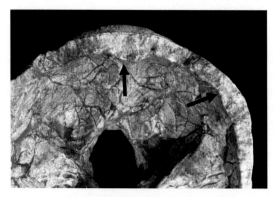

FIGURE 7.17 **Cranial vault thickness in the Sangiran 4 H. erectus from Java, showing the frontal bone at bregma (above, *arrow*) and at the occipital and parietal margins (below, *arrows*).** *(Source: Courtesy of Professor Freidmann Schrenk, Senckenberg Museum)*

'hair-on-end' condition applies to WLH 50 confined to a small area around the pre-bregmatic region and even that is minimal. Many regions in Africa have had endemic malaria for a long time because monkeys and anthropoids suffer from it and formed suitable vectors for transfer of the disease to other susceptible creatures such as modern humans. I cannot explain the thickened vaults of WLH 50 and Singa in any other way than emanating from a pathological process. Although it seems reasonable to suggest WLH 50's thick vault indicates a form of palaeohaemoglobinopathy, by today's diagnostic criteria, this individual must have been heterozygous for the condition to reach adulthood, because homozygous subadult sufferers do not survive without regular blood transfusions. But few osteological changes occur in heterozygous individuals. However, vault thickening and symmetrical osteoporosis are normally associated with homozygous sufferers and WLH 50's uniform cranial vault thickening is not diagnostic of the presence of a haemoglobinopathy among modern populations.

While it does not account for thickening among other vaults in the Willandra series or elsewhere among Late Pleistocene and early Holocene Australians, I know of no other individual with the same vault thickness as WLH 50 and indicates a certain amount of uniqueness so far. Nevertheless, it could be suggested that, although WLH 50 and Singa are different ages, they may reflect an early biological adaptation among hominins, for coping with malaria. And while this cranium does not display skeletal changes conforming *exactly* to those found among recent populations suffering chronic anaemias, the morphology of thin tables and thick diploe strongly indicates the body's need to produce extensive amounts of haemapoetic tissue as a reaction to something like a blood dyscrasia. But how and when might this condition have emerged?

Although life threatening in its homozygous form, these haemoglobinopathies in there heterozygous form probably emerged as an adaptive protection against malaria, probably selected for in areas of endemic malaria. According to the modern distribution map of malaria issued by the *Centres for Disease Control and Prevention*, means virtually everywhere along modern human dispersal routes to Australia. The origin of one of the deadliest malaria parasites (*Plasmodium vivax*) has been traced back to 300 ka in Africa, originating among wild apes. That means there was plenty of time for some form of palaeohaemoglobinopathy to emerge among early modern humans for alleviating the effects of malaria. The Singa cranium, therefore, might represent the emergence of such a condition in northeast Africa. However, while it is difficult to imagine WLH 50 was an indigenous African that carried it to Australia, there are several possibilities to consider. First, he could have been an unwitting recipient of associated genes via their flow from African ancestors, inheriting a heterozygous form in Australia from others who brought it in. They were then transmitted between generations during their dispersal and maintained in the population as they passed through endemic malarial areas in South East Asia and Sahul. The second possibility is that a similar polymorphism to that in Africa emerged in Asia to combat malaria there, if that is what Singa signifies. It could then have been passed on to modern humans during episodes of contact between those people that mixed with dispersing heterozygous or non-carrier modern humans. We might also consider how a palaeohaemoglobinopathy might present tens of thousands of years ago in terms of its bony morphological appearance. Could we be facing bones changes that no longer occur in modern populations? Some examples of such bony mysteries of that kind do exists (E. Trinkaus, pers. comm.) The enigmatic thickening of WLH 50's vault might represent one of these and an unknown form or precursor palaeohaemoglobinopathy, or an original emergent type unknown today. Was it similar to those of today or was it different, and, moreover, how would it affect people already having a rather thickened cranium as a normal developmental condition derived from an archaic ancestor. In the absence of a selective agent for the balanced polymorphism in Australia and its comparative rarity, after an initial introduction, it could disappear quickly, perhaps within a few thousand years. Indeed, if WLH 50 had the condition it could mean that he was probably an early entrant on the continent and very few of those can be expected to have been carriers displaying such changes. Any loss or lack of offspring would compound the rarity and so there should be no reason to expect other individuals to be found with a similar vault construction.

My interpretation of WLH 50's vault structure and specifically its uniform hyper-cranial thickening is that it was the result of a mixed archaic modern ancestry. Its thickened vault is just one trait among a suite of robust features showing that association, by direct or indirect genetic linkage with robust archaic hominins. Its vault structure is in addition to these traits. It represents further thickening of a naturally thickened vault as an archaic feature brought about by an adaptive pathological phenomenon associated with an haemoglobinopathy of some kind or one in a precursor stage of development. The suite of characteristics WLH 50 possesses is impossible to explain in any other way. I note also that it possibly represents the thickest fossil hominin vault found so far. However, as a mixed ancestry person, it either inherited a palaeohaemoglobinopathy that had evolved among modern people that brought it from Africa or the Middle East, or he received it from a local adaptation among archaic groups coping with malaria with whom he mixed. Moreover, I would suggest that because WLH 50 has retained a high proportion of morphological traits as well as a very robust general appearance associated with an archaic population, he was temporally not far from the point of his original mixed common ancestors.

I emphasise that vault thickening seen in many Late Pleistocene and Holocene Australian crania is not necessarily a pathological phenomenon, only extraordinary thickening is. Therefore, it is likely that this individual would normally have had a thick vault in keeping with other robust features in the same manner that many other Late Pleistocene and Holocene Australians have (Figs. 7.18 and 7.19). Little or nothing is known of the origins and palaeo-epidemiology of many of our commonest diseases or of human responses to them if encountered tens of thousands of years ago. The evidence presented here requires further thought about the age, distribution and effect of malaria

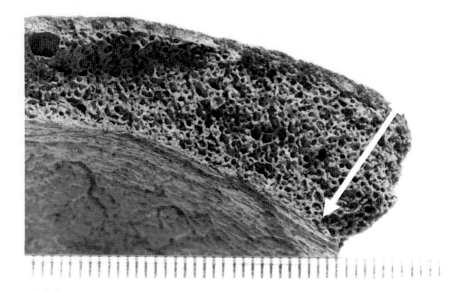

FIGURE 7.18 A Holocene Australian with a thickened vault (16 mm on extreme right, *arrow***).** Scale in millimetres. *(Source: S. Webb)*

FIGURE 7.19 A thickened section of frontal vault in a Kow Swamp individual held by A. G. Thorne. *(Source: S. Webb)*

on all hominins in the past, their responses to it and to appreciate the antiquity of adaptive human biology that must have taken place not only during the Late Pleistocene and Holocene but also in the Lower and Middle Pleistocene among hominins moving into new environments and exploring new regions.

CONCLUSION

When I first proposed vault thickening could be a pathological condition it was pointed out that such pathologies could not exist in Australia because we did not have endemic malaria which would select for them (Webb, 1990, 1995). I did mention, however, that those who had lived in places where endemic malaria existed could have brought it into Australia and without selection and suitable vector the trait could be eliminated from the population. If we look at the origins of cranial thickening in WLH 50, it can only come from a haemoglobinopathy. Perhaps it was a forerunner of the haemoglobinopathies we know today, which evolved by the late Middle Pleistocene when Singa lived and who possessed a polymorphism in its early stages of development. It is possible that the condition did not display bone-changing characteristics in the same way then that it does today.

From the differential diagnosis presented above, only one reasonable conclusion can be made. WLH 50 displays a pathological alteration of an already thickened cranium, but its other traits have come from an ancestral association with archaic people that occurred through differentially mixed modern/archaic populations that spread from South or Eastern Asia into Australia. After they arrived, the polymorphism causing the hyper-thickened vault was selected out because of the few carriers who arrived and because malaria, as a selecting agent for the condition, was not encountered. Further mixing of those people with modern humans that continued to arrive, particularly in the north where 'frontline' encounters were made in greater frequency, reduced robusticity there. That pattern then than trickled down through the continent but reached the south last where it persisted albeit in a much greater frequency. Some was retained through local selection factors as well as isolation and/or small populations living in refuge-like areas in different parts of the continent during the height of the last glaciation 40–20 ka. Relic populations in those refuges later displayed those traits to somewhat lesser extent and populations living along the Murray River, especially those of Kow Swamp and Coobool Crossing that lived in the very late Pleistocene and early Holocene who may have represented them. The presence of WLH 50 in the Willandra suggests that it was brought there from the north or was one of the first of its 'type' in the region. Either way, it would suggest it was part of a progenitor population: one of the first arrivals of its kind.

References

Angel, J.L., 1964. Osteoporosis: Thalassemia? American Journal of Physical Anthropology 22, 369–371.

Angel, J.L., 1967. Porotic hyperostosis or osteoporosis symmetrica. In: Brothwell, D., Sandison, A.T. (Eds.), Diseases in antiquity. C.C. Thomas, Springfield, USA, pp. 378–389.

Brown, P., 1981. Artificial cranial deformation: A component in the variation in Pleistocene Australian Aboriginal crania. Archaeology in Oceania 16, 156–167.

Caddie, D.A., Hunter, D.S., Pomery, P.J., Hall, H.J., 1987. The ageing chemist – can electron spin resonance (ESR) help. In: Ambrose, W.R., Mummery, J.M.J. (Eds.), Archaeometry, further Australasian studies. Department of Prehistory, Australian National University, Canberra, pp. 167–176.

Cooley, T.B., Lee, P., 1925. Series of cases of splenomegaly in children with anaemia and peculiar bone changes. Journal of the American Pediatric Society 37, 29.

Hawks, J., Oh, S., Hunley, K., Dobson, S., Cabana, G., Dayalu, P., 2000. An Australasian test of the recent African origin theory using the WLH-50 calvarium. Journal of Human Evolution 39 (1), 1–22.

Coppens, Y., Tseveendorj, D., Demeter, F., Tsagaan, T., Giscard, P.-H., 2008. Discovery of an archaic *Homo sapiens* skullcap in Northeast Mongolia. Human Palaeontology and Prehistory 7 (1), 51–60.

Hrdlicka, A., 1914. Anthropological work in Peru in 1913 with notes on pathology of the ancient Peruvians. Smithsonian Miscellaneous Collections 61, 57–59.

Rightmire, G.P., 1990. The evolution of *Homo erectus*. Cambridge University Press, Cambridge.

Stringer, C.B., 1979. A re-evaluation of the fossil human calvaria from Singa, Sudan. Bulletin of the British Museum of Natural History (Geology) 32 (l), 77–83.

Stringer, C.B., Cornish, L., Macadam, P.S., 1985. Preparation and further study of the Singa skull from Sudan. Bulletin of the British Museum of Natural History (Geology) 38 (5), 347–358.

Tseveendorj, D., Gunchinsuren, B., Gelegdorj, E., Yi, S., Lee, S.-H., 2016. Patterns of human evolution in Northeast Asia with a particular focus on Salkhit. Quaternary International 400, 175–179.

Webb, S.G., 1990. Cranial thickening in an Australian Hominid as a possible palaeoepidemiological indicator. American Journal of Physical Anthropology 82, 403–411.

Webb, S.G., 1995. Palaeopathology of Australian Aborigines. Cambridge University Press, Cambridge, UK.

Weidenreich, F., 1943. The skull of *Sinanthropus pekinensis*; a comparative study on a primitive hominin skull. Palaeontologia sinica. Lancaster Press (New Series D. No.10.), London.

Weidenreich, F., 1951. Morphology of Solo man. Anthropological papers of the American Museum of Natural History 43 (3), 205–290.

Williams, H.U., 1929. Human paleo-pathology with some observations on symmetrical osteoporosis. Archives of Pathology 7, 839–902.

Chapter 8

Impenetrable Obscurity

WHEN AE1 AND AE2 MET AE4

So, who were the first people to arrive in Australia? Are we any closer to answering that one? Answers to 'how many journeys' and 'when' are important in this regard but one thing we do not have to contemplate for AE1 as we do for AE4 is the complications arising from contact with those already living outside. But for modern humans such contact brought another element into dispersals and outcomes and possibly introduced sudden changes to morphological variation among some of the earliest modern human populations wandering outside Africa. We know modern humans met archaic humans living in various parts of the Old World, in effect the AE4 explorers met up with the AE1 and 2 explorers by meeting the latter's descendants. Those archaic groups had forged their own histories and evolutionary pathways living in scattered pockets stretching from Europe to East and South East Asia; all regions modern humans transitioned. With so few over such vast distances there were times when they missed each other. Occasionally they interacted and sometimes they fought, perhaps eliminating each other. They may even have unwittingly travelled in similar directions but the possibility of meeting, fighting and interbreeding depended on timing and routes taken. The cultural or biological influence of one on another varied depending on group size, the power of one over the other and sometime deliberate avoidance. One or other could be overwhelmed, particularly if they found themselves isolated or in the minority and no doubt that happened to modern humans moving through the Levant 120,000 years ago and in southern China about the same time or a little later.

Made in Africa. http://dx.doi.org/10.1016/B978-0-12-814798-6.00008-X

When modern humans entered a region already occupied by archaics they were imposing on people who had thrived in 'their' region, they knew it intimately and had adapted to local conditions. Such contact experience over several thousand years and in different places could change modern humans biologically and/or culturally, make them think differently through the realisation they were not alone and looked different. Such encounteres could have also altered their plans, direction of travel and even stalled their dispersals. Another factor to consider is the extinction of archaic groups and the subsequent isolation of mixed modern-archaic people who looked somewhat like but yet different from unmixed modern people. These encounters as well as the effects of *9EM* during these long journeys would also effect morphological changes. Such issues, for people travelling at least 20,000 km to Australia taking many generations, were a very real possibility. Moreover, the longer the dispersal the greater the opportunity for those changes to occur. Therefore, it is possible to imagine the groups that entered Australia being more a product of their journey and encounters that changed them from their African forebears, and that could have played out elsewhere among other modern humans on long dispersal events. However, it all depended on the vicissitudes of their journey, the presence or absence of archaic populations along the way and if mixing took place. The later modern human journeys occurred the less likely it was that archaics had any influence on them because they gradually were disappearing.

The presence of modern humans living in Fuyan Cave 120 ka leaves open the possibility of an early modern human entry into Australia. They lived there when archaic humans were around but we do not know how many archaic types there were, their distribution and exactly when and how they disappeared although 80,000 years later Neanderthals, Denisovans and *floresiensis* were still around as well as the possibility of others but only just. The timing of archaic extinctions is important for evaluating the possibilities of modern/archaic mixing particularly in the Far East. But, with an increasing cast of archaic humans living in the region it cannot be ruled out that even archaics themselves entered Australia before modern humans, perhaps pushed by the Toba eruption around 74 ka or for some other reason. It is possible that event may have pushed both groups onto the same path and then towards Australia. There are so many possibilities. Here's another, unmixed modern humans as well as other mixed or nonmixed archaics arrived here at a similar time or travelling together unphased by any differences in appearances. They could arrive within 500 years of each other or less and we might never be able to separate those arrivals with present geochronological methods.

CONSIDERING THE EARLIEST AUSTRALIANS

The title of this chapter comes from a quote by the 19th century Polish explorer Sir Paweł Edmund Strzelecki (1845), who, my Polish wife informs me, is pronounced Shcheletski not Strezleki. Anyway, what he was referring to was the baffling origin of the first Australians in his journal *Physical Description of New South Wales* and van Diemens Land. Unbeknown to him, he was addressing the hardest of questions that would be pondered and argued about for the next 150 years and more and one that has both haunted and puzzled many anthropologists and anthropological study for many years. In the light of the subject of this book it is worth expanding the paragraph of Strzelecki's observations:

> *Their history has no records, no monuments; but consists mostly of traditions, which, in common with their language, customs, moral, social, and political condition, seem, ever since their discovery, to have been regarded as a subject unworthy of European study.... Their origin, like that of most things in creation, is involved in impenetrable obscurity:....... (1845:333).*

Things have vastly changed over the last century or more and may be becoming a less impenetrable task to unravel that mystery than it seemed 170 years ago. The common thread of bafflement has been the robust element seen in fossil Australian skeletons. With the background of discoveries over the last decade, it is now more likely than ever that it came from a mixed or unmixed archaic hominin population whom entered Australia. The goal now is to find out if that is, indeed, where the robusticity came from and who that archaic population was. That might help resolve the puzzle of where the first Australians originated, at least some of them.

It was all so simple 50 years ago, when comparatively few hominin fossils existed. What there was seemed to form a logical if incomplete sequence that made human evolution comparatively easy to piece together and understand, the only thing was, it was wrong. With magnificent truth of hindsight, we can say it was a time of many beliefs most of which have been overturned. One thing that has not changed, however, is that it was believed then that *Homo erectus* (in one form or another) was the founder of all modern humans as its generic name implies. However, it was also thought that *erectus* first left Africa only 1 million years ago at the earliest, ventured into Europe and crossed the Old World settling in China and Java. That was true. They then flourished and evolved in their various regions and in so doing gave rise lineages that slowly evolved into modern people with increased brain size and capabilities to match. That was not true although we all had one *H. erectus* ancestor at some point. It was assumed that natural evolutionary and adaptive processes acted locally on each population to produce distinctive modern humans regionally that exist today. That is true. Only 50 years ago, all biological anthropologists and palaeoanthropologists accepted that story and that was the Regional Continuity or Multiregional hypothesis and as you can see, it was partly correct.

The Regional Continuity model was first proposed by Franz Weidenreich 70 years ago. He had taken over study of the Choukoutien *H. erectus* site outside Peking (now Beijing), following the death of the Canadian anatomist Davidson Black. Black had been leading the work until his unexpected death at his laboratory bench one evening in 1934. Weidenreich would also die suddenly in

1951 and in mid-sentence as he wrote his classic monograph *Morphology of Solo Man* describing *Homo soloensis*, his last words ending in mid-sentence as he described: *The distance from the foramen fissure....* (p 286). Weidenreich's model used four main populations or lineages he termed: *Australian, Mongolian, African* and *Eurasian* (Weidenreich, 1946:30). A column beneath each heading was cross linked by diagonal lines representing gene flow between them that kept them thriving and prevented speciation. Each population was derived from *H. erectus* and had evolved through different stages noted on the left-hand side of his illustration. These different levels of evolutionary development he termed Archanthropines (*H. erectus*), Paleoanthropinae (*Homo heidelbergensis* and Neanderthals) and Neoanthropinae (modern humans). This explained the origin of modern humans as he saw it with the progressive rise of all regionally separated populations. Although terms were dropped, his model formed the basis for the Regional Continuity hypothesis which by then had the benefit of more fossil discoveries to draw on. We now evaluate hominin differences in terms of what we know about how they lived, their environment, climate and genetic history. It was only recently that we realised such monumental climatic and environmental changes brought about by Ice Ages throughout *Homo* history, were probably the greatest catalyst in making us who we are.

The puzzle of the origin and meaning of robusticity in Australia has persisted, however. But what was about to occur was a total rethink concerning modern human origins and that would require an explanation about the origin of the First Australians and incorporate an explanation for the widely contrasting morphologies found here, robusticity something anthropologists have mused upon for over 100 years. The question has always been: *who* were the people who brought it here and from where did *they* originate? It cannot be denied that there is a wide range of skeletal variability among Australia's very Late Pleistocene and even some Holocene people but where did it come from. A single population with such characteristics has no equal among modern people, so who lived here and later mixed together or who lived just outside and mixed together out there? Who brought the robusticity in and by what migratory strategy did it arrive. Much speculation has followed this trail and the subject has fuelled its fair share of argument.

Robust Aboriginal crania was noted by Paul Topinard (1872) and so did Charles Darwin's friend and avid supporter, Thomas Huxley (1863:165), who had studied the craniology of skulls from many parts of the world including Australia. When he published *Evidence as to Man's Place in Nature,* Huxley suggested that Australian skulls had several characteristics in common with Neanderthal crania. What he was suggesting was that robusticity was part of the general morphology of Australian skulls as it was in Neanderthals. His suggestion is interesting for his time because *soloensis* and *H. erectus* were not known then, the only robust hominin fossils in Huxley's time were Neanderthals and there were very few of them. All he could do when it came to making a comparison was to suggest Neanderthal. Others, much later described the Solo crania as tropical Neanderthals (Howells, 1973). I suggested the same thing to Alan Thorne when I was still an undergraduate. He just laughed and said Howells and I were both wrong. Not long after Ter Haar's discoveries of the Solo crania, Weidenreich made a link between them and Australians when he was studying *H. soloensis*:

> *...the ancient Javanese forms of, Pithecanthropus and* Homo soloensis, *(sic) agree... with certain fossil and recent Australian types of today so perfectly that they give evidence of a continuous line of evolution leading from... Javan forms to the modern Australian bushman (Weidenreich1946:83).*

Support for *soloensis* as the source of Australian robustness was embraced from the late 1960's to the early 1990's because as long-term neighbours, they were the best candidates and although undated were believed to be Middle Pleistocene. Therefore, they could have arrived in Australia before 150,000 years ago, perhaps during the MIS 6 glacial when sea levels were low, or even before that. From 1960 onwards, researchers outside Australia also noticed Australian cranial robustness like physical anthropologist Don Brothwell (1960):

> *... it is apparent that there is a general acceptance that the early populations of South-east Asia included both a robust fossil type and a more slender (?later) form.... (1960:341).*

His observation was repeated a couple of years later by N.W.G. Mackintosh, Professor of Anatomy at Sydney University, who supported Weidenreich's suggestion of a connection between fossil Australians and *soloensis* people. The case was building through a series of shared robust cranial characteristics between recent Australians and the Javan fossils that included thick cranial vaults, prominent brow ridges, postorbital constriction and a sloping frontal bone. Mackintosh's noted that, even as far back as 1965, none of then presumed oldest Australian crania found up to that time had been properly described:

> *All these crania need complete morphological description and re-examination, including comparison with one another, with more recent man in Australia and with earlier fossil man elsewhere. The mark of ancient Java is on all of them, but that can be seen in modern Aboriginal crania too.... (Mackintosh 1965:59).*

The crania he was referring to were, Keilor, Talgai, Mossgiel, Tartanga (a juvenile) and Cohuna. Sadly, except for the description of the Willandra Collection included here, those Australian fossil crania continue to lack a detailed published description including the unpublished Cobool Creek (CC), WLH 1 and Kow Swamp populations that are described only in unpublished PhD theses. However, Mackintosh's second sentence regarding *the mark of ancient Java* stuck, and became the flagpole on which many nailed their colours, including me at one time, but it was all we had and it seemed to make sense. The *mark of ancient Java* was initially the sloping forehead in individuals like Cohuna but later that was expanded to a set of features that amounted general robusticity. It was always assumed Australian robusticity was derived from the template that earlier Javan *H. erectus* populations had laid down and *soloensis* had inherited. The Indonesian lineage went back to Sangiran 17 *H. erectus* cranium (1.2 Ma, a date unknown at that time). The key to temporally as well as evolutionarily placing fossil hominins is good dating and nowhere is without problems in that regard. Indonesia certainly has its problems and we are still not sure of the correct age of *H. soloensis* with dates of 150–50 ka proposed a couple of decades ago (Swisher et al., 1996) and then more recently 546–143 ka (Indriati et al., 2011). Using the latter range with an average of around 350 ka would mean *soloensis* could have reached Australia long after its supposedly earlier *erectus* ancestors had moved to other Indonesian islands like Flores and Sulawesi over half a million years before *soloensis* appeared. I would question, however, if *soloensis* was a descendent of earlier *H. erectus* types. It is only assumed they were, but there is no direct proof of it (see Chapter 3).

The proposed linkage between Javans and early Australians was strengthened by the discovery in 1967 of the Kow Swamp fossil population uncovered by workman building a levee in the central Murray River (Thorne, 1975). Regional Continuity made very good sense after these discoveries. Because the Kow Swamp people possessed an extraordinary set of robust features, among males and some females. Their morphology matched the previously discovered Cohuna cranium found on the north-western edge of Kow Swamp 42 years earlier (Fig. 7.10). It shared many cranial features with Kow Swamp and there is little doubt it was part of that population. Cohuna and Kow Swamp displayed robustly constructed crania but a date of between 16–9 ka obtained for Kow Swamp placed them in a special place in Australian prehistory. The earliest archaeological evidence for people in Australia then was 16 ka obtained at Kenniff Cave in central Queensland by John Mulvaney in 1962. Therefore, Kow Swamp people, with a similar age, were thought to represent the first or some of the first people to enter Australia and they had the expected morphological connection to the robust archaic people of Java to prove it! So, the connection between Indonesia and Australia was largely cemented and the idea of Regional Continuity was also. Kow Swamp showed that more modern people had resulted from what was believed to be an ancestral stock in Java thus showing a descent group that had begun with *H. erectus* a million years before, as the date was then for *erectus* in Java.

Much later another fossil population morphologically similar to Kow Swamp and jst upstream from Kow Swamp supported those conclusions. The set of fossil crania from Coobool Creek, sometimes called Coobool Crossing, was discovered in early 1980 in the Anatomy Department of the University of Melbourne by Peter Brown. They were originally discovered in 1950 as part of a group of 126 crania found eroding out around the Wakool River adjacent to the Murray River. George Murray Black, a grazier from Gippsland in eastern Victoria, found them. He did not take any field notes of exactly where the remains had been found or any depositional circumstances. Over the decade prior to this finding, he had regularly been distributing hundreds of Aboriginal skeletal remains, either found by him or given to him, to the Anatomy Department of the University of Melbourne and the Australian Institute of Anatomy in Canberra. He believed he was contributing to science by his activities and he continued his collecting mainly from other graziers who lived along the Murray river and elsewhere either side of the river. Much later, research carried out on the Coobool Creek sample by Loring Brace led him to remark *…that in both cranial and post-cranial form the Coobool Crossing material displays a degree of robustness well beyond that of most recent Aborigines.* (Brown, 1981:157). Both Kow Swamp and Coobool Creek people shared the same robust features which included: broad faces, prominent brows, thick cranial vaults, broad deep palates, large teeth, sloping foreheads and prominent bony places of facial and occipital muscle attachment. A uranium/thorium date from the CC 65 cranium provided a date of around 14 ka, placing it contemporary with Kow Swamp dated between 14.0–9.5 ka. For Alan Thorne and Milford Wolpoff, the biggest proponents of Regional Continuity, the Coobool Creek population further reinforced the idea of an archaic connection with Java as well as supporting the idea of Regional Continuity (Thorne and Wolpoff, 1981).

The Kow Swamp date was the first properly dated and widely accepted Late Pleistocene skeletal evidence found in Australia. Then as now theories were built on the best evidence of the time, but over time, theories are either confirmed or refuted by new evidence whether it be empirical or experimental. The Coobool Creek 'discovery' 15 years after Kow Swamp by Brown, who also led the restoration and research of the collection, seemed to support Java connection. At that time, modern humans were just modern humans, they were people that had evolved from earlier types and the most famous type specimen was Cro-Magnon from France. It was a time when modern Europeans like Cro-Magnon were thought to have evolved in Europe from Neanderthals, an idea that seemed required a rapid skeletal modernisation, particularly of the skull, over a very short time.

OTHERS EMERGE

In 1969 WLH 1 was found, only 2 years after Kow Swamp. Then in 1974 WLH 3 was found less than half a kilometre from the WLH 1 site. WLH 1 and 3 were initially dated to 25 and 32 ka, respectively. Their redating in 2002 resulted in an age of 40–41 ka (Bowler et al., 2003). In 1999 a claim was made that WLH 3 was around 60 ka, a date later retracted (Thorne et al., 1999). Both WLH 1 and 3 looked totally different from the Kow Swamp and Coobool fossils They were fully modern with gracile features and not a hint of robusticity and they were of course much older. This begged the old question but with a new theme: where did these *gracile* people, who looked so different from Kow Swamp people come from? It was obvious they had nothing to do with Java or Kow Swamp.

They were Cro-Magnon-like but were 25,000 km from home, so could not be related to him, and anatomically modern humans from Africa were not a concept at that time. The other problem they brought was their age. They were twice as old as any other fossil humans in Australia. The Mungo people were, therefore, another puzzle to be explained.

The solution seemed to be another link, not to Europe or Java, this time with China 7500 km away across the Equator. While the nearest archaic people to Australia were Middle Pleistocene Javans, other Middle Pleistocene people had been living in China and they now became contributing contenders to the earliest Australians. Followers of the Regional Continuity model believed that somehow Chinese *H. erectus* had evolved into a gracile hominin over time, as they were assumed to have done in Indonesia, but much more so than *H. soloensis* who had remained robust albeit with increased encephalisation. But why it was thought these two groups with similar ancestors had undergone such different evolutionary processes in their respective regions was never explained, but there again neither is why *H. rhodesiensis* turned into modern humans and *H. heidelbergensis* did not if they were closely related.

In Australia at that time, Chinese gracilisation was thought to be the key to the Mungo people. It was exemplified by a fossil cranium discovered in 1958 in Tongtianyan Cave located in Liujiang County in southern China (Woo, 1959). The Luijiang cranium was a fully modern, a lightly built human possibly male around 40 years old. The earlier discovery of seven people in the Upper cave at Choukoutien also showed an unmistakable modernisation although some individuals displayed an element of cranial robusticity. However, they were lost with the *erectus* remains in 1941 and subsequently with only casts to examine, have been largely sidelined in the story and the Liujiang specimen became the unofficial 'type' specimen' of Chinese gracilisation or modernity presumed, then, from its very distant 'Peking Man' *erectus* ancestor. For Alan Thorne, who had removed hundreds of fragments of the WLH 1 skeleton from its entombment in a large lump of calcium carbonate and who also excavated WLH 3 (Fig. 8.1), the Liujiang cranium, was a dead ringer for these people, particularly WLH 3, in terms of its gracile appearance. He then set about proposing a model that explained the presence of two quite separate ancestral populations in Australia (Thorne, 1976). His later comparison of the gracile Mungo morphology with that of Liujiang and the Kow Swamp people with Javan *erectus*-like *soloensis* fossils, prompted him to suggest, with others, Java and China were the two regions from where the first Australians had come. All was now explained; and Regional Continuity explained it. The then date of 60 ka for Liujiang also confirmed that the people it represented had enough time to move from China to Australia. They were believed to have reached Australia first because their remains at Lake Mungo were the oldest on the continent, then placed at 32 ka. It was a good idea but it did not really fit. The gracile people should have arrived after the robusts because they had been there longer and lived much closer to Australia and so had the opportunity to reach Australia first. It was only logical that this was so.

FIGURE 8.1   Alan Thorne excavating WLH 3 in 1973. The picture shows the skeleton laying on its right side, knees slightly bent with its hands in its lap.

However, the very nature of Thorne's work, both at Kow Swamp and Lake Mungo, changed Australian palaeoanthropological debate in a couple of ways and took it forward. No longer was the human record in Australia being argued using a few undated and largely undescribed crania that came from different parts of the continent as in Macintosh's time. There was now fresh, empirical, in situ skeletal evidence that included contrasting morphologies, a good sample size, and they were securely dated to the Late Pleistocene. It all seriously challenged previous arguments of the phylogenetic homogeneity of modern Aboriginal people and made for a good logical model that could be presented as the story of the first Australians. Previous arguments were often based on shaky morphometric data gathered from surveys of limited numbers of regionally restricted modern Aboriginal people, heavily impacted for decades by nontraditional nutritional and medical problems, and based on the idea that people had not been in Australia very long (Abbie, 1968, 1976; Howells, 1973). Some continued to argue that the 'two groups' (gracilles and robusts) actually represented contrasting ends of a single population that had a very wide range of variability (Abbie, 1976; Macintosh and Lamach, 1976; White and O'Connell, 1982; Habgood, 1986; Brown, 1987). Thorne's dual origin idea also reduced by one the 'tri-hybrid' theory of Jo Birdsell (1967, 1977), that had been the mainstay of 'origins' here. Birdsell had formulated his tri-hybrid ideas in the late 1930's following a similar proposal by Huxley and others who believed Australians were a homogeneous group, although Tasmanians were different from mainland people. Birdsell identified three 'types' of people, *Carpentarians*, who were tall and slender and derived from western Indian and Sri Lankan Veddas;

Barrineans, a small people similar to the *Negritos* that were scattered throughout South East Asia, and *Murrayans*, a stockier, thickset group that lived along the Murray River and other parts of Southeast Australia and thought to be related to the indigenous Ainu people of Japan. The theory did not consider the long environmental adaptation of people living on this continent that had been shaping them for many millennia, although to be fair, Birdsell did not know about that long tenure and the adaptive contours imposed on human form over thousands of years. He had formulated his ideas while travelling around Australia, often with his co-investigator Norman Tindale. Between them they wanted to biologically and culturally quantify Aboriginal populations to show the Aboriginal nation was neither culturally nor biologically homogenous, as many perceived it to be. He was influenced by variations in hair type, skin colour and body build, and it was assumed that these variations were derived from different migrations entering Australia in the last few thousand years. Birdsell's 'tri-hybrid' theory remained a strong force for a long time in Australian anthropology, and it did seem to make sense of the evidence at the time just as Regional Continuity did later.

Outside Australia, emerging fossil discoveries were added to the story although some did not make sense in the Regional Continuity paradigm. Fit or not, the skeletal evidence was squashed into the multiregional pattern of lineages often with the idea that they will eventually make sense when more fossils turn up and we learn more about evolutionary processes and human transitional types that lived in this part of the world. It was a nice idea, simple and logical, but we can now look at the new evidence and see that the conclusions were wrong. It was nobody's fault or lack of judgement, as I have said before, that's how science works. It is often said that simple ideas are the best, but in this case that was not so. Human evolution is not simple, as we are constantly reminded. There were, indeed, several fossils and dated sites found that were a puzzle but that puzzle was not put together until DNA research joined in the detective story. What followed changed anthropology and the story of recent human evolution as it became the biggest gun in the analytical arsenal of palaeoanthropology. One of those was the story of how humans had not only emerged from archaic people living in different parts of the world but also had instead emerged from the Dark Continent and eventually overlapped and mixed with those formative people.

THE OLD CHESTNUT: HUMAN VARIATION IN LATE PLEISTOCENE AUSTRALIA

The story unfolding in South East Asia and China not only continues to be very important it probably holds the key to Australia's human story that will unlock some of the palaeoanthropological mysteries here. The fact is that story starkly shows how modern humans shared the landscape with archaic hominins living in the region in the same way that modern humans and Neanderthals did in Eastern Europe and the Levant. Modern humans and archaic people mixed together in those places so it seems likely they did the same in East Asia, Sunda or even Australia? Genetic research has shown that Denisovans and Neanderthals mixed with modern humans somewhere in Asia and East Asia by 70 ka, so the idea of two different people, mixed or unmixed, entering Australia is not a fantastic idea. Mixing could produce morphological outcomes such as robust traits in modern humans in the following generations and equally they could have inherited/developed their own evolutionary responses to disease some of which were passed down to us. Our pathogens developed along with us and we developed responses to some of them that could have been passed both ways between modern and archaic people after all, they had developed their own pathogens as they travelled their evolutionary path. These two groups had evolved and lived in different regions with different ecological surroundings and different diseases. Modern human dispersals would have encountered these as they moved to where they were encountered and some responses were selected for when pathological vectors were met with, whereas archaic groups had lived longer in their respective regions and had had more time to develop adaptive biological responses to them. Those adaptive responses could have been beneficial not only to themselves but to any modern humans they mixed with, and WLH 50 was one of them (see Chapter 7).

It is possible that either archaic humans arrived here and then mixed with modern humans or more likely they mixed externally. Denisovans and other archaics moved around in the Far East and had plenty of time to do so, but where they moved to exactly we are not sure. Their genes certainly reached Australia and Melanesia by some manner, but does that mean they reached Sahul themselves, it could mean mixed modern/Denisovan people brought it. If mixing took place on Sahul, did modern humans arrive first or after mixed modern/archaic groups? The next question is: who were the archaic people in the mix, Denisovans or another archaic hominin we know nothing about, the DNA does not tell us and cannot extract DNA from WLH 50 which is our most likely candidate for such mixing. But what about other robust individuals in the collection, what is their story? It is easy to talk of 'mixing', but the actual mixing process and its outcomes would have been much more complex than it might seem. The genetic and, thus, the morphological outcomes of mixing at dispersal endpoints like China, South East Asia and Australia could be viewed as a dammed river channel, where water terminates and swirls back on itself remixing and causing eddies, whirlpools and countercurrents back flowing along the oncoming stream. Various genetic and morphological consequences of such processes in Eastern and South East Asia and Sunda could present a very confusing picture of morphology in those places during the terminal Pleistocene. Indeed, the whole argument of dispersal is always seen one way, but back flow even back into Africa from various points along dispersal routes was always possible (Garcia, 2016).

The Willandra collection shows gracility and robusticity among early Australians that must have been a factor of migration. Further migration into Australia, between 40–20 ka, carrying a myriad of mixing histories may have resulted in rapid change producing the variety in Australian morphology subsequently and that has been a hard story to reconcile from the archaeological and fossil record so far. It is still the case that among Australia's oldest fossils, gracile people are the oldest firmly dated. We know that they are modern humans but we do not know exactly when the very first of them arrived, although for now we assume 80–70 ka is the minimum time. It seems very likely that a time of first arrival could very well 100 ka or even earlier. What we do know is that there is a definite archaic component among early Australians but we do not know where it came from or when it arrived.

Even the modern human component in the Australian story may not be easily explained. For example, some gracility observed in the Willandra collection might be allometric or evidence of a smaller type of person similar to modern 'Negrito' people or others of small stature and displays the natural gracility that goes with it. That idea is not new and gracile morphology is seen close to home in the Niah cranium (45–39 ka) from Sarawak (Barker et al., 2006). Birdsell (1979:421) described WLH 1 as '*somewhat reminiscent of the skull from Niah and which certainly is to be classified as Negritoid in derivation*'. Besdies Birdsell's classification, Niah, WLH 1 and other Willandran graciles certainly indicate the arrival of modern people with a smaller stature. Whether early arrivals included a population of generally small people remains in the realm of speculation. However, WLH 1 is only 151 cm tall, WLH 3, while gracile, is not particularly short (173–175 cm) although their different statures could indicate sexual dimorphism. However, WLH 6 is 158 cm if male and 153 if female, a similar size to WLH 1. The possibility of morphological variation among migrating modern humans is an interesting thought, whether representing one or several populations, and should be considered because there must have been such variations. Modern humans were not all homogeneous and neither were ancient hominins and later archaic groups, all had morphological variation as humans do today. One variation was probably stature and general body size. Another was steatopygia, a form of enlarged buttocks and hips among recent hunter-gatherers as a way of storing body fat. It is an adaptive trait among some recent hunter-gatherers against lean times and occurs in Koisan, Bantu and Andaman Islanders. Rock paintings in South Africa and carved Palaeolithic Venus figurines from Europe, together with hair sculpting of small, tight curls, show these adaptive traits, going back at least 25 ka in Europe. Steatopygia could then be a body form marker originally transported by modern human dispersals to places like the Andaman Islands that could be accessed during low sea levels. People of small stature that were the precursors of the Andamanese could have spread through Sunda and Sahul mixed with or as separate dispersals of modern humans. Eventually giving rise to other small phenotypes such as the Aeta (Philippines), Semai or Semang (Malaysia), and some groups in the Papua New

Guinea Highlands and other places in Melanesia. Perhaps Australia's northern rainforest regions were where some ended up giving rise to people of small stature that lived there. However, explaining the early peopling of Australia based on one or two separate waves of migration is no longer viable. Much later, Australian populations may have replicated the natural complexities of population movement and composition within the continent after 70 ka.

An initial population of both mixed and unmixed modern and archaic groups could have entered Sahul first and moved far across the continent probably moving around the coasts. Their inevitably small, discrete populations experienced long periods of genetic isolation with respect to the size of the continental mass and the number of people inhabiting it. Continuing trends towards gracility operating externally and the demise of archaic people externally brought infrequent pulses of more gracile people into Australia which eventually added to and 'watered down' Australia's internal mix. After the last glacial maximum (20 ka) it was likely that only gracile forms of modern human arrived in Australia. Essentially, they would have been part of a continuous but sparse process of gene flow from Sunda to Sahul. These processes resulted in the archaeological 'detection' of two completely separate groups. There seems little doubt that at some time around 45–30 ka, people bearing the two sharply contrasting morphologies lived in the Willandra. The likelihood of WLH 1 and 50 being 'sister' and 'brother', WLH 3 and WLH 19 being brothers' or WLH 1 and WLH 45 'sisters' is difficult to believe and always has been (see also Chapter 10). The terminal Pleistocene populations across Australia reflected the wide range of morphology produced by the intermixing of these types and, no doubt, the addition of further arrivals, but dominated overall by a larger phenotype. In addition, there has been local adaptation and drift of varying frequency and direction operating temporally and spatially over millennia selecting for retention of a more robust morphology in some and gracility in others. Finally, once the mixing of gene pools began, differential mixing could take place involving different gene frequencies across the continent. Such a process would confuse the morphological picture further in sorting out origins, particularly if these populations were joined by later migrations over thousands of years bearing a modern or intermediate morphological appearance. We know such migrations took place elsewhere during the terminal Late Pleistocene and early Holocene in Eurasia, the Caucasus and Eastern Europe (Pugach et al., 2013).

POSTSCRIPT: A LEGACY OF MODERN HUMANS

Thus, from the war of nature, from famine and death, the most exalted object which we are capable of conceiving, namely the production of the higher animals, directly follows. There is a grandeur in this view of life, with its several powers, having been originally breathed by the Creator into a few forms or into one; and that, whilst this planet has gone cycling on according to the fixed law of gravity, from so simple a beginning endless forms most beautiful and most wonderful have been, and are being evolved. (Charles Darwin 1859, pp 669–670)

I have unashamedly written this book perhaps using an element of grandiose to tell the story of modern human migrations and their journey to Australia almost as a heroic tale of courage and fortitude that I believe it was. Indeed, those anonymous modern explorers are, to me, heroes of a sort. We have inherited an intriguing story invested with tremendous toil, adventure, resilience and strength woven into a fascinating tapestry that our modern human ancestors left us and some of us are gradually uncovering. It is a story worth telling and it will be told in much greater detail, probably better and with more detail, in the future, by many authors from many perspectives. The journeys of modern humans show us how clever and adaptive we were. We were tenacious and bold and worth the title of 'hero' to have undertaken them and, to coin a phrase, boldly go where no modern human had been before. That image is now under severe threat through our poor treatment of our planet and its creatures.

In 2013 I published *Corridors to Extinction* that discussed Upper Pleistocene megafauna's extinctions, particularly in Australia, They had occurred on many continents seemingly as modern humans entered them. I defended humans from the accusation that as they made their journeys outlined in this book, they slaughtered many of the largest mammal species that that they had encountered and were common at that time. That accusation never sat well with me not because I couldn't accept that we would do such a thing but because many issues did not make sense, could not be resolved and there was little evidence for the supposed slaughter, at least in Australia. Recent dates for those dispersals now seem to confound the 'Blitzkreig' theory even more than then. In Australia we now know there was a 40,000 year overlap, at least, between the first arrivals and the last megafauna. Blitzkreig arguments were largely based on the well documented cases of species extinctions caused by humans during the last 1000 years. Salient examples include the nine species of Moa in New Zealand; the Dodo in Mauritius; ferocious hunting by seafarers of the Great Auk (*Pinguinus impinis*) and Steller's Sea Cow (*Hydrodamalis gigas*) in the early 19th century, the needless slaughter of the Tasmanian Tiger (*Thylacinus cynocephalus*) and many bird species in Hawaii and elsewhere the list goes on. Whaling was another onslaught that depleted the world's whale stocks to dangerously low levels and brought many whale species, including the largest mammal that has ever lived on Earth, the Blue Whale, to the verge of extinction. Unfortunately, there are many, many other examples. The heroes of this book, modern humans, no longer live in groups and bands or are reliant for survival on killing wild animals. Indeed, I would argue, as I did in *Corridors*, they probably did not kill as many terrestrial mammals as we might give them credit for, but perhaps we judge them by our own voracity.

We are no longer a world population of a few hundred thousand or even 1–2 million, we are closing in on 8 billion of us. The result is that we have expanded into places that bring numerous pressures on hundreds of species with whom we share this planet and depended on in the past. We are forcing them towards rapid extinction mainly by taking their land and wrecking their environments. The number of species included in the *International Union for Conservation of Nature (IUCN) Red List of Vulnerable, Endangered and Critically Endangered Categories* are not only in their thousands but the numbers are increasing exponentially (http://www.iucnredlist.org/). Animal icons like the gorilla, all African elephant species, all rhino species and big cats, bears, frogs, bees, fish stocks and even sharks are heading for extinction down a path we have constructed. Some we will not be able to save in the long run and our grandchildren will never see them in the wild. Many causes are not wilful but how we treat the environment, what we take from it and put in it.

We are no longer the pioneering hero, the stalwart modern human worth praising because of its strength and tenacity – perhaps we became too strong and arrogant so that we could no longer see the consequences of our activities. We are killing planetary life through anthropogenic climate change, plastic polution and consciously by stealth. We poach, kill and traffic in horn, body parts and in the live creatures themselves, we also kill each other to do it. We do it for greed, profit, to fund armament sales and wars. It is done for quack potions, false aphrodisiacs, meaningless medicines, artless carvings and ornaments and to supply foreign currency for corruption and to traffickers of all nationalities and creeds. Animals belong to all of us as our natural heritage and they are supposed to be in our care. They are already under pressure, but we kill them for sport: one more death won't make any difference! Our hero has become so greedy and thoughtless about his fellow animal travellers and the Earth on which he stands (men are overwhelmingly, if not totally, to blame). He has become bereft of common decency, humility, humanity and the ethical and moral consideration that made us different from animals. But once, we were as close to them as family because we relied on them and sometimes worked with them. The saddest thing is that we cannot see we are now not only a menace to the welfare of the planet but also to ourselves. Why do we do it; because we have become removed from nature.

'Modern humans': a term that denotes and separates us from what… the archaic hominins of the past, those who were doomed to extinction because they were inferior or looked different? Perhaps, on occasion, what we did then to archaic people was our first attempt at the 'extinction business'? Of course, sometimes it worked out between us, but they looked malevolent; they were less than us and did not even have language, and some anthropologists still purvey that argument. Of course, they were a barbarous people, less intelligent than us, who had taken the wrong evolutionary road. We are now the hominin that will kill off the planet's animals for profit, rip out their environments, pollute the oceans with plastic and other modern dross and destroy ecosystems to graze one or two domestic livestock species and plant monocultures like palm oil trees replacing thousands of plants and animal species specifically orang utangs 100,000 have been killed in 16 years because of palm oil plantation spreading. They had lived and evolved, as did many other species did, in complex, diverse rainforest ecosystems that are now being destroyed. Similarly, modern humans destroy the marine equivalent of rainforests, coral reefs, with dynamite to catch a few fish, and indirectly by coral bleaching caused by anthropogenic climate change. And we still make plans to open enormous coal mines in Australia...... Our hero is a hero no longer.

The modern humans in this book have departed this Earth and this story and others like it are all that is left of them apart from a few archaeological remains. We have also departed but unlike them our archaeological footprint will be for all who visit the planet to see. It will not take painstaking work to piece together lifestyle and legacy. Unlike those modern humans, we have deliberately separated ourselves from the Earth. Most of us find ourselves living in cities, suspicious of and cringing away from open spaces and fearing the wide outdoors. Cities are places where it is easy to think we no longer need the Earth, to be close to it or need to know anything about its workings and recognise our interconnectedness to it. Most modern humans are totally ignorant of how it works, nor do they care because food appears in the supermarket. Our hero knew all about interconnectedness and being part of the rich tapestry of life that surrounded them, if they did not have that knowledge and understanding, they died. Today, we live in an electronic bubble that separates us from the real world and teaches us all we need to know from the comfort of an armchair, desk or while we walk in an overcrowded street and step in front of vehicles. We hope the electronic highway we tread will take us where we want to go, we don't really have to go and see or experience the world for ourselves. We don't have to touch it, smell it, listen to its sounds, taste it or feel its temperature like our ancestors had to in order to survive. We hope our highway will serve us well for all our wants and needs, carry us into the future, whatever that might look like, and save us from ourselves.

Our forefathers lived close to the environment but like them our future depends on the world around us. We need the organisms that live in it, their biological pyramids and the interconnections they have between them and between them and us. The vast gamut of those organisms both macro and microscopic made our planet fit for human evolution to move along its pathway, we would not be here without it and the desperate will to understand it. It stands to reason that as organisms become fewer, the fewer structures and networks there will be that support all life including ours. So far as we know, they are what make this planet unique in the cosmos. Our electronic world will not make us better keepers of the planet unless we want to be better keepers but we need to know our world and look after it. If we do not begin to once again intimately touch and see our world, care for it and use our ingenuity to reinforce and intimately understand our connection to it, in the way those modern human travellers understood it, humanity will suffer, possibly terminally.

History is not the prerogative of the human species. In the living world there are millions of histories. Each species is the inheritor of an ancient lineage. (E.O. Wilson, 2016, p155)

We humans have an ancient lineage but most of theirs are much older. We can only hope that modern humans everywhere will realise we have no right to terminate the history of any species by turning inwards on ourselves, moving deeper into our electronic cave and looking only at the electronic screen wall, rather than reaching out of that cave to embrace the reality of the world outside and use again the ingenuity that took at least 3 million years to evolve to keep us all alive and provide us with a future.

PLUS POSTSCRIPTS

Abbie, A.A., 1968. The homogeneity of Australian Aboriginals. Archaeology and Physical Anthropology in Oceania 3, 223–231.

Abbie, A.A., 1976. Morphological variation in the adult Australian Aboriginal. In: Kirk, R.L., Thorne, A.G. (Eds.), The origin of the Australians. Australian Institute of Aboriginal Studies, Canberra, pp. 211–214.

Barker, G., Barton, H., Bird, M., Daly, P., Datan, I., Dykes, A., 2006. The 'human revolution' in lowland tropical Southeast Asia, the antiquity and behaviour of anatomically modern humans at Niah Cave (Sarawak, Borneo). Journal of Human Evolution 52, 243–261.

Birdsell, J., 1977. The recalibration of a paradigm for the first peopling of Greater Australia. In: Allen, J., Golson, J., Jones, R. (Eds.), Sunda and Sahul, prehistoric studies in Southeast Asia, Melanesia and Australia. Academic Press, New York, NY, pp. 113–167.

Birdsell, J.B., 1979. Physical anthropology in Australia today. Annual Reviews in Anthropology 8 (1), 417–430.

Bowler, J.M., Johnston, H., Olley, J.M., Prescott, J.R., Roberts, R.G., Shawcross, W., 2003. New ages for human occupation and climatic change at Lake Mungo. Australia. Nature 421, 837–840.

Brothwell, D., 1960. Upper Pleistocene human skull from Niah Caves, Sarawak. Sarawak Museum Journal 9, 323–349.

Brown, P., 1981. Artificial cranial deformation: A component in the variation in Pleistocene Australian Aboriginal crania. Archaeology in Oceania 16, 156–167.

Brown, P., 1987. Pleistocene homogeneity and Holocene size reduction: The Australian human skeletal evidence. Archaeology in Oceania 22, 41–67.

Garcia, E.A., 2016. Dispersals out of Africa and back to Africa: Modern origins in North Africa. Quaternary International 408, 79–89.

Habgood, P.J., 1986. The origin of the Australians. Archaeology in Oceania 21, 130–137.

Howells, W.W., 1973. Evolution of the genus *Homo*. Addison-Wesley, Reading, MA.

Huxley, T.H., 1863. Evidence as to man's place in nature. D. Appleton & Co, New York.

Macintosh, N.W.G., Lamach, S.L., 1976. Aboriginal affinities looked at in a world context. In: Kirk, R.L., Thorne, A.G. (Eds.), The origin of the Australians. Australian Institute of Aboriginal Studies, Canberra, pp. 113–126.

Pugach, I., Delfin, F., Gunnarsdóttir, E., Kayser, M., Stoneking, M., 2013. Genome-wide data substantiate Holocene gene flow from India to Australia. PNAS 110 (5), 1803–1808.

Strzelecki de, P.E., 1845. Physical description of New South Wales and Van Diemen's Land: Accompanied by a geological map, sections and diagrams, and figures of the organic remains. Longman, London.

Swisher, III, C.C., Rink, W.J., Anton, S.C., Schwarcz, H.P., Curtis, G.H., Widiasmoro, S., 1996. Latest *Homo erectus* of Java: Potential contemporaneity with *Homo sapiens* in Southeast Asia. Science 274, 1870–1874.

Thorne, A.G., 1975. Kow Swamp and Lake Mungo, towards an osteology of early man in Australia (Unpublished PhD Thesis). Australian National University, Canberra.

Thorne, A.G., 1976. Morphological contrasts in Pleistocene Australians. In: Kirk, R.L., Thorne, A.G. (Eds.), The Origin of the Australians. Australian Institute of Aboriginal Studies, Canberra, pp. 95–112.

Thorne, A.G., Wolpoff, M.H., 1981. Regional continuity in Australian Pleistocene hominin evolution. American Journal of Physical Anthropology 55, 337–349.

Thorne, A.G., Grün, R., Mortimer, G., Spooner, N.A., Simpson, J.J., McCulloch, M., 1999. Australia's oldest human remains, age of Lake Mungo 3 skeleton. Journal of Human Evolution 36 (3), 591–612.

Topinard, P. (1872) Etude sur les Tasmaniens/par Paul Topinard. Mémoires de la Société d'anthropologie de Paris vol 3, fasc 4.

Weidenreich, F., 1946. Apes, giants and man. The University of Chicago Press, Chicago, IL.

White, J.P., O'Connell, J.F., 1982. A prehistory of Australia. New Guinea and Sahul Academic Press, Sydney.

Woo, J, 1959. Human fossils found in Liukiang, Kwangsi, China. Paleovertebrata & Paleoanthrpology 1, 97–104.

POSTSCRIPTS

Darwin, C., 1859. Origin of species. John Murray, London.

Wilson, E.O., 2016. Half Earth. Our planet's fight for life. Liversight Pub, New York.

FURTHER READINGS

Birdsell, J., 1972. Human evolution an introduction to the new physical anthropology, 1st ed. Rand McNally, Chicago, Illinois.

Indirati, E., Swisher, III, C.C., Lepre, C., Quinn, R.L., Suriyanto, R.A., Hascaryo, A.T., 2011. The age of the 20 meter Solo River terrace, Java, Indonesia and the survival of *Homo erectus* in Asia. PLoS One 6 (6), e21562. doi: 10.1371/journal.pone.0021562.

Macintosh, N.W.G., 1965. The physical aspect of man in Australia. In: Berndt, R.C., Berndt, C.H. (Eds.), Aboriginal man in Australia. Angus and Robertson, Sydney, pp. 29–70.

Part III

The Willandra Lake Collection: A Record

Chapter 9

A Descriptive Analysis of the First Australians

Lake Mungo lunette with residuals and ancient living floors.

THE WILLANDRA LAKES COLLECTION

The Willandra Lakes Collection is the largest and oldest of its kind in Australia and is probably one of the largest of its time frame in the World and WLH 1, 3 and 50 are its prize individuals (Fig. 9.1). It is also the biggest set of human remains from the times of modern human dispersals. It not only marks the time that humans have been in Australia, it also presents a picture of the new arrivals, their physical attributes and culture. The collection consists of remains recovered from around the lakes since 1969 but the bulk come from Lakes Garnpung, Leaghur and Mungo and areas between these lakes (Fig. 9.2). The collection gradually grew as intermittent surveys were carried out by Peter Clarke and Mike McIntyre as well as by Alan Thorne and myself during short visits from the late 1970s till 1982. The WLH prefix used in the original numbering system stood for 'Willandra Lakes Hominin' and was instituted by Alan Thorne and Peter Clark in 1982. Thorne sought to use 'hominin' for the collection because of its early chronological placement in Australia but also because

Made in Africa. http://dx.doi.org/10.1016/B978-0-12-814798-6.00009-1

FIGURE 9.1 **Left lateral view of WLH 3 (left) and WLH 1 (right).** *(Photo S. Webb)*

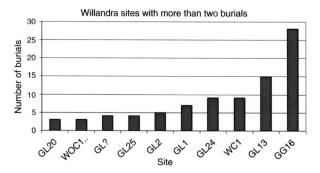

FIGURE 9.2 WLH burial distribution.

it aligned the collection with older fossil human remains found elsewhere in the world. In that way, it added to Australia's prestige in palaeoanthropological study. Ignoring taxonomic nomenclature and recognising the wide range of chronological age of these individuals, and taking into account propriety and the comparatively recent age of the collection, it is better to make 'Hominin' redundant and replace it with 'Human' in the WLH prefix and it is used that way here.

The WLH collection was assembled to study what it could contribute to our understanding of who we are as a species; what it might say about our shared history as a species and who the first Australians were and how they lived. However, such study has not always been welcomed by Aboriginal Australians because of the poor history of colonialism in Australia and the treatment of Indigenous people that resulted from that. Therefore, some Aboriginal people have seen anthropological study, as another form of oppression. The reasons for such work have not always been understood, believed or accepted by Indigenous people and is often seen as another way of proving they may not have owned their traditional lands or they were not the original people of the continent. Unfortunately, those Indigenous sentiments miss the point of recent research. The Willandra remains were collected because of the natural desire to establish who the first Australian arrivals were, their place in world history, what they could tell us about their culture and origins and show how long people have been living here. The collection was also gathered and studied because of concern for the loss of extremely valuable, ancient information about the Indigenous heritage of this continent and the depth of human occupation that would otherwise be totally unknown.

Some fragmentary bone was not collected and it is certain the remains of many individuals are yet to be discovered. That is obvious by the constant exposure of fossil human bone, most usually in a fragmentary condition. Some individuals were not formally included in the series although new discoveries were normally given numbers, whether they were collected or not. One example of this is a cremated individual found in 2008 south of the WLH 1 site and because of its identical gracile cranial morphology reflecting that of WLH 1 is estimated to be contemporary. It was not collected on an Indigenous request to leave it where it was found.

Since the 1980s Lake Mungo has been a focus for tourism with over 30,000 visitors annually. That brings the danger of tourists removing bone, not realising its significance or even that it could be human. An example of that emerged several years ago. A person entered the Queensland Museum quickly depositing a box of fossil human remains at the Information desk with the words to the effect: I collected these at Mungo years ago when on holiday and I am returning them. The person quickly disappeared leaving a bemused Information Officer holding the battered cardboard box of fossil bone that was covered in calcium carbonate. I was informed of this event by the museum and after inspection to confirm their status I returned the remains to the custodianship of the recognised traditional owners at Lake Mungo where it was placed in a Keeping Place, a cellar under the floor of the Ranger Station office.

Peter Clark first listed the collection in 1982–1985 during his tenure as Regional Archaeologist. That listed 103 individuals collected or left in the field since the WLH 1 discovery. The collection extends from Lake Mulurulu in the north to the Arumpo/Chibnalwood Lakes in the south. Lakes north of Mulurulu have not been surveyed. In November 1984, I took a group of students to the region to carry out an archaeological survey of the Middle Willandra Creek, particularly two small lakes, Pan Ban and Baymore, between Mulurulu and Garnpung. During the survey five individuals were found as well as stone flakes, grindstones and hearths, freshwater muscle shell scatters and middens (Dowling et al., 1985). Human remains were discovered but not collected in keeping with the wishes of the local Aboriginal communities but numbers WLH 93–96 were allocated to them. The main Willandra collection was not researched or reviewed until 1985 when I compiled the first descriptive record. During the work, two or more individuals were discovered in some boxes under a single number. Separating these resulted in an increase in individuals expanding the listing to 132 individuals. From 1986 to 2002 further discoveries were made in the field but none were collected. Some individuals were allocated numbers if they were deemed significant or consisted of a significant portion of the skeleton that extended the collection from WLH 133 to WLH 151.

In May 2002, the Traditional Owners requested help to recover and manage burials eroding out near Lake Garnpung. This was a break with the long-standing moratorium on research but it was important that the work was initiated by them. It provided an opportunity to work in partnership with the local communities which was something they had always wanted. Those burials are described in detail at the end of Chapter 10 because of the large and complete skeletal elements present; their morphological and palaeopathological significance and a detailed scientific record was required providing as much information as possible both for scientific reasons and the Aboriginal community who were genuinely interested in what we could find out about the people. Those individuals are numbered WLH 152–154.

Continuous and rapid erosion around the lakes has uncovered remains, removed them, often taking them out of context, hence the rapid examination for WLH 152–154 described above. Unfortunately, however, that was a rare occurrence. The WLH series lacks whole skeletons, complete skeletons. It contains a few partial crania consisting mainly of vault sections and in rare cases parts of the face or mandible. Long bones are generally represented only by sections of diaphysis occasionally with parts of the proximal or distal ends, but most are fragmentary. Taphonomic processes in the region are responsible for the condition of the bone with surface erosion and/or carbonate surface wash. Parts of some lunettes erode quicker than others, rapidly and completely exposing bone in a comparatively short period. Speedy exposure reduces fragmentation if the bone if discovered in time. Frequent survey is important to reduce exposure time, so areas of severe erosion are targets for regular monitoring. Rain and wind exacerbate losses, the latter having been observed to move fragments of bone some distance, particularly downslope. The steeper the slope, the quicker its removal and its loss of context, but as erosion continues lunette height and slope are reduced, degradation slows and so does the removal of bone. Small fragments are more easily scattered and reburial by wind driven sand storms that are not uncommon in the area. Variation in the speed of bone loss depends on where the bone emerges, its condition, size, and its length of exposure. In the past, lengthy bone exposure subjected it to damage by domesticated animals when sheep farming occurred in the area for 120 years till 1982.

Cortical bone lasts longer both in and out of the ground than trabecular bone. Usually, cancellous bone is scoured from the centre of long bones and is usually found only in the neck of the femur and intracranial vault sections. Variation in cortical bone thickness and bone size selects for differential preservation of skeletal parts, such as long bones, with thick cortices. In descending order of frequency, sections of femora, tibiae and humeri are well represented in the collection. There are vast differences in the thickness of cortical bone between individuals. Metatarsals, metacarpals and phalanges are found almost as regularly as long bones because of their compact shape and construction reduces fragmentation. Overall, smaller and lighter bones have a greater chance of being broken or lost. Flat bones, such as the scapula and sections of the ilium, as well as ribs are vulnerable to breakage and regional environmental conditions. They include hot summers and cool to cold winters. Exposure to high winds is a factor in lunette erosion and the generally flat surrounding country allows strong gusty winds to not only remove sediments but also sand blast bone with air-borne quartz grains and remove it from its original context.

Any assessment and interpretation of the WLH collection is severely limited by its poor condition. Moreover, very few individuals have been or can be accurately dated and most of those have only an approximate date. The fossilised and general condition of the bone as well as its sedimentary origin in lunettes, almost certainly places all remains in the Late Pleistocene (50–20 ka). However, while the use of such material is limited its antiquity and regional importance requires recovery of all or any possible information. Most individuals are missing important diagnostic features degenerated by long-term interment followed by exposure. Cranial and postcranial bone is missing preventing metrical and non-metrical data being recorded. Many individuals are represented by small or single pieces of bone, others consist of only a few pieces but these are often sufficient to reveal the general morphology of the individual albeit limited.

Metrical assessment was undertaken where possible on cranial remains particularly cranial thickness, supra-orbital size and malar morphology. These bony regions often survive while the mandible, dentition and lightly built facial and cranial structures, are almost always missing. Those features have been used separately and in conjunction to provide a glimpse of individual morphology. They are important in any discussion of origins of the first Australians but their true value will be realised when DNA studies are carried out. There may be as many as 38 individuals suitable for such study but that will not take place with reburial of the collection. It is worthwhile providing whatever measurements are possible to assemble a comparative physical appearance. I believe this to be important now that the collection has been returned to Willandra to be either kept in a Keeping Place or, as some wish, reburied.

Stature

Except for WLH 4, which is a recent skeleton, all long bones in the series are fragmentary. The original lengths of the whole bones were usually estimated

TABLE 9.1 Willandra Lakes Series Stature Measurements

WLH No	Humerus (mm)	Ulna (mm)	Radius (mm)	Femur (mm)	Tibia (mm)	Height male (cm)	Height female (cm)
1	–	–	180l	297r	262l	–	151
3	352r	297	265re	480le	405le	173–177	–
6	–	–	–	–	330re	158	153
25	310re	260re	–	463le	360le	–	161–165
45	310re	–	–	–	–	–	160
67	–	–	–	504le	–	177–180	–
72	305re	–	–	450re	–	–	159–162
106	340re	260re	–	485re	388re	171–173?	171–173?
107	–	–	–	473re	–	170	–
109	328re	–	270re	–	–	170–174?	170–174?
110	312re	–	–	–	–	164?	164?
117	–	–	–	465le	–	169?	166?
152	–	310r	290r	505re	–	176–180	–
153	320r	–	267r	–	–	173	163

e, Estimated length of an incomplete bone; r, right; l, left; ?, sex unknown.

using corresponding bones from a reference collection and stature calculated from those (Table 9.1). Stature estimates were enhanced using other long bones belonging to the same individual or from the same section of bone from other individuals. Preserved bone length was used as the basis of estimation, and often the gender of the individual. Thus, some general assessment of stature was provided where otherwise none was available. Differences in stature estimates using different bones occurred. Those made on longer bones like the femur or tibia was preferred, but bones were incomplete. Estimates on all major long bones available were included so that some idea of stature by averaging stature estimates could be given even when preferred bones were absent. The term 'actual length' used in the record refers to the length of a long bone as preserved. Minimum long bone shaft circumference was calculated using average antero-posterior and medio-lateral diameters at the narrowest part of the diaphysis.

Sexing

Sexing fossil remains in Australia has traditionally relied on a technique developed on recent Aboriginal skeletal remains from New South Wales (Larnach and Freedman, 1963). The method assessed the degree of development of a set of cranial features that included the glabella, superciliary ridges, zygomatic trigone, malar tuberosity, mastoid process, sub-occipital rugosity and palate size. The method has been useful for general identification of recent Aboriginal skeletal remains. However, the method was developed on recent regionally limited samples (coastal New South Wales and Queensland). Therefore, using the same features on Late Pleistocene skeletal remains cannot reasonably be expected to fit the same sexing criteria. Developments in our understanding of modern human evolution and its history now make that criteria difficult to use. Therefore, sex identification using skeletal robusticity, particularly cranial robustness, as well as pelvic bones when available, was used for separating males and females. Sexual ambiguity arises for Late Pleistocene individuals when using criteria developed on Holocene materials. This is well illustrated by the fact that by using the suite of cranial features listed above, the male WLH 3 would, in the absence of the pelvis, have been pronounced female. The example underlines the truth of the statement by Shipman et al. (1985:249) that the 'overall accuracy of … determining sex is improved by a thorough knowledge of the general population from which the skeletal remains are drawn'. However, that is not always available when dealing with Late Pleistocene remains.

In the absence of knowledge regarding the Willandra people other broad criteria were used. There is no doubt that very robust individuals in the series, with strongly developed areas of muscle attachment, are male. The difficulty arises when identifying those who may not fit into accepted morphological categories and, because of the generally fragmentary and limited nature of the material, may not be correctly identified without standard sexual dimorphic parts of the skeleton. At present, there are no remedies for these problems, although they should be borne in mind. In the event, male and female individuals have been recognised and the reasons for each decision are set out in the following record. In all but one case, female status has been attributed solely on the gracile morphology of cranial and/or postcranial remains. One example of this is WLH 3 and WLH 45, for which sections of the pelvis are present. Cranially, the former is more gracile than the latter, but generally, more males than females can be expected among gracile individuals and more males than females among robust individuals.

Ageing

Again, the fragmentary nature of the collection presented similar problems to those that arose for sexing. The lack of dental remains precluded attrition rate and eruption sequences and only WLH 105 has a deciduous dentition. The sandy environment of the Willandra always caused gritty inclusions in food causing high rates of dental attrition unlikely to correspond with crown reduction due normal tooth wear and personal age. For the bulk of the collection even basic age grades, such as 'old' or 'young' adult, could not be fixed with any certainty. Most individuals, therefore, were identified simply as adults, unless other factors brought greater precision to the aging process. Some include cranial suture fusion, the presence of third molar alveoli, alveoli resorption, pre-mortem tooth loss and, in a few cases, remnant molar tooth roots present in sockets. Use of post-cranial epiphyseal fusion was not possible, because all long bones lacked proximal and distal ends. Osteoarthritic degeneration was also used if present. Long bone lacking epiphyseal parts reduced this method to very few individuals.

Other Cranial Remains

Comparatively few cranial measurements are available from the collection. Only WLH 19 and WLH 130 yielded a significant information, but the degraded surfaces and poor condition of WLH 19 make its measurements shaky at best and possibly inaccurate. All figures range within both Late Pleistocene and modern comparative groups, but the lack of fundamental information, such as cranial height and length, prevents basic comparisons between the Willandra group and others. The frontal curvature index of WLH 19 is 18.9, the only individual for which an index can be taken. That is slightly higher than a combined mean (18.3) for the Coobool Creek/Kow Swamp populations (inclusive of individuals with possible head-binding), but far less than the 22.6 for more recent Australian crania (Thorne, 1976; Brown, 1981). There is, however, no obvious sign of cranial deformation in WLH 19 as there is among the Coobool Creek and Kow Swamp groups.

Mandibular metrical data is extremely limited; however, the lower end of the range of symphyseal height among the Kow Swamp and Coobool Creek populations barely overlaps with the upper end of the Willandra range. As well as being completely outside the female range of the Coobool Creek population, WLH 11, 20 and 1 are three, four and almost five standard deviations (S.D. 3.14 mm), respectively, smaller than the mean (37.2 mm) of the Coobool range (WLH 20 is unsexed). In comparison with a recent female population from the Murray Valley, WLH 1, 11 and 20 are over two, three and four standard deviations (S.D. 2.48 mm) below the Murray mean (32.5 mm). The corpus height of WLH 11 is almost four standard deviations (S.D. 2.43 mm) below the mean (31.2 mm) for the same measurement in the Coobool females. WLH 11's small size supports conclusions analyses of the supra-orbital, cranial thickness and malar morphology that match it with other gracile individuals like WLH 1. Both males in the sample (WLH 3 and 22) have a symphyseal height two standard deviations (S.D. 2.62 mm) smaller than the mean (39.1 mm) for Coobool Creek males.

The WLH 1, 11 and 20 mandibles are small compared to other Late Pleistocene and more recent Aboriginal skeletal samples. In fact, the great contrast in size between the three Willandra specimens and the two samples from the Murray Valley strongly show a fundamental difference in morphology. On the other hand, could females have subsequently become more robust. I would find that hard to accept unless there was an input of more robust genes into the population at some time in the past or female robustness was selected for. An increase in robustness also goes against the general trend of reduction in robusticity among humans during the last 20,000 years, world wide.

Thick Cranial Vaults

I have already discussed this feature in Chapter 7, however, Willandra crania with thick vaults reflect other Australian Late Pleistocene crania from elsewhere. The thickest yet measured from outside the Willandra is 13.5 and 16.3 mm at mid-frontal in a Kow Swamp and Coobool Creek individual, respectively (see angular and supra-mastoid tori Chapter 7). However, that may have been due to head binding that was practised among some individuals in those populations. Head binding changes vault structure resulting in thickening of posterior areas of the frontal bone, particularly anterior to bregma hence the 'bregmatic bulge' often seen in head-bound crania, however, head binding is not apparent among Willandrans. More recent Australian crania are thickened somewhere on the vault, usually at lambda, asterion or inion (13–16 mm). None, however, reach the stage of development of WLH 19 and WLH 50.

Postcranial Remains

The poor condition of post cranial bone among all individuals confines discussion only to the larger long bones of the humerus, femur and tibia.

Humeri

Only 14 individuals have humeri that yield measurements. The difference in mid-shaft diameter and minimum circumference between left and right arms highlights the overall bias towards right handedness. For so few representative bones the size range of both the antero-posterior and medio-lateral diameters is large, with 10 mm between minimum and maximum sizes. Minimum humeral circumference lies at the robust end of the late Holocene range. The 56.6 mm measurement obtained for WLH 45 is interesting, because it puts this female well within the modern male range and two standard deviations above the recent female mean (48.5 mm), which further supports her robust features. The robusticity index reflects the overall bone development of Willandra humeri, although the six available are a small sample. However, the estimated maximum and minimum humeral length is 305–361 mm, which lies within the late Holocene range of 276–366 mm.

WLH 11 has a slim left humerus producing the smallest robusticity index of the series in keeping with its other gracile traits already discussed. In contrast, the right humeral shaft of WLH 110 has the thickest cortex in the series, measuring 7–8 mm, which is similar to the cortical thickness of the Shanidar 4 Neanderthal femur (Trinkaus, 1984: 267). It is difficult to believe that the cortical morphology displayed by WLH 110 has nothing to do with its ancestral derivation, but the limited remains of this individual prevents further discussion of its overall morphology. The thick cortical development is also seen in the left humerus although slightly thinner (5–6 mm) but that suggests handedness in someone with thick cortices. The humeral robusticity of WLH 110 is unusual compared with others in the series and among other Australian humeri. The cortical thickness is not associated with pathology or other possible origins and must be attributed to normal developmental factors than malformation of any kind.

Femora

The sub trochanteric and midshaft portions of the Willandra femora lie towards the robust end of the late Holocene range in both their anteroposterior and medio-lateral diameters (the late Holocene ranges are 18–29 mm and 24–34 mm respectively, (Davivongs, 1963)). The medio-lateral diameter of the midshaft, moreover, exceeds that of the modern range by about 4 mm, which suggests a transversely flatter bone. The platymeric indices, however, lie well within the modern range and indicate that the bone has a normal configuration higher up on the shaft. Again, sample size must be taken into consideration when making statements concerning morphological trends, although some generalisations do seem valid. The range of the robusticity index for the Willandra group is almost exactly that for modem Aboriginal femora. The female WLH 45 once again demonstrates her robust character by being 0 8 below the top of the male robusticity index and 0.1 below the maximum for females. The range of estimated femoral length in the Willandra series is close to that for recent Aboriginal femora and, using this bone alone, indicates a wide range of stature, from 149 cm for WLH 1 to 177 cm for WLH 67.

An outstanding feature of the femoral development is the very thick cortex of some bones around the mid-shaft and lateral to the *linea aspera* thickness ranges from 3 mm on WLH 3 and 121 to around 12 mm and up to 16 mm on certain parts of the shaft of WLH 107 (Plate 1), which is far in excess of the 8 mm for a Neanderthal femur noted above after Trinkaus (1984:267). Cortical bone composition is solid and consists entirely of compact bone, without trabecular networks or pathological formations of any kind. There is a basic, positive association of cortical bone thickness with cranial thickness, but because of the poor condition of many of the remains and the consequent inability consistently to replicate measurements for a particular area of the cortex, accurate statistical assessment of this association cannot be undertaken.

Tibiae

The smallest tibia, in both mid-shaft diameter and overall length, is that of WLH 6 and lies at the low end of the tibial range in the Willandra series. It is exceptionally short indicating an estimated personal height of around 158 cm. It is also small in both antero-posterior and medio-lateral dimensions, particularly when compared to recent Aboriginal tibiae. Because of the prominent lines at points of muscle insertion, this bone could be male. Other tibiae in the series are somewhat larger in both the antero-medial and medio-lateral mid-shaft dimensions, although the right tibia from WLH 45, a robust female, is slightly larger. Hyperplatycnemia, or extreme lateral flattening of the shaft, occurs only in the right tibia of WLH 67, but there is no obvious sign of a diaphyseal curvature ('sabering') that can accompany this tibial shape. The small sample spans the normal range of the platycnemic index and, like the other long bones, includes a wide range of stature.

PATHOLOGY

This report excludes WLH 152–156 that are described separately at the end of Chapter 10.

We know little about the types of disease encountered by Late Pleistocene human migrations. New suites of parasitic, zoonotic and environmentally determined diseases could have affected modern human dispersals travelling through new regions for the first time that acted as virgin soil groups. Looking at Willandra series bone surface erosion and fragmentation tends to hide subtle pathologies and because of missing articular ends of major long bones assessment of osteoarthritic conditions is impossible. However, that is not the case with WLH 152–154 (see Chapter 10). Besides those individuals only one other (WLH 106) shows osteoarthritic changes in the appendicular skeleton. The distal articular surfaces of the right humerus of WLH 106 show minor osteophytic lipping (osteophytosis) and surface bone erosion. There is erosion of the mid-trochlear ridge, whose normally sharp border has been removed, leaving a worn, granulated surface and the articular surfaces either side unaffected. Minor osteophytic lipping has proliferated around the rims of the coronoid and radial fossae, accentuating the olecranon fossa. Post-mortem damage has removed bone but the base of osteophytic outgrowth remains.

There is a moderate, medially directed bow in the right humerus of WLH 106 involving the distal three-quarters of the shaft, the proximal quarter is missing. However, the well-developed deltoid tuberosity suggests the arm was used normally. Bowing can result from congenital deformation; nutritional inadequacy; a metabolic disorder (rickets); 'endemic' non-specific bowing of the limb, or as a response to a shoulder or arm accident during growth. No other bones have this shape, nor do they exhibit signs of nutritional inadequacy. In fact, the cortical bone is thick and quite normal and shows none of the osteoporotic resorption and lack of mineralisation typical of rickets, osteomalacia and other nutritional disorders. Rachitic bones are also light and rather brittle; these are not. The well-known bowing that is often symptomatic of rickets occurs during childhood; is usually bilateral and is more common in the weight-bearing bones of the legs, weakened by demineralisation. All long bones of WLH 106 are normal and robust. Congenital deformities of the skeleton usually cause distinct osteological changes which continue into adulthood. Radiographic examination indicates the humerus is normal, with a uniform cortical thickening. There is the possibility of 'endemic' non-specific bowing of the type found in Cro-Magnon. In Australia, it takes the form of an antero-posterior bend in the tibia often accompanying platycnemia. It was originally associated with endemic treponematosis but is more commonly seen, at least in dry bones, without any form of associated infection of the bone. It occurs in Aboriginal skeletal remains from most parts of Australia, but the greatest frequencies are in the tropical north and the Murray River region, 4.8% and 3.7%, respectively, and occurs in all long bones. Although bilateral involvement is common, asymmetrical bowing is not rare. No aetiology is offered here for this bowing but there are no other changes in the general appearance of the bones of WLH 106. It might be similar to a condition observed in many other, more recent individuals. If this is so, the Late Pleistocene age of the individual makes it one of the oldest examples of the condition in the world.

Osteoarthritis of the Temporomandibular Joint

Five individuals (WLH 27, 44, 67, 72 and 101) have osteoarthritic degeneration of the glenoid fossa. The condition varies in degree and involves different parts of the mandibular condyle. Only the condyle of WLH 27 remains and that has lost most of its articular surface, advanced osteophytic growth occurs on the anterior edge. The anterior half of the left glenoid fossa of WLH 44 has been almost destroyed. The fossa was originally shallow, with little development of the articular tubercle contrasting with the normally deeper fossae among recent Aboriginal crania. The damage would have prevented proper function of the mandibular condyle and the fossa aggravating the condition.

The left glenoid fossa surface of WLH 67, and the articular tubercle have worn away. The fossa was wide but shallow, but new bone deposited on the surface masks its exact depth. New bone suggests loss of articular cartilage and the articular disc resulting in bone-on-bone contact between the mandibular condyle and fossae surface. New bone has itself then been scoured out as the degenerative process continued, leaving deep antero-posteriorly directed grooves etched into the surface caused by lateral excursion of the mandibular condyle with continued jaw use resulting in a jagged, uneven surface. Well-emphasised muscular markings on the medial and lateral surfaces of the gonial angle and gonial eversion on the mandibular body indicate strong chewing behaviour.

The right glenoid fossa of WLH 72 is pitted together with the posterior surface of the articular tubercle. Fossa morphology is different from those already discussed. It is very deep (12.5 mm) and narrower in the coronal plane, which gives it a distinct oval appearance. Mechanically, it seems to be far more adaptive for retaining placement integrity of the mandibular condyle during strong chewing than the shallower fossae of WLH 44 and 67. Its design would have prevented the antero-posterior excursion of the mandibular condyle that would sit deeper than the fossa of WLH 67, making it less likely a complete anterior displacement overriding the articular tubercle would take place as in WLH 44 and WLH 67, causing complete degeneration of the tubercle. Incisor chewing and using the anterior tooth complex for gripping objects encourages the condyle to **override the articular tubercle** through an antero-inferior movement. The WLH 72 fossae would have been highly adaptive for maintaining joint integrity and helping prevent osteoarthritic degeneration.

Both glenoid fossae of WLH 101 show osteoarthritic changes. The floor of both fossae and articular tubercles are extensively eroded, together with the inferior surface of the zygomatic processes indicating lateral excursion of the mandibular condyle during chewing. Fossa depth is eight millimetres, despite this, the fact that lateral movement occurred is due to its triangular shape and roof shape that ends in an apex. New bone has been deposited in the fossae, but while this is a natural process for joint protection, it reduces fossa depth, while allowing natural movement of the mandibular condyle. Thus, the scouring continues and the deposited new bone itself becomes scoured, as in WLH 67.

In conclusion, it seems that the pattern of osteoarthritic erosion in the few individuals examined highlights two different temporomandibular joint morphologies. The small, shallower glenoid and smaller mandibular condyle of some individuals has succumbed to behavioural or dietary change encountered in Australia. Another more robust and deeper glenoid fossae seem to cope much better with those same jaw stresses. This tentative pattern may reflect general morphological differences of people who entered the continent.

Cranial Trauma

Cranial trauma consists of small depressed fractures on the vaults of WLH 22 and 63. It is minor in both cases affecting only the outer cranial table. On WLH 22 there is a shallow, slightly oval depression, 20 mm in diameter, situated posteriorly on the left parietal, close to the sagittal suture. WLH 63 has two elongated oval dents (36 and 21 mm long), orientated sagittally and positioned well forward on the frontal bone. These appear to be more severe than that on WLH 22, and have only penetrated the outer bone table probably because of its thick construction (7.5 mm). In fact, WLH 63 is one of the few individuals whose combined cranial table thickness exceeds 60% of the whole vault thickness. Bone remodelling has taken place in the centre of both fractures leaving uneven scar tissue.

Post-Cranial Trauma

Five individuals have some form of postcranial trauma. These are WLH 27, 72, 106, 117 and 127. The right third metacarpal of WLH 27 has suffered a complete mid-diaphyseal oblique fracture in an infero-superior direction. The shaft has telescoped, shortening the diaphysis, but it healed well with very little callus formation around the break. This sort of fracture usually stems from a fall onto outstretched hands or a blow to the knuckles. A section of the left fibula of WLH 72 has a healed diaphyseal fracture. The shaft is not misaligned indicating the *tibialis posterior* and the *flexor halluces longus* muscles acted to maintain relative anatomical position of the fibula to the tibia, preventing complete discontinuity of the shaft. There is a complete break of the shaft of the left fibula of WLH 106. The edges of the bone at the break site are jagged with no sign of healing and only a minimum amount of new bone growth just below the break on the distal section. The bone itself is in two pieces and the broken ends are stained and/or covered by the same calcium carbonate deposit as other surfaces suggesting pre-mortem non-unification and the trauma occurred close to the time of death.

Both WLH 117 and 127 have typical parrying fractures of the ulna positioned in the distal one third of the shaft. These are normally on the left or shield arm which is used to block a blow from a weapon held in the right hand of an opponent. The parrying fracture in WLH 117 was a complete break of the diaphysis between the distal and middle third of the diaphysis. Bone ends are jagged, slightly misaligned and did not unite. The broken edges are slightly swollen, indicating that a collar of sub-periosteal bone began to form marking pre-mortem healing with the individual possibly dying within weeks of the injury. The ulna fracture in WLH 127 is long and oblique and the ends of the bone united completely in almost exact anatomical position.

Normally little misalignment of the shaft occurs because of the presence of soft tissue. Both the strong interosseous membrane between the ulna and radius and the presence of the radius itself brace the fracture. There is substantial callus formation indicating the arm was used, in a limited fashion, during healing. Healing was good, without any sign of secondary infection, thus attesting to the individual's general good health.

Infection

Only two individuals in the series, WLH 25 and 117, have distinct lesions caused by bone infection. WLH 25 has a localised periostitis on a section of the right clavicle. Changes to internal cancellous bone structures suggest that it may have suffered a previous fracture followed by a localised infection of the outer bone surface (periostitis). A patch of periostitis occurs on the upper medial surface of the right tibia consisting of a thin layer of new bone five centimetres along the diaphysis and causing minor cortical thickening. The infection is more likely to have resulted from a blow to the leg rather than a systemic cause. An antero-proximal section of the left tibia of WLH 117 has a long (7.2 cm) strip of new bone formation on its lateral surface reminiscent of periostitis and contrasting with normal cortical bone structures beneath. As with WLH 25, this person probably received an infection from a blow, but the structure of the new bone is distinct from that derived from trauma. The woven architecture of the cancellous-like formation is different from the new bone of WLH 25 and conforms to a more extensive and virulent type of infection.

Harris Lines

Only a trace of one Harris line was found in any of the long bones, on the posterior wall at the proximal end of the right tibia of WLH 69. Its position corresponds to an age of about 3–6 years in the growing bone. From the common occurrence of a line in this position in tibiae of more Aboriginal people, it has been proposed that it may be caused by a metabolic disturbance during weaning (Webb, 1995). This single example of Harris lines is more a reflection of the poor condition of the long bones in the collection due to bone breakage and exposure of medullary areas during interment.

Dental Hypoplasia

Only one upper central incisor of WLH 133 has any evidence of dental hypoplasia. Dental Hypoplasia normally results from disturbance of normal ameloblastic activity during enamel formation on the growing tooth. It is caused by any stress or metabolic disturbance that interferes with the tooth's natural growth processes during bud formation and growth. The most likely way this occurs in hunter–gatherer children is at times of systemic illness or nutritional deprivation. The result is the formation of a transverse line, groove or, in severe cases, row of pits across the dental enamel, the position of which indicates to some degree the age at which the metabolic insult took place. In WLH 133 the stress occurs 1.7 mm from the cemento-enamel junction, indicating an age at formation around 1.5–2.0 years. Normally weaning does not occur among hunter–gatherers at this early stage of a child's development, so the stress is more likely to have been a moderately severe illness or period of nutritional inadequacy before weaning.

Cyst

WLH 55 has a small, oval dent just below the left temporal line. The internal surface is smooth and there is no sign of infection, stellate fracturing or healed bone tissue. It appears to be too deep in relation to its width to be a depressed fracture and looks more like a scar left by a dermal cyst.

Pre-Mortem Tooth Loss

Four individuals show pre-mortem tooth loss: WLH 3, 11, 20 and 22. WLH 3 has been described in Chapter 6 so I begin with WLH 11 whose left mandibular body has complete resorption of the first and second molars sockets. The latter has almost completely disappeared, with redeposited bone filling the socket. The third molar alveolar bone is largely missing, but from the little that remains it seems likely that this tooth might also have been lost pre-mortem. No evidence of periodontal disease can be seen.

The mandible of WLH 20 is incomplete but retains the lateral part of the socket of the third molar which was lost pre-mortem. The lower central incisors of WLH 22 are missing. Bone resorption around the gingiva suggests these teeth were lost some considerable time before death. Nothing remains of the sockets, which have been filled with new, spongy bone. Resorption processes have narrowed the breadth of the mandible at this point, which is further evidence of elapsed time since the incisors were in place. The presence of the lateral incisors at death indicates only involvement of the central incisors that could only be lost either through accident or attrition. A third option is tooth avulsion that I believe is the case here. Tooth avulsion has been suggested as causing the missing lower incisors in WLH 3 (see Chapter 6).

SITE DESIGNATION AREAS (SDAS) AND THEIR CODING

Before defining each site designation area (SDA) in the Willandra Lakes system an explanation of their coding, which is included at the beginning of each individual description, is required. The SDA was used to pinpoint individual burial sites in the Willandra. It was devised by Peter Clark before the advent of GPS or laptop computers and is a way of dividing the 3600 km^2 Willandra into 37 areas within which all archaeological sites and skeletal remains were found. Originally, the widely scattered sites required a unique system of recording unlike that used anywhere else in Australian archaeology. The area had to be divided into workable areas using a coding system involving natural and man-made geographical features, but without GPS only an approximate site placement could be made within those. This is a record of that system and the only means of relocating the burial place of the Willandra Lakes collection when it was collected. The SDA system placed them in relation to major lakes and associated features and positioned a site to within 50 m or so of the original. Today it would be impossible to relocate sites without using this system and a new survey program should be instigated to match those original sites using this system and then record them using GPS. Landscape change over time has been comparatively rapid making features associated with sites unfamiliar to the infrequent visitor and in some cases obliterating the original site altogether. The SDA system is now the only way of finding the original burial places of individuals in the series and so is an important part of this record.

The SDA's system was invented by Peter Clark and refers to sections of discrete land systems around each lake, its surrounding area and the Willandra Creek. They include lake beds, channels, lunettes, and tree covered inter-lake areas. Each SDA has a clearly defined boundary comprising different landforms and old property boundary fences that existed at the time. Some of those properties no longer exist and their boundary fences and other station installations have been removed or are no longer featured. Management of the area required abandoned stations to be dismantled including fence lines, wells, windmills and other minor buildings and sheds that were functional when the SDA system was constructed. Alphanumeric codes match the geographic position on both lake and inter-lake areas. Most sites can still be relocated using old pastoral property maps of the area.

The example of SDA 9 refers to a location enclosing all the Garnpung East Sites prefixed GE. Similarly, the Walls of China (WOC, Mungo lunette) are in SDA 25 and the sites there are prefixed WOC. It is worth noting the GG in SDA 11 is the only reference to sites situated around or near the small Lake Gogolo positioned between Lakes Garnpung and Leaghur.

TABLE 9.2 Major Willandra Lakes Sites with SDA Areas and Site Codes

Lake/area/Creek	SDA	Sites
Mulurulu		
East	1	ME1–8
West	2	MW1,2,4,6
Lake bed	4A	WCM
Baymore		
East	5	BE1,2
West	5	BW1–3
Pan Ban		
East	6	PBE1–3
South	6	PBS1–4
Garnpung		
North	7	GN1–5, 7–9, 16–18, 19–25, 10–13
West	8	GW1, 5–7, 12
East	9	GE8,9
South	10	GS1–3, 9–11
L. Garnpung/L. Gogolo	11	GG1–5, 7–9, 16–25
Garnpung/Leaghur	12	GL1,2,4,5, 13–15, 20, 23–30
Lake bed	4D	WCG1,2
Leaghur		
North	13	LN7,8–11
West	14	LW1,4–10,20
Peninsular	15	LP1–3
Leaghur/Garnpung	16	LG1,2
Lake bed	4C	LLB1,8
Mungo		
Mungo/Leaghur	17	ML1–4,20,50,51
North	18	NM1–5

(Continued)

TABLE 9.2 Major Willandra Lakes Sites with SDA Areas and Site Codes (*cont.*)

Lake/area/Creek	SDA	Sites
Mungo/Arumpo	24	MA1–4,11,100–113
Walls of China	25	WOC1–3,145–155
East	26	MLE100–106
Lake bed	28	MLB1–21
Arumpo		
Outer Arumpo	29	OA1–15
Arumpo West	32	AW1
Arumpo South	33	AS1–4
Chibnalwood, Durthong and Prungle Lakes		
Chibnalwood Lakes	30	CL1–5
Durthong L. (all areas)	20	DTL4–6
Prungle North Basin	22	PN20–29,35–42,85,90–96
Prungle South Basin	23	PS1–19,30–32,43–52, 60–89,97–107

TABLE 9.3 WLH Series Skeletal Listing with Site of Origin

WLH no	Sex	CR	P-C	SDA
1	F c	Y	Y	WOC1
2	M c	Y	Y	WOC1
3	M	Y	Y	WOC1
4	F	Y	–	WOC1
5	NC	NC	NC	WOC1
6	M b	–	Y	WOC1
7	NC	NC	NC	WOC1
8	NC	NC	NC	WOC2
9	? c	Y	Y	WOC1
10	? c	Y	Y	WOC1
11	F	Y	Y	GL1
12	M	Y	Y	GL2
13	F	Y	Y	GL2
14	M	–	Y	GL2
15	F	–	Y	GL2
16	?	Y	Y	GL2
17	?	Y	Y	GL?
18	M	Y	Y	GL?

TABLE 9.3 WLH Series Skeletal Listing with Site of Origin (*cont.*)

WLH no	Sex	CR	P-C	SDA
19	M	Y	–	GG16
20	?	Y	Y	GG16
21	?	Y	–	GG16
22	M b	Y	Y	GG16
23	F	Y	Y	GG16
24	F b	Y	Y	GL13
25	F	–	Y	GL13
26	?	Y	Y	GL13
27	M	Y	Y	GL13
28	M c	Y	Y	GL13
29	F	Y	–	GL13
30	NC	NC	NC	GG20
31	NC	NC	NC	GG22
32	NC	NC	NC	LW9
33	NC	NC	NC	LG12
34	NC	NC	NC	NM4
35	NC	NC	NC	MLB14
36	NC	NC	NC	PSB19
37	NC	NC	NC	GG19
38	NC	NC	NC	MA109
39	?	Y	–	GL13
40	NC	NC	NC	GE8
41	NC	NC	NC	WOC3
42	?	Y	Y	GL24
43	?	Y	Y	GL20
44	F c	Y	–	GL24
45	F	Y	Y	GL24
46	?	Y	Y	GG16
47	?	Y	Y	GG18
48	?	Y	–	GL25
49	?	Y	–	GL25
50	M	Y	Y	GL20
51	F	Y	–	WOC145
52	M	Y	Y	GS10
53	F	Y	Y	WOC145
54	NC	NC	NC	GG8

(*Continued*)

TABLE 9.3 WLH Series Skeletal Listing with Site of Origin (*cont.*)

WLH no	Sex	CR	P-C	SDA
55	?	Y	Y	WOC152
56	?	Y	Y	WOC152
57	?	Y	–	WOC1
58	?	Y	Y	MA1
59	?	–	Y	WOC1
60	NC	NC	NC	MLE105
61	NC	NC	NC	MLE105
62	?	Y	Y	GL20
63	? b	Y	–	LW4
64	?	Y	Y	LW4
65	?	Y	Y	ML3
66	?	Y	Y	WCW6
67	M	Y	Y	GL5
68	F	Y	–	GL25
69	M	Y	Y	LP1
70	NC	NC	NC	LN7
71	NC	NC	NC	LN10
72	F	Y	Y	ME1
73	F	–	–	ME2
74	?	Y	–	GL24
75	?	–	–	GL25
76	NC	NC	NC	GS3
77	NC	NC	NC	ME2
78	NC	NC	NC	MWC1
79	NC	NC	NC	GN10
80	NC	NC	NC	GG1
81	NC	NC	NC	OA14
82	NC	NC	NC	GG7
83	NC	NC	NC	OA15
84	NC	NC	NC	GL1
85	NC	NC	NC	GL1
86	NC	NC	NC	GL1
87	NC	NC	NC	GL1
88	NC	NC	NC	GL1
89	NC	NC	NC	GL1
90	NC	NC	NC	LME106

TABLE 9.3 WLH Series Skeletal Listing with Site of Origin (*cont.*)

WLH no	Sex	CR	P-C	SDA
91	NC	NC	NC	OA2
92	NC	NC	NC	MWC2
93	NC b	NC	NC	BW1
94	NC	NC	NC	BW3
95	NC	NC	NC	GN12
96	NC	NC	NC	GN13
97	NC	NC	NC	LW10
98	?	Y	Y	GL?
99	?	Y	Y	GL?
100	F	Y	–	GG16
101	M	Y	–	GG16
102	M	Y	Y	GG16
103	?	Y	–	GG16
104	?	–	–	GG16
105	JUV	–	–	GG16
106	M	–	Y	GG16
107	M	–	Y	GG16
108	?	–	Y	GG16
109	?	–	Y	GG16
110	M	–	Y	GG16
111	?	–	Y	GG16
112	?	–	Y	GG16
113	?	–	Y	GG16
114	?	–	Y	GG16
115	? c	Y	Y	GG16
116	?	–	Y	GL13
117	?	–	Y	GL13
118	?	–	Y	GL13
119	?	Y	Y	GL13
120	? b	Y	–	GL13
121	?	–	Y	GL13
122	F	Y	–	GL13
123	F	Y	–	GL24
124	F	Y	Y	GL24
125	?	Y	–	GL24
126	? b	Y	–	GG16

(*Continued*)

TABLE 9.3 WLH Series Skeletal Listing with Site of Origin (*cont.*)

WLH no	Sex	CR	P-C	SDA
127	?	Y	–	WOC152
128	?	Y	–	GL24
129	?	Y	–	GG16
130	F	Y	–	GL24
131	?	–	Y	GG25
132	? c	–	Y	GG16
133	?	Y	–	GL13
134	F	Y	–	GL26
135	?F	NC	NC	WOC1
136	NC	NC	NC	GL13
137	NC	NC	NC	GL13
138	NC	NC	NC	Unknown
139	NC	NC	NC	Unknown
140	NC	NC	NC	Unknown
141	NC	NC	NC	Unknown
142	NC	NC	NC	Unknown
143	NC	NC	NC	Unknown
144	NC	NC	NC	Unknown
145	NC	–	Y	GL13
146	NC	NC	NC	Unknown
147	NC	NC	NC	Unknown
148	NC	NC	NC	Unknown
149	NC	NC	NC	Unknown
150	NC	NC	NC	Unknown
151	NC	NC	NC	Unknown
152	M	Y	Y	GG16
153	M	Y	Y	GG16
154	F	Y	–	GG16
155	M?	Y	Y	GL3

Key: *b*, Burnt bone; *c*, cremated; *CR*, cranial remains present; *NC*, not collected; *P-C*, post-cranial remains present; *SDA*, site designation area; *Y*, representative bone present.

SDA Definitions (See also Table 9.2). Italics refer to sheep station names (Table 9.3).

SDA 1 – Mulurulu East (ME)

Lake Mulurulu lunette and an area to the east. Northern limit at Mulurulu/Spring Hill boundary fence. Southern limit at northern Willandra Creek shoreline upstream of Lake Mulurulu.

SDA 2 – Mulurulu West (MW)
An area west of Lake Mulurulu including the western shoreline. The northern limit is the *Mulurulu/Spring Hill* boundary fence. The southern limit is the northern edge of Willandra Creek floodplain and the *Pan Ban/Garnpung* boundary fence.

SDA 3 – Willandra Creek, Upstream (WCM)
The Willandra Creek flood plain upstream of Lake Mulurulu lake bed.

SDA 4a – Mulurulu Lake Bed (WCM)
The Mulurulu Lake bed below high water strand lines that mark the edge of the lake bed.

SDA 4b – Willandra Creek Middle (WCU)
Willandra Creek floodplain between Mulurulu lake bed and Garnpung lake bed excluding the Lakes Baymore or Pan Ban.

SDA 4c – Garnpung Lake Bed (WCG)
The area within high water strand line of Lake Garnpung.

SDA 4d – Leaghur Lake Bed (LLB)
An area within high water strand line on Lake Leaghur.

SDA 5 – Baymore Lake east (BE) or west (BW)
Lake bed, shorelines and lunettes of Lake Baymore.

SDA 6 – Pan Ban Lake east (PBE) or south (PBS)
Lake bed, shorelines and lunettes of Lake Pan Ban.

SDA 7 – Garnpung North (GN)
Includes: Willandra Creek floodplain, upper beach strand line of Lake Garnpung, *Balmoral/Baymore* boundary fence and part of northern end of the Garnpung lunette.

SDA 8 – Garnpung West (GW)
The area west of the upper beach strand line on the western Garnpung shoreline. Northern limit is *Garnpung/Pan Ban* boundary fence and southern limit a line running west of the southern edge of Garnpung soak sand sheet where it cuts the western lake shoreline.

SDA 9 – Garnpung East (GE)
Area west of high water each strand line on eastern shore of Garnpung. The northwestern limit is the *Baymore/Balmoral* boundary fence and the southern edge of the Willandra Creek floodplain. The southern limit is an internal *'Gol Gol'* fence line which runs due east of Stitz's tank.

SDA 10 – Garnpung South (GS)
An area between Lake Garnpung high water beach strand lines and the same feature on Lake Gogolo, including the Gogolo lunette except northern end. Northern limit is the Garnpung/Gogolo interconnecting channel and southern limit a line running southeast from the end of the Gogolo lunette and the *Zanci/Leaghur* boundary fence.

SDA 11 – Garnpung/Gogolo (GG)
Area between high water strand lines on Lakes Garnpung and Gogolo. All of Gogolo lunette except for the northern end. Northern limit defined by Garnpung/Gogolo interconnecting channel; southern limit is a line running southeast from end of Gogolo lunette to the *Zanci* and *Leaghur* boundary fence.

SDA 12 – Garnpung/Leaghur (GL)
An inter-lake area between high water marks on Garnpung and Leaghur lakes. Northern limit is the edge of the irregular parabolic dune field west of the G/L interconnecting channel which marks the southern limit. The eastern limit is the northern end of Gogolo lake bed and the *Leaghur/Zanci* boundary fence.

SDA 13 – Leaghur North (LN)
This area is West of Leaghur lake bed with a northern limit defined by a line running west of Garnpung shoreline where it is cut by Garnpung soak dune sheet. Southern limit is the western edge of the Willandra Creek floodplain and northern edge of Durthong Lake complex.

SDA 14 – Leaghur West (LW)
An area along eastern edge of Willandra Creek, southwestern shoreline of Leaghur lake bed and northern end of the Mungo lunette.

SDA 15 – Leaghur Peninsular (LP)
Peninsular area along Mildura/Ivanhoe road running north of *Leaghur* homestead.

SDA 16 – Leaghur/Gogolo (LG)
Area between Leaghur and Gogolo lake beds. Northern boundary is *Leaghur/Garnpung* boundary fence and southern limit is Leaghur/Zanci fence line.

SDA 17 – Mungo/Leaghur (ML)
Area south of Leaghur lake bed and north of the Mungo lunette. Eastern limit is a fence separating 'Gilbey's' and 'Paradise' paddocks on old Zanci Station. The *Mungo/Leaghur* interconnecting channel defines the western limit.

SDA 18 – North Mungo (NM)
A small area between the northwestern end of the Mungo lunette to the *Top Hut/Leaghur* road. The eastern limit is the Mungo-Leaghur interconnecting channel.

SDA 19 – Willandra Creek West (WCW)
The Willandra Creek floodplain between Leaghur lake bed and Outer Lake Arumpo lake bed without Durthong Lake.

SDA 20 – Durthong Lake (DTL)
Lake bed, lunette and shorelines of Lake Durthong.

SDA 21 – Gogolo Lake Bed (GLB)
Defunct

SDA 22 – Prungle North Basin (PN)
Lake bed to strand lines and lunette of Prungle lakes north basin.

SDA 23 – Prungle South Basin (PS)
Lake bed to strand lines and lunette of Prungle lakes south basin.

SDA 24 – Mungo/Arumpo (MA)
Interlake area between Mungo and Outer lake Arumpo. Western limit is Willandra Creek floodplain and eastern limit is a line running south of the southern tip of the Mungo lunette to the *Joulnie/Turlee* boundary fence.

SDA 25 – Walls of China (WOC)
The Lake Mungo lunette.

SDA 26 – Mungo East (MLE)
East of WOC, Lake Leaghur and Outer Lake Arumpo. Southern limit is *Joulnie/Turlee* boundary fence and north western limit the internal *Zanci* fence line.

SDA 27 – Moonlight Lake (MLL)
No sites recorded.

SDA 28 – Mungo Lake Bed (MLB)
Mungo lake bed within high water strand lines.

SDA 29 – Outer Arumpo (OA)
Lake bed, shorelines, and lunette of Outer Arumpo excluding Chibnalwood lake beds shorelines and lunettes.

SDA 30 – Chibnalwood Lakes (CL)
Lake bed, shorelines, and lunette of both north and south Chibnalwood Lakes, Lake Benenong and Inner Arumpo.

SDA 31 – Lower Willandra Creek channel (LWC)
The Willandra Creek floodplain and associated lake basins between Outer Arumpo lakebed and the Prungle lakes.

SDA 32 – Arumpo West (AW)
An area west of Outer Lake Arumpo and the Willandra Creek floodplain.

SDA 33 – Arumpo South (AS)
An area south of Outer Lake Arumpo and east of the lower Willandra Creek floodplain. Northern limited defined by Joulnie/Turlee boundary fence.

REFERENCES

Brown, P., 1981. Artificial cranial deformation: a component in the variation in Pleistocene Australian Aboriginal crania. Archaeology in Oceania 16, 156–167.

Davivongs, V., 1963. The Femur of the Australian Aborigine. American Journal of Physical Anthropology 21, 457–467.

Dowling, P., Hamm, G., Klaver, J., Littleton, J., Sanderson, N., Webb, S.G., 1985. Middle Willandra Creek archaeological site survey. Australian National University, Canberra, (Unpublished report to the New South Wales National Parks and Wildlife Service).

Larnach, S.L., Freedman, L., 1963. Sex determination of Aboriginal crania from coastal New South Wales. Records of the Australian Museum 26 (11), 295–308.

Shipman, P., Walker, A., Bichell, D., 1985. The human skeleton. Harvard University Press, Cambridge, MA.

Thorne, A.G., 1976. Morphological contrasts in Pleistocene Australians. In: Kirk, R.L., Thorne, A.G. (Eds.), The origin of the Australians. Australian Institute of Aboriginal Studies, Canberra, pp. 95–112.

Trinkaus, E., 1984. Western Asia. In: Smith, F.H., Spencer, F. (Eds.), The origins of modern humans. Alan R. Liss, New York, pp. 251–293.

Webb, S.G., 1995. Palaeopathology of Australian aborigines. Cambridge University Press, Cambridge, UK.

Chapter 10

Willandra Lakes Skeletal Collection: A Photographic and Descriptive Catalogue

Made in Africa. http://dx.doi.org/10.1016/B978-0-12-814798-6.00010-8
Copyright © 2018 Elsevier Inc. All rights reserved.

Willandra Lakes skeletal collection serial numbers run from WLH 1 to WLH 154, although not all are included here. WLH 5, 7, 40, 41, 79 and 90 are vacant numbers created during collection curation in 1986 or are missing. WLH 30–38, 54, 60, 61, 70, 71, 76–78, 80–89, 91, 97 and 138–151 are numbers allocated to skeletal elements that were not collected and remain in the field or in a Keeping Place at Lake Mungo. WLH 135 was discovered early in 1986 at the southern end of the Lake Mungo lunette, close to the WLH 1–3 site.

NOTE

WLH1, 2, 3 and 50 are described in Chapters 7 and 8. WLH 4 is a late Holocene skeleton excavated in 1974 and not included here. WLH 5 is missing, therefore, the following descriptions begin with WLH 6.

The following photographs are attributed to S. Webb.

WLH 6, Site: WOC 1, Lake Mungo. Found: 1980. Sex: Male?

WLH 6 is a mineralised right tibia without proximal and distal ends. The distal half has surface scorching and sites of muscular attachment are prominent and deep. The tibial tuberosity is prominent and the *soleal* oblique line is well developed. The origin of *tibialis anterior* is deep, the inter-osseous border is sharp, well defined and runs to the level of the tibial tuberosity. Bone surface at the *popliteus* insertion is roughened, cortical bone thickness varies from 4 to 9 mm and is thickest at midpoint of the anterior border. The bone is smaller than the tibia of WLH 1 and normally the estimated stature and bone thickness might suggest a female person. However, strong muscle insertion markings could indicate male muscularity. WLH 6 could be between 153 and 158 cm depending on gender which suggests it may be from a population of small stature. Bone gracility may, therefore, be an allometric characteristic associated with small stature and build and that is the conclusion drawn here. This tibia was found in the same general area of the Mungo lunette as WLH 1, 2 and 3.

Metrical data (mm)
Actual length – 285   Stature – male 158 cm and female 153 cm
Estimated length – 330
Mid-shaft diameter
Antero-posterior   32.0
Medio-lateral   21.0
Platycnemic index   65.6

WLH 9, Site: WOC 1, Lake Mungo. Found: 1978. Sex: Undetermined (Female?)

WLH 9 consists of fragmented and burned, cranial and postcranial bone. Colour varies from chalky white calcination through grey to blue-black typical of high temperature burning. Post-cranial cortical bone fragments have transverse, crescentic surface cracking identical to WLH 1 caused by burning. This individual was cremated with bone smashing as in the case of WLH 1. The extremely gracile cranial fragments suggest a similar morphological type to WLH 1 and 3 with a maximum cranial thickness of 5 mm, thin (<1 mm) cranial tables and little diploe. Non-fused sutures and unworn molar enamel suggest an older juvenile or young adult. Gracile postcranial bone fragments are also consistent with these conclusions.

WLH 10, Site: WOC ?, Lake Mungo. Found: 1978. Sex: Undetermined

WLH 10 also shows cremation with bone smashing. Pieces of long bone, with transverse, crescentic cracking, have chalky calcination, coated with orange-coloured calcium carbonate. Thick cortical bone (7 mm) indicates femur, another piece with a thinner cortex is an antero-distal portion of the right humerus superior to the olecranon fossa. The single piece of vault is thin (5 mm) with 1 mm tables.

WLH 11, Site: GL 1, Lake Garnpung. Found: 1977 Sex: Female

All bone of WLH 11 is mineralised, postcranial pieces have carbonate deposition and are white with minor surface erosion. The fragmented cranial remains are from the left parietal, left the occipital, left zygomatic process and a superior portion of the left malar with a thin vault and tables 1 mm thick. The left parietal boss is moderately developed and the internal surface shows the bony impression of the middle meningeal artery and a section of the middle cerebral artery near the frontal border. The superior occipital fossa is moderately deep. A portion of the lambdoid suture between the posterior left parietal and adjacent occipital has two gaps indicating lost lambdoid Wormian bones. Incomplete suture fusion may indicate a younger adult.

A small section of frontal includes a zygomatic process lacking a trigone. A short anterior section of the temporal crest disappears as it moves posteriorly and vault thickness at that point is 6 mm. A large supra-orbital foramen lies medial and slightly behind the superior orbital margin. On the medial edge are the superior and lateral walls of a large frontal sinus. The supra-orbital region is not developed although the orbital margin is rounded making a smooth transition from the orbital plate of the frontal to the brow. The internal opening of the supra-orbital foramen lies on the orbital rim and laterally there is smooth descent from the zygomatic process across the fronto-zygomatic suture and onto the malar at its marginal process. The malar is small but the temporal process and almost all the inferior sections of the body are missing. A small foramen lies just below the inferior orbital rim.

A section of left mandibular body extends from the *sulcus extramolaris* to the right side of a small mental trigone which lacks mental spines. The body is short and gracile with a short tooth row and small alveoli indicating a small dentition all of which emphasise general gracility. Bone resorption around the first and second molar alveoli indicates pre-mortem tooth loss indicating an older adult. Postcranial bone consists of the middle two-thirds of the left humeral diaphysis. It is small and slim with thin cortical bone (3.5 mm maximum). There is a feint trace of a deltoid tuberosity.

This seems to be an older female of the gracile group and contrasts strongly with robust individuals. Although unfused cranial sutures suggest a young adult, the loss of molars with socket resorption indicates the remains are those of a middle-aged or older adult.

Metrical data (mm)
Cranial thickness

Opisthocranion	9.0
Left parietal boss	7.5
Inion	7.5

Brow thickness

Medial (est.)	7.0
Middle	7.0
Lateral	7.0

Mandible

Symphyseal breadth	12.5
Symphyseal height	28.0
Body breadth	11.0

Humerus
Mid-shaft diameter

Antero-posterior	16.0
Medio-lateral	17.0
Least circumference	51.8

WLH 12, Site: GL2, Lake Garnpung Found: 1977 Sex: Undetermined

The postcranial remains of WLH 12 are a yellow-white colour, very fragmented, eroded and only two small pieces of ulna can be identified. All bones are mineralised, but completely free of calcium carbonate. Cranial remains show open sutures, indicating a young adult. The vault bone is moderately thick (10 mm), suggesting general cranial robusticity. A small section of occipital bone has a section of the superior nuchal line developed to a slight crest. The internal occipital surface has a marked superior occipital fossa. Inferior is a wide lateral sulcus. Maximum occipital thickness is 20 mm.

A middle section of the left mandibular body reflects the robust nature of this individual. Worn cusps on a surviving upper left first molar indicate a greater age than minimal suture fusion might suggest. The lower left first and second molar crowns associated with this individual have little wear on any of their cusps (below, right). Interestingly, the second molar shows more wear than the first. It is possible that the differential wear may indicate the molars are from different individuals.

Metrical data (mm)
Mandible
Body height
Body breadth
Dental measurements

	Upper left		Lower left	
	B-L	**M-D**	**B-L**	**M-D**
Ml	11.4	10.6	13.0	9.7
M2	–	–	11.5	10.0

WLH 13, Site: GL26, Lake Garnpung Found: 1977 Sex: Undetermined

This individual consists of only a few highly fragmented and heavily eroded cranial and postcranial pieces. Bone surfaces are severely eroded removing all features. Inner and outer cranial tables are 1 and 2 mm, respectively, and maximum thickness is 7 mm. A single crown of an upper left third molar is unworn and may not have fully erupted but it may not be associated with the other fragments.

WLH 14, Site: GL26, Lake Garnpung Found: 1977 Sex: Male

All bone is highly fragmented and covered in orange-brown calcium carbonate. The poor quality of the postcranial material renders description unreliable. Cranial remains consist only of six maxillary molars, an upper left lateral incisor, a canine and a lower left first molar but measurements were possible only on four teeth. Some are cracked with calcium carbonate infill holding them together. Most occlusal surfaces have heavy wear, suggesting a mature age (30–40? years) and their size indicates a male person, something supported by the robust construction and thick cortex of post-cranial fragments. A single third molar shows little cusp wear inconsistent with the wear pattern of other possibly indicating a second, younger individual.

Metrical data (mm)

	Maxillary dental measurements			
	Left		Right	
	B-L	*M-D*	*B-L*	*M-D*
M1	13.3	10.2	13.8	11.0
M2	14.2	11.3	14.1	11.5

WLH 15, Site: GL 26, Lake Garnpung Found: 1977 Sex: Female

WLH 15 consists of fragmentary cranial and postcranial remains showing extreme surface erosion that has removed surface bone. Despite this, the bone displays a degree of 'freshness' although it is mineralised with minor carbonate encrustation. Identifiable bone comes from the cranial vault (maximum thickness 7 mm), maxilla, mandible, axis, various vertebral bodies, clavicle, pelvis, phalanges, femora, patellae, tibia and humerus. There are also tooth fragments. The bone's poor condition prevents further analysis (Figs. 10.1–10.8).

WLH 16, Site: GL 26, Lake Garnpung Found: 1977 Sex: Undetermined

With exception of a small mid-shaft section of the (left?) humerus, all WLH 16 remains are cranial, extremely fragmented, eroded and heavily mineralised. Cranial suture fusion of a section of lambdoid suture extends from asterion to near lambda, indicating an older adult, perhaps around 50 years. The maximum thickness of the only remaining section of cranial vault is 10 mm (Fig. 10.9).

Metrical data (mm)
Humerus
Mid-shaft diameter:

Antero-posterior	21.0
Medio-lateral	17.0

FIGURE 10.1 Partially burnt tibia of WLH 6.

FIGURE 10.2 Cremated cranial fragments from WLH 9 showing burning.

FIGURE 10.3 A section of the thin cranial wall of WLH 9.

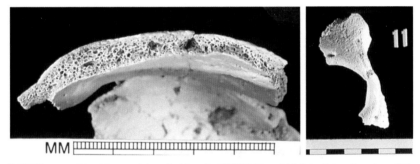

FIGURE 10.4   Long-bone fragments subjected to high temperature burning resulting in transverse cracking, calcination and colouration typical of cremation.

FIGURE 10.5   The thin cranial vault (above) of WLH 11 and (right) a lateral section of the right frontal and malar around the left eye socket. The light construction of these pieces as well as the small left body of the mandible (below) indicates a gracile individual. The loss of the molar teeth with total resorption is visible.

WLH 17, Site: GL20, Lake Garnpung Found: Unknown Sex: Undetermined

WLH 17 consists of small, yellow-white, highly mineralised fragments with silicification. Vault bone is gracile with a maximum thickness of 7.5 mm. Suture fusion has begun indicating an individual in their late 20s or early 30s and two molar fragments have only moderate attrition which supports that age range. The remains of a second tooth (molar?) have considerably more attrition and have lost all its enamel. Postcranial remains consist entirely of small heavily weathered and damaged fragments, unidentifiable to any particular bone.

WLH 18, Site: GL20, Lake Garnpung Found: Unknown Sex: Male

This individual consists of cranial remains mostly white, with a shiny surface typical of surface silicification and some dark, patchy manganese staining. The

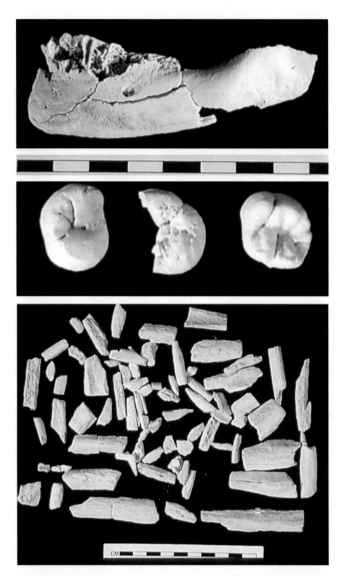

FIGURE 10.6 Post-cranial bone fragments (below) from WLH 12. Minimal wear on the molar tooth crowns (top) indicates a young adult.

largest piece of vault is a section of the left side of the frontal bone extending from near the coronal suture anteriorly to the left brow that includes an anterior portion of the left orbital plate. The frontal bone rises smoothly from the brow region without an ophryonic groove. It is robustly constructed and although inner and outer bone tables are not thick, 1 and 2.5 mm, respectively, the average diploeic bone thickness is 10 mm with a maximum of 13.5 mm. That pattern occurs in other Willandra individuals. Cranial thickness suggests a male person which is supported by its supra-orbital development.

FIGURE 10.7   The molar teeth of WLH 14 showing advanced attrition.

FIGURE 10.8   Post-cranial fragments WLH 15. A section of the pelvis (bottom right) indicates a female.

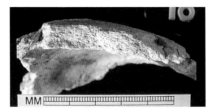

FIGURE 10.9   A section of cranial vault from WLH 16.

Marked postorbital constriction is emphasised by a prominent temporal crest and temporal line. The zygomatic trigone is well developed and the superior orbital margin is rounded, with a small supra-orbital foramen on the upper surface and a wide, deep supra-orbital notch medially. The right brow is not overly developed and its superciliary ridge is small. A small, round depressed fracture lies close to the temporal line, two-thirds of the way to the distal border.

Metrical data (mm)
Cranial thickness
Left frontal boss	12.0
Mid-frontal	13.0

Brow thickness
Medial	17.0
Middle	9.0
Lateral	13.5

WLH 19, Site: GG16, Lake Garnpung Found: 1977 Sex: Male

This individual consists of large sections of the occipital, both parietals, the right temporal around the mastoid and external auditory meatus, and frontal bone less temporal surfaces. Bone condition is good although some surface features have been partially eroded away and the parietals are somewhat distorted so they do not meet at the sagittal suture. That gives the impression of a high vault which is not the case. Parts of the external surface have manganese staining beneath a fine wash of red-brown calcium carbonate and the cranial sutures are not fused.

The frontal bone consists of a small, oval section with well-developed super-ciliary ridges almost forming a small torus meeting above the glabella in a shallow fossa. The torus is reduced laterally, but missing bone prevents further supra-orbital assessment. The glabella is accentuated by marked postorbital depression. A shallow ophryonic groove provides a noticeable demarcation of the supra-orbital from the frontal. The frontal curve is smooth but low and the short glabella-bregma chord suggests low cranial vault height at bregma despite bone distortion presenting otherwise. This gives the effect of a rather sloping frontal with a distinct fullness. The bone is uniformly thick (max. 15 mm) with thick diploe and thinner inner and outer tables 2 and 3 mm, respectively. Small portions of the orbital plates remain. The right plate has a small, shallow pit in the lacrimal fossa like those caused by chronic trachoma (Webb, 1995). Such lesions are believed to stem from a long-term lacrimal gland infection causing a porotic lesion on the adjacent bony plate. A localised periostitis continues, eventually developing into a small osteitic infection causing limited destruction of the bony plate.

The posterior third of the right temporal line crosses the bone, curving infe-riorly and ending near asterion. The occipital has a prominent nuchal torus fol-lowing the contour of the superior nuchal line that reaches both occipito-mastoid borders. The central point of the torus at inion is inferiorly displaced giving a distinct 'M' shape to the torus. Vault thickness at inion is 19 mm. A small supra-iniac fossa is flanked by a supra-nuchal sulcus partly derived from the distinct morphology of the torus itself. The maximum vertical width is 15 mm and its height 7 mm. The upper half of the occipital is rounded when viewed laterally and from the base although the nuchal plane is flattened somewhat giving a stark basi-occipital angulation. There are prominent muscular markings, par-ticularly at the origins of *semispinalis capitis* and *rectus capitis posterior minor.* Flattening is emphasised towards the anterolateral sections of the occipital that sits infero-posteriorly to the mastoid process and medial of the occipital torus.

A small portion of the posterior rim of the foramen magnum, at opisthion, suggests that this point was tilted slightly upward relative to the nuchal plane.

Internally, both the superior and inferior occipital fossae are deep and well demarcated. The inferior portion of the internal occipital crest at endinion is shifted inferiorly and to the left when compared to inion. This adds an odd discontinuity to the crest. The internal cruciform ridges are prominent and provide a definite separation of the left and right superior and inferior occipital fossae, and there is a slight depression on the right superior side of the internal protuberance for the lateral sulcus.

The right temporal extends from the mastoid border at asterion to the anterior wall of the mandibular fossa. The tip of the large mastoid process is missing exposing a number of air cells. Anterior to the mastoid lies the roof of the external auditory aperture. Internally, near the posterior wall, is a small example of the notch of Rivinus (*incisura tympanica*) and medially is a shallow but distinct digastric fossa. The squamo-mastoid suture runs inferiorly and follows the anterior line of the mastoid process until it turns medially. The mandibular fossa is shallow (4 mm) but well defined and takes the form of a narrow oval 20 mm wide and 11 mm long. Internally, there is a deep impression for the sigmoid sinus and there is a large mastoid foramen at a point where it turns inferiorly. A sharp demarcation occurs between the sigmoid sinus and the small middle fossa.

WLH 19 is interesting in several respects. The frontal bone is small and the apex of the vault seems to be low, although an estimated cranial length suggests it is long. When these features are added to the development of the supra-orbital region, the arc-shaped superciliary ridges, prominent occipital torus and thick vault, the appearance of WLH 19 is very robust and archaic looking. Moreover, its supra-orbital morphology and occipital tori is unique in the series although reflects the very robust WLH 50. This is a male person.

Metrical data (mm)
Cranial thickness

Midfrontal	12.0
Frontal bosses (av)	12.5
Bregma (near)	10.0
Vertex	10.0
Obelion	10.0
Lambda	10.0
Right parietal boss	10.0
Left parietal boss	11.0
Right asterion	11.5
Left asterion	11.5
Inion	19.0
Brow thickness	
Medial	16.0
Middle	16.0
Lateral (est.)	7.5
Craniometric data	
Glabella-bregma	113.0
Glabella-metopion	64.0

Nasion-bregma	111.0
Bistephanion	114.0
Interorbital width	27.0
Maximum width of frontal	106.0
Maximum width of parietals	126.0
Lambda-asterion	87.0
Lambda-opisthion	94.0
Bregma-lambda	122.0
Bregma-asterion	145.0
Inion-lambda	47.0
Inion-bregma	156.0
Inion-opisthion	59.0
Asterion-inion (left) (est.)	68.0
Asterion-inion (right)	66.0
Asterion-bregma	145.0
Asterion-opisthion	69.0
Bi-asterion	112.0
Lambda-asterion (left)	80.0
Lambda-asterion (right) (est.)	78.0
Estimated length of cranium	205.0
Frontal curvature index	18.9

WLH 20, Site: GG 16, Lake Gogolo Found: 1977 Sex: Undetermined

Cranial remains consist of a section of the left parietal, just anterior to the boss, and part of the right mandibular body. Among the few highly fragmented and eroded postcranial fragments are mid-shaft portions of the left ulna and humerus. All remains are covered in red-brown carbonate.

The parietal is thick (max 12 mm) and inner and outer cranial tables are 2 and 3 mm, respectively. Erosion of the inner surface has removed features there, but grooves for branches of the middle meningeal artery are evident. The mandibular section is normal size and extends from the right side of the mental trigone to the lateral part of the third molar socket. The latter may have been lost pre-mortem although two roots of the second molar are in place. The digastric fossa is shallow without a genial tuberosity or spines.

Metrical data (mm)
Mandible

Symphyseal height	24.0
Symphyseal breadth	14.0
Body height	31.5
Body breadth	15.5
Left humerus	
Actual length	93.0
Mid-shaft diameter	
Antero-posterior	18.0
Medio-lateral	14.0

WLH 21, Site: GG16, Lake Gogolo. Found: Unknown Sex: Undetermined

This is a single, small piece of highly mineralised, grey-white, cranial vault, probably parietal. It is thin (max. 7.5 mm) and the inner and outer tables measure 1 and 1.5 mm, respectively. It gives the impression of a more gracile individual.

WLH 22, Site: GG16, Lake Gogolo Found: 1977 Sex: Male

Both cranial and postcranial remains are light in colour, some grey-white and others yellow-orange. Postcranial remains consist of long bone fragments while cranial remains represent large sections of both parietals, less the temporal surfaces. There is a section of frontal bone and two pieces of mandibular body. Some bone is burnt but not as thoroughly as those described as 'cremated'. The internal cranial surface has patches of blue-black scorching. Bone surfaces have probably been scorched because the diploe and outer tables have escaped discolouration. The calvarium has a sagittal keel with slight parasagittal depressions near obelion. The parietals slope evenly from the midline to just below the temporal line then the contour becomes rounded crossing the parietal eminence and straightening as it descends to the temporal surface.

All sutures are fused with some parts of the sagittal suture obliterated. A large, shallow, depressed fracture is located on the left parietal near obelion. Four small parietal foramina lie medial of the fracture and straddle the sagittal suture. The outer table is 3 mm thick while the inner table ranges from <1 to 1 mm thick, a condition noted in other individuals. Maximum vault thickness is 12 mm at lambda and 14 mm at inion. In some places the diploe has large intra-trabecular spaces similar to those noted for others in the series.

Internally, two large, deep depressions lie either side of the sagittal suture. The one on the right contains two smaller depressions reducing vault thickness to 5 mm in those places. One depression is probably associated with the passage of the superior sagittal sinus, the origin of the second is uncertain. The endocranial surface has many arachnoid foramina. Only the upper half of the occipital consists of the occipital plane above the superior nuchal line. The line is moderately developed at its centre although the right half is missing, together with all bone inferior to it. The plane is rather flattened when viewed laterally and there is a slight depression at lambda.

A central section of the frontal above the supra-orbital area includes the posterior walls of the centrally located frontal sinuses. The sinus was large compared to others in the series with the exception of WLH 50. A sharp, strongly developed frontal crest lies on the internal surface with 12 mm vault thickness on either side. There are fragments of the supra-meatal region from both sides of the head bearing the *incisura tympanica*. The internal surfaces are missing exposing the mastoid air cells that extend beyond the immediate area of the mastoid as in WLH 50.

A thick anterior section of the mandibular body displays a prominent mental trigone with a strong mental protuberance. The mandible was robust and contains second and third molar sockets and an anterior root of the first molar. The genial spines are broken, but their stumps remain prominent. Inferiorly, the digastric fossa is not as deep as might be expected of a large mandible. The size of the dental sockets, the diameter of the root impressions and socket depth indicate large teeth. The lower central incisors were lost pre-mortem and resorption and closure of the right canine socket indicates loss early in adult life. The tip of the root of the right lateral incisor is in situ. A small section of the right gonial angle displays a roughened surface typical of strong insertion of the pterygoid muscle but bony eversion is unexpectedly slight at this point.

Postcranial remains consist only of an 84-mm-long section of the proximal right ulna and the proximal half of a phalange. The ulna has well-marked of muscular insertion points at the supinator ridge and ulna tuberosity, as well as a prominent interosseous border. The general robust morphology of the remains and closed suture fusion suggest an older male person

Metrical data (mm)
Cranial thickness

Frontal	12
Obelion	9
Lambda	12
Asterion (left)	8
Parietal (right)	10
Inion (close to)	14
Mandible	
Symphyseal height	33
Symphyseal breadth	18
Body height	39
Body breadth	17

WLH 23, Site: GG16, Lake Gogolo Found: 1977 Sex: Female

WLH 23 consists of very eroded, fragmentary and heavily mineralised bone with a high gloss and some manganese spotting. Many cranial pieces have sections of suture along one margin and some are fused on the endocranial surface. Few have anatomical detail but all pieces are gracile with a maximum vault thickness of 8–9 mm at asterion and 7–8 mm on the vault. Vault thickness consists of diploeic bone with very thin inner and outer tables.

The largest cranial piece is an anterior section of frontal with a central portion of the supra-orbital. The area around the glabella is missing exposing the posterior and lateral walls of the frontal sinus. The brow extends laterally enough to show small superciliary ridges and very sharp superior orbital margins each with a very shallow supra-orbital notch. The internal surface of the vault has a small section of frontal crest that originally was high, long, fairly wide and with a sharp border. The thickness of the vault is 6 mm on the left and 7 mm on the right of the crest. The frontal rises acutely from the superciliary ridges without an ophryonic groove. The generally gracile cranial construction, particularly in the supra-orbital region, suggests a female person.

WLH 24, Site: GL 13, Lake Garnpung Found: 1977 Sex: Female

WLH 24 consists of highly mineralised cranial and postcranial fragments. Inner and outer vault surfaces have a high (silica?) gloss similar to that noted on other individuals. All remains are light grey-blue typical of high temperature burning with some bone covered in a pale brown carbonate wash. Broken edges show internal charring of the bone. Sections of vault extend posteriorly from the left zygomatic process to the anterior third of the left parietal. Little of the frontal bone remains, while just behind the coronal suture a section of vault extends irregularly from the left inferior squamous border of the temporal to the right parietal just lateral of the sagittal suture and posterior to bregma. An anterior view of the vault shows it was small and very oval.

The zygomatic trigone is smooth and the temporal crest is barely developed. A faint temporal line can be traced for about 3 cm from its posterior ending abruptly at a break. The coronal, sagittal and lambdoid sutures are completely fused. A small section of the orbit indicates little development. Although the cranium is thin medial to the temporal crest, vault thickness increases in the sagittal plane. Thickness is variable but marked (10 mm maximum) for an otherwise delicately featured individual. The inner cranial table is often less than 1 mm, contrasting with the outer cranial table which varies from 1 mm to 3.5 mm. The left side of the occipital shows part of a minimally developed superior nuchal line. A small section of the left temporal has extensive groups of mastoid air cells occupying a large portion of the endocranial surface. To the right of the superior sagittal sulcus, the vault is 4 mm. Other internal features include deeply etched sulci for cerebral vessels that are prominent for both the frontal and parietal branches of the middle meningeal and cerebral arteries. Postcranial remains consist mainly of sections of the larger long bones. They are highly mineralised and smooth, with thin cortices.

Suture fusion and general gracility suggest an old female. However, the overall gracility is not in keeping with the rather robust dimensions of the central areas of the cranial vault. Although bone fragmentation might indicate deliberate smashing the pieces are not as small as normally seen in such cases.

Metrical data (mm)
Cranial thickness

Bregma	9
Asterion	8
Lambda	10
Porton-asterion	46
Bi-parietal	144

WLH 25, Site: GL 13, Lake Garnpung Found: 1978 Sex: Female

These mineralised remains are greyish white with patches of yellow carbonate stain and some surface gloss. Cranial fragments are few and small with a maximum vault thickness of 8 mm. Reconstruction produced large sections of the femora and tibiae and sections of the diaphyses of the left humerus, right ulna and left radius. Other bones include sections of metacarpals and phalanges of the right hand and a left second or third phalange.

Left Femur

The *linea aspera* and gluteal tuberosity are well defined and a small exostosis occurs on the lateral supra-condylar line. All bones above the gluteal tuberosity and below the divergence of the medial and lateral supra-condylar lines are missing.

Right Femur

The *linea aspera* and gluteal tuberosity are prominent on both left and right femurs. The femoral head, neck and both trochanters are missing, together with almost the whole distal third of the shaft. Marks consistent with rodent gnawing or other small animals are present lateral to the *linea aspera* at mid-shaft.

Left Tibia

This consists of the diaphysis below the tibial tuberosity and above the distal end. It is not robust and lacks the thick cortical bone often seen on other tibiae in the series.

Right Tibia

Only part of the proximal half of the diaphysis remains. It has an oval patch of periostitis, about 21 mm long, on the medial side of the diaphysis at about the level of the nutrient foramen. The cortical bone is thickened there by new bone formation. The cause of this pathology is not obvious but could have been the result of an overlying soft tissue infection.

Right Humerus

Only the distal half of the diaphysis remains and is gracile in keeping with the rest of the postcranial skeleton. There is some development of the lateral supra-condylar ridge, probably in response to extensive use of the *pronator teres* muscle during lower arm pronation and supination. Use of a spear thrower could have produced this kind of wear (see also Chapter 7). At its distal end, the shaft presents a marked triangulation in cross-section. The thickness of the shaft cortex, measured anteriorly just below the level of the deltoid tuberosity, is 4.4 mm, which could be regarded as proportional to the 7.6-mm-thick cortex of the anterior border of the left tibia. The right ulna consists of the whole shaft less both ends.

Pathology

Besides the periostitis noted on the right tibia, a portion of tubular bone (clavicle?) shows a similar non-specific infection. Some bone remodelling and lay-down of very porous new bone have occurred over the thin cortical bone. The small fragment makes further comment unjustified but the osteitic infection is likely to have covered a large area. A connection between the two sites may be possible in a wider systemic pathology such as yaws, but that cannot be proven. Internally, trabecular bone is thickened diagonally across the medullary cavity. The formation is consistent with a fused fracture, but, again, the size of the fragment prohibits further analysis. A tentative guess at the origin of this pathology is an open wound infection following a partial (greenstick) fracture of the right tibia.

Stature and general skeletal gracility suggest that this individual is female. The incidence of infection is interesting, particularly if the remains date to within the Late Pleistocene. It is generally thought that hunter-gatherer demography at this time (low populations, constant movement and small groups) created an environment where the chances of infectious disease would have been low (Figs. 10.10–10.24).

Metrical data (mm)
Left femur
Estimated length   463
Stature   165 cm
Subtrochanteric diameter

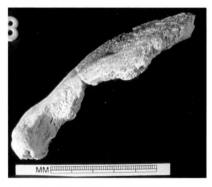

FIGURE 10.10   **Lateral view of left frontal bone WLH 18.**

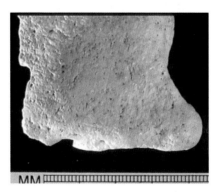

FIGURE 10.11   **Superior view of right frontal WLH 18 showing prominent post-orbital constriction and prominent zygomatic trigone.**

FIGURE 10.12   The vault construction in WLH 18 composed of thick cancellous bone and thin cranial tables.

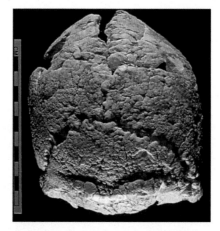

FIGURE 10.13   WLH 19 posterior view showing sagittal keeling, almost vertical parietals and nuchal torus.

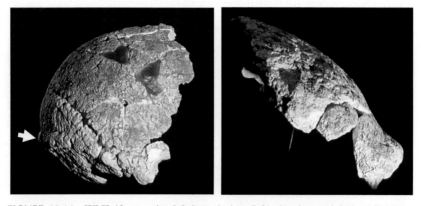

FIGURE 10.14   WLH 19 posterior left lateral view (left) showing occipital nuchal torus (*arrow*), and frontal bone (right) with prominent brow.

FIGURE 10.15   The sub-occipital region is detailed above. The prominent double-arc of the nuchal torus is prominent and flat external occipital surface is broad and flat to accommodate a well-developed nuchal muscle group.

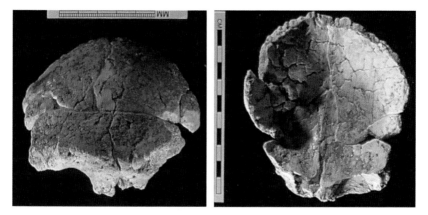

FIGURE 10.16   Anterior (left) and internal (right) views of the frontal bone of WLH 19 show its robust character.

FIGURE 10.17   Mandibular, cranial and post-cranial fragments belonging to WLH 20.

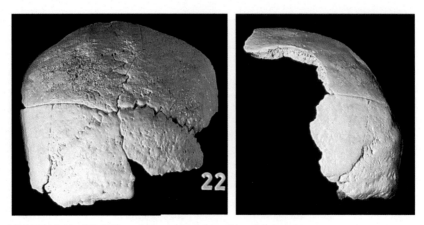

FIGURE 10.18   WLH 22 posterior and left lateral cranial view.

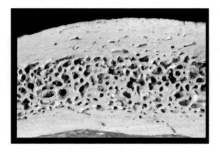

FIGURE 10.19   Thick cranial vault of WLH 22 composed thick trabecular bone.

FIGURE 10.20   Cranial fragments and part of the right brow of the female WLH 23 (top).

FIGURE 10.21   The developed superciliary ridge of WLH 23.

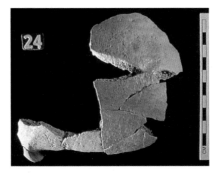

FIGURE 10.22   Partial vault of WLH 24.

FIGURE 10.23   Section of cranial vault from WLH 24.

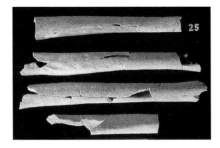

FIGURE 10.24   Sections of long bone from WLH 25 (top) and metacarpals and phalanges (below).

Antero-posterior	25.5
Medio-lateral	36.0
Mid-shaft diameter	
Antero-posterior	29.4
Medio-lateral	30.7
Mid-shaft circumference	92.4
Platymeric index	72.2
Robusticity index	13.0
Right femur	
Actual length	286
Subtrochanteric diameter	
Antero-posterior	25.0
Medio-lateral	35.0
Mid-shaft diameter	
Antero-posterior	27.5
Medio-lateral	28.5
Mid-shaft circumference	86.4
Platymeric index	71.4
Left Tibia	
Actual length	239
Estimated length	360
Actual length	313
Diameter at nutrient foramen	
Antero-posterior	34.4
Medio-lateral	24.4
Platycnemic index	70.6
Right Humerus	
Actual length	170
Estimated length	310
Stature	164 cm
Mid-shaft diameter	
Antero-posterior	22.5
Medio-lateral	19.0
Minimum shaft diameter	
Antero-posterior	19.1
Medio-lateral	18.0
Least circumference	58.1
Robusticity index	22.1
Right ulna	
Actual length	212
Estimated length	260
Stature	161 cm
Mid-shaft diameter	
Antero-posterior	12.3
Medio-lateral	12.0
Least circumference	37.7

WLH 26, Site: GL13, Lake Garnpung Found: 1977 Sex: Undetermined

WLH 26 consists of a few cranial and postcranial fragments covered in a thick layer of calcium carbonate. The largest piece of cranial vault is from a posterior section of the left parietal with a maximum thickness of 11 mm at asterion (Fig. 10.25). Both cranial tables are thin, with the inner table <1 mm and the outer 1 mm. Impressions of the branches of the middle meningeal artery are prominent endocranially. The only postcranial piece is the distal one-third of the left radius.

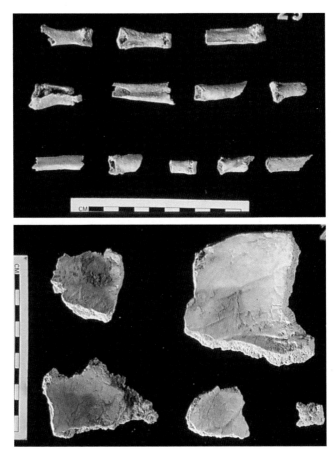

FIGURE 10.25   Cranial fragments from WLH 26 (above) and a three long bone fragments (below).

WLH 27, Site: GL 13, Lake Garnpung Found: 1977 Sex: Male

Both cranial and postcranial remains are highly mineralised, yellowish white in colour. Much of this material has a familiar surface shine identical to that on other individuals. Cranial remains include large sections of the vault. There is a left zygomatic bone and a mandible. Part of the left parietal is 14 mm thick, but the whole vault seems to be uniformly thick. Both cranial tables are thin (1.5 mm), with the rest comprising diploeic bone. A section of occipital includes the superior nuchal line while the nuchal plane reflects a rugosity originating from muscular insertions for the *rectus capitus* muscles. There is also moderate development of the nuchal torus. The occipital inner table is missing although a prominent internal occipital protuberance is present with crests on either side.

Part of the mastoid portion of the right temporal is present but the tip of a large mastoid is missing. Medially, there is a deep but narrow digastric fossa together with part of a well-formed supra-mastoid crest, below in which is a large pre-mastoid foramen. Another sits directly infero-laterally on the surface of the mastoid. Enlargement of these foramina probably occurred post-mortem. The internal surface of the mastoid portion of the temporal has a large, deep sigmoid sulcus. Inferior is an extensive mass of mastoid air cells that extend anteriorly beyond the auditory canal and upward to the lambdoid suture.

The left malar includes the temporal process minus an inferior border and most of the orbital surface. There is a large zygomatic foramen and part of a moderately developed malar tuberosity. The body of the malar is robust and thick obliquely through the temporal surface to the orbital margin. Two pieces of the supra-orbital have extremely large superciliary ridges. The left-hand section also has a large but shallow supra-orbital notch. The ridges themselves are so strongly developed that they impinge on the orbital margins, filling them out and causing them to rise almost vertically, 20 mm on the left. Although the lateral surfaces of these ridges are missing, it is likely that they did not extend as far as the zygomatic process. Nevertheless, it is suggested that the expression of this trait in this form be termed a torus or semi-torus. The right brow of WLH 50 has the same morphology as the left brow of this individual, the two being the only examples of this type of superciliary development in the series. Robust development of the supra-orbital area is impressive and similar to WLH 50. The thickened vault is consistent with brow ridge development, but the malar and sub-occipital region are not as robust as might be expected. All features, however, emphasise the maleness of the individual.

The mandible includes the left mandibular condyle, less the superior pterygoid *tubercle, and* a small portion of the *planus triangulare*. Osteophytic lipping on the anterior part of the condyle indicates a chronic condition possibly from consistent stress on the temporo-mandibular joint through using the jaws in the preparation of vegetable fibre. An antero-inferior section of the right ramus has a pre-coronoid fossa and buccinator crest. The left distal portion of the mandibular body has a prominent mental protuberance, a small mental fossa and an extremely shallow digastric fossa. The roots of the lateral incisor, canine and first and second premolars are in situ. A large mental foramen lies halfway between the alveolar rim of the second premolar and the inferior border.

Postcranial remains are few and very fragmentary. Pieces of the left femur represent the middle section of the diaphysis with a strong *linea aspera* and thick cortices up to 11 mm. A right third metacarpal has a healed complete fracture with a small callus midway along its diaphysis. The injury has shortened the bone by drawing the anterior section posteriorly towards the wrist which then set in this position. X-ray of the bone confirms the orientation and exact nature of the fracture.

Metrical data (mm)

Cranial thickness

Left parietal	14.0
Inion	17.5

Mandible

Symphyseal height (est.)	33.0
Symphyseal breadth	15.6

Left femur

Mid-shaft diameter

Antero-posterior	35.0
Medio-lateral	25.3

WLH 28, Site: GL 13, Lake Garnpung Found: 1977 Sex: Male

All these fragmented cranial and postcranial remains have been completely burnt, with some fragments showing the chalky white of calcination. Apart from this, the colour ranges from an off-white/light grey through various darker shades of grey and blue-black. The broken edges of the bone fragments, smooth and well worn, are coloured in an identical way to external bone surfaces. Many fragments are extremely small something suggesting the mortuary practice of cremation with bone smashing.

The largest piece of cranial bone is a robust mid-section of the occipital with a prominent nuchal crest extending to the mastoid suture on both sides. The nuchal plane is rugged at points of muscle insertion, particularly *semispinalis capitis*. The thickest part of the vault (21 mm) is at inion, comprising thick inner and outer tables of 4 and 8 mm, respectively. There is high gloss on both the inner and outer surfaces of the occipital bone which displays a large internal occipital protuberance and well-defined, broad sulci for the sagittal, transverse and lateral sinuses.

Surface bone on the medial surface of the mastoid has been removed exposing extensive pneumatic cells. Little remains of the mastoid itself, but enough suggests it was large. In contrast to the occipital, a piece of the left supra-orbital has a small superciliary ridge indicating minimal development raising doubts that it is from the same individual. (The remains of a second, burnt but much more gracile individual was found with WLH 28, see WLH 122.) There is, however, a clear separation between the supra-orbital region and the rise of the frontal bone, although without an ophryonic groove. Pieces of the cranial vault show extensive thickening, with a maximum of 12 mm.

A portion of the left malar has most of the inferior border and orbital surface missing. It shows the differential pattern of dark blue-grey and light grey colouration, with heavier burning on the orbital surfaces. The malar tuberosity is small, but, as for WLH 27, the bone is ruggedly constructed and thick through the body of the malar from the temporal surface to the orbital border (see Fig. 6.1).

A piece of the left mandibular condyle lacks the superior pterygoid tubercle. The external pterygoid fossa is present but all bone inferior to this is missing. The condyle has posterior and medial sections missing. Minor pitting on articular surface may have been caused by erosion and burning rather than osteoarthritic degeneration.

All postcranial remains are fragmentary and only a few pieces of the long bones, such as the tibiae and femora, are present. The largest of these is a proximal section of the right femoral diaphysis, situated just below the lesser trochanter burnt grey-blue. Cortical thickness is 9 mm and the periosteum has been stripped away leaving a pitted surface, which is typical of severely burnt bone (see also WLH 68). The gluteal tuberosity is long and deep adding to the robust appearance of this bone.

WLH 28 is the only cremation of a robust individual. Since the robust construction of the cranium indicates a male, it is also the only firmly identified male cremation. Although WLH 2 is tentatively identified as male ait is a gracile form.

Metrical data (mm)
Cranial thickness

Parietal	12
Right asterion	9
Inion	21
Right **Femur**	
Subtrochanteric diameter	
Antero-posterior	27.0
Medio-lateral	34.0
Platymeric index	79.4

WLH 29, Site: GL 13, Lake Garnpung When found: 1977 Sex: Female

This individual consists only of cranial remains. They are mineralised and covered in a moderately thick, dark brown layer of carbonate. Most of the posterior section of the calvarium is present represented by medial sections of both parietals and a right superior portion of the occipital. The right parietal extends as far as the temporal surface and shows the radiating lines of temporal articulation. Overall, the vault is gracile, suggesting a female individual. The vault is large and rounded, with a slight sagittal keel and no rugged features. The vault is generally thin (3–5 mm) with a maximum thickness of 8 mm. There is relatively much less diploeic bone than in many other individuals in this series, but the inner and outer cranial tables have similar thicknesses, 1 and 2 mm, respectively. Much of the diploeic bone and inner table has been lost, however.

A section of the right lambdoid suture has not fused, but almost all the sagittal suture, with the exception of a small section near lambda, has disappeared. Two large lambdoid wormian bones are missing from the left side of the suture, as indicated by two indentations in the lambdoid suture. Degree of suture fusion suggests a mature adult, probably over 45 years.

There are fragments of the occlusal surfaces of two lower molars (Ml and M2) showing substantial attrition. This wear pattern adds some support to the tentative age suggested above. There are two small sections of bone originating from almost the same part of each temporal. They show a small part of the meatal area and a moderately developed mastoid crest.

Metrical data (mm)
Cranial thickness: Obelion 7, lambda 8, left parietal 7, right parietal 8, right asterion 7

WLH 39, Site: GL 13, Lake Garnpung Found: 1979 Sex: Undetermined

Six highly mineralised white cranial fragments only. Three are covered in a thick grey-brown coating of calcium carbonate.

WLH 42, Site: GL 24, Lake Garnpung Found: 1976 Sex: Male?

Only a few fragmented and eroded cranial and postcranial remains. Maximum cranial thickness is 10 mm and the inner and outer tables are thin. A left lower canine shows extreme attrition exposing a large area of dentine.

Metrical data (mm)
Lower left canine

Labio-lingual	9
Mesio-distal	7

WLH 43, Site: GL 20, Lake Garnpung Found: 1979 Sex: Undetermined

WLH 43 consists of one medium and one very small piece of cranial vault as well as a thin section of an ulna diaphysis. The medium-sized piece of vault is from an area around right asterion. The suture on the occipital border is unfused. Maximum thickness is 10 mm and both cranial tables are very thin.

WLH 44, Site: GL 24, Lake Garnpung Found: 1978 Sex: Female

WLH 44 consists of a few small and highly mineralised fragments. Most are burnt and covered in a pale orange carbonate wash. Complete carbonisation of the bone and its state of fragmentation put this individual into the category of cremation with bone smashing. Most pieces come from the left temporal and left parietal, near asterion. A large portion of the crown of an upper second molar has moderate attrition. There is also the root of a small upper left second premolar. The mandibular or glenoid fossa is small and shallow.

Osteoarthritic degeneration appears on the articular surface of the fossa and the articular tubercle has been eroded and flattened because of extensive and consistent joint stress. The fossa has allowed the mandibular condyle to override the articular tubercle indicating stress applied to the anterior dental complex, when the incisors were used for chewing, biting, or gripping an object while being pulled forward in the mouth (Shipman et al., 1985). A large, shallow, pitted hollow lies anterior to the articular tubercle, but it is not known whether this was caused by the above stress or by injury. It closely resembles osteoarthritic degeneration when constant stress is applied to the temporo-mandibular joint.

The cranial remains are gracile and internally, there is a well-defined curved depression to accommodate the middle meningeal artery. The maximum measurable thickness of the vault is 7 mm, with 1 mm inner and outer tables. The gracile appearance and moderate molar attrition suggests a mature age female.

WLH 45, Site: GL 24, Lake Garnpung Found. 1977 Sex: Female

WLH 45 is a partial but fragmented skeleton. Cranial remains include a large part of the calvarium, the right maxillary arcade and a small section of the left temporal and mastoid region. The calvarium has patches of erosion on the inner and outer occipital surfaces. A pale brown carbonate wash covers most of the external vault beneath which is white, shiny bone suggesting silicification. Most of the frontal and supra-orbital regions are missing. The remaining lateral half of the right orbit has a rounded but not thick superior margin but from the little that remains it may not have been well developed. The frontal rises smoothly from the brow with little post-orbital constriction. The vault is rounded coronally, with minimal keeling, some minor parasagittal grooving and rather perpendicular parietals. Maximum cranial breadth occurs low down on the temporal. The roof of the vault is thick, measuring a maximum of 14 mm. The inner and outer cranial tables are moderately thick, 1.5 and 2.5 mm, respectively. The coronal, sagittal and lambdoid sutures are fused, indicating a mature adult probably over 50 years of age. A small piece of the left temporal includes part of a mastoid crest and the roof of the external auditory meatus. A large cluster of mastoid air cells are exposed on the endocranial surface.

A section of the inner wall of the right maxilla indicates a deep palate and the roots of all teeth from the lateral incisor to the third molar indicating no pre-mortem tooth loss. The occlusal surfaces of M1 and M3 show moderate attrition, wearing down all cusps but not exposing dentine. Some alveolar bone has been resorbed, but no pathology is evident suggesting aging as the cause. There are fragments of other teeth, but these are in very poor condition. All surfaces are heavily worn.

Left Femur

Only the proximal half of a poorly preserved femoral diaphysis remains. The surface is eroded and pitted and parts of the bone are covered in light brown-orange calcium carbonate. Both the *linea aspera* and the gluteal tuberosity are prominent. Cortical bone thickness is 9 mm.

Right Femur

This bone is represented only by the proximal third of the diaphysis. The gluteal tuberosity and the *linea aspera* are prominent, as on the left femur, together with the spiral and pectinial lines. The brown carbonate layer is thicker on this bone than on the left femur.

Right Tibia

The right tibia is represented only by a small mid-section of the diaphysis at the level of the nutrient foramen.

Left Humerus

Only the proximal half of the left humerus is present, less the head. Like other postcranial remains, it is covered in a thick coating of calcium carbonate. Beneath this is some blue-black manganese spotting. The bone is rather slender and delicately built, with thin cortical bone (4 mm). The inter-tubercular sulcus is deep and about 5.1 cm long. Inferiorly, there is a small, underdeveloped deltoid tuberosity.

Right Humerus

The right humerus consists of the distal two-thirds of the diaphysis. Its condition is very like that of the left humerus, with a substantial covering of calcium carbonate and an underdeveloped deltoid tuberosity. The diaphysis is straight, slim and rather gracile. The morphological similarities between the left and right humeri preclude an assessment of handedness.

Pelvis

The pelvis consists of a section of the left ilium. Part of the auricular surface shows deep notches, originally filled with a large quantity of calcium carbonate. Removal of the carbonate revealed a deep, 28 mm long pre-auricular sulcus. The superior rim of the greater sciatic notch is shallow and the shallow curve into the inferior border of the notch suggests a wide angle. These traits indicate a female person. A small area of the same section of the right ilium lacking a lateral surface is present.

Cranial vault thickness (14 mm) is interesting in this female and usually associated with males. The femaleness of the remains is supported by ill-defined muscle markings; the shape of the vault, which is well rounded and with a full frontal; the lack of supra-orbital development, particularly of the supercili-ary ridges and zygomatic trigones; pelvic morphology; and the lack of deltoid tuberosity development on both humeri.

Metrical data (mm)
Cranial thickness

Right frontal boss (near)	11
Bregma (near)	9
Obelion	10
Left parietal boss	13
Right parietal boss	12
Lambda (est.)	10
Brow Thickness	
Medial	-
Middle	6
Lateral	7
Biasterion (est.)	120
Lambda-bregma (est.)	142
Maximum breadth	134
Maximum length of maxilla	63
Endomolare-ectomolare	13
Left Femur	
Actual length	290
Estimated length	435
Stature	159 cm
Subtrochanteric diameter	
Antero-posterior	23.5
Medio-lateral	34.2
Mid-shaft diameter	
Antero-posterior	29.0
Medio-lateral	26.0
Platymeric index	70.6
Robusticity index	12.6
Right femur	
Actual length	139
Subtrochanteric diameter	
Antero-posterior	24.0
Medio-lateral	34.0
Platymeric index	70.6
Right tibia	
Actual length	140
Mid-shaft diameter	
Antero-posterior	33.0
Medio-lateral	21.5
Left humerus	
Actual length	125
Surgical neck diameter	
Antero-posterior	21.2
Medio-lateral	21.8

Mid-shaft diameter

Antero-posterior	16,0
Medio-lateral	22.0

Right humerus

Actual length	187
Estimated length	310
Stature	160 cm

Mid-shaft diameter

Antero-posterior	20.6
Medio-lateral	19.0
Minimum circumference	56.6

Right maxillary dental measurements (mm)

	Ml	M2
Bucco-lingual	11	10
Mesio-distal	11	12

WLH 46, Site: GG16, Lake Garnpung Found: 1977 Sex: Undetermined

WLH 46 consists of a small fragment of long bone (bone unknown), a piece of right zygomatic trigone, a section of the right maxilla containing the roots of the canine and first premolar and some small segments of cranial vault. The maximum thickness of the vault is 10 mm, with an outer table of 3.3 mm and the inner table <1 mm. All these remains are mineralised, eroded and many surfaces have manganese spotting and patches of brown carbonate. No further description is possible.

WLH 47, Site: GG18, Lake Garnpung Found: 1977 Sex: Undetermined

These remains consist of eroded and mineralised cranial and postcranial fragments. Little carbonate deposition is visible, but the inner surface of a fragment of vault is shiny, possibly due to silicification.

The mastoid portion of the left temporal contains the parietal notch, a large section of the parieto-mastoid suture and the posterior wall of the mastoid. Most of the inferior section is missing, but from what remains it would have been small. Internal features show a section of the sigmoid sulcus and an extensive area of pneumatisation. Both the distribution and size of air cells far exceed their limited development as seen in other remains. Two pieces of cranial vault include a section of the temporal surface and part of the vault proper. Maximum thickness is 7 mm, with 1 mm tables. There are only a few small fragments of the postcranial skeleton.

WLH 48, Site: GL 25, Lake Garnpung Found: 1980 Sex: Undetermined

Only two small fragments of vault comprise WLH 48. They are very mineralised and have a coating of reddish brown carbonate over almost all the inner and outer surfaces. The pieces are very thin, with a maximum thickness of 5 mm and negligible cranial tables (<1 mm).

WLH 49, Site: GL 25, Lake Garnpung Found: 1978 Sex: Undetermined

This individual consists mostly of postcranial fragments, particularly of long bones. Some are covered in a pink-white carbonate crust and all are mineralised. The only cranial piece is a small portion of vault lacking most of the inner table and some diploeic bone are missing. Maximum thickness is 8 mm and the inner and outer bone tables are thin. There are no diagnostic characteristics on any of this material.

WLH 51, Site: GL 13, Lake Mungo Found: 1978 Sex: Female

The largest bone is a section of the left frontal close to the zygomatic process. The rest consists of a few heavily eroded, small fragments. The vault is thin, with a maximum thickness of 5 mm, and lacks all other features. It rises almost vertically from the supra-orbital region with almost no indication of separation between the two. The section includes part of the supra-orbital region from the fronto-zygomatic suture to a point medial of the supra-orbital notch. The notch is wide but shallow and the superior orbital margin is sharp. A section of the superciliary ridge has a small tuberosity and the zygomatic trigone is almost non-existent. Posterior to the zygomatic process is a well-defined temporal crest which extends into a pronounced temporal line. Non-closure of the fronto-zygomatic suture and the extreme gracility of the frontal bone and supra-orbital region indicate a young adult probably female.

WLH 52, Site: GS10, Lake Garnpung Found: 1980 Sex: Undetermined

Both cranial and postcranial bone is fragmented, with surface erosion. Some is covered in a dense, red/brown calcium carbonate that has coloured the bone. The vault is moderately thin, with a maximum thickness of 9 mm, although in many areas it is only 6–7 mm thick. Despite much of the outer cranial table is missing, enough remains to indicate both tables were thin. A small section of the left frontal bone has part of the temporal line. A small oval hollow, resembling a dermoid cyst, sits on and below this line. Postcranial remains consist mainly of long-bone fragments. The largest piece is from the middle third of the diaphysis of the right ulna, which has a strongly developed inter-osseous border. There is also a section of the acromial neck from the right scapula.

The highly mineralised, red stained bone fragments of WLH 52 suggest it originates from the deep red Gol Gol sediments, the oldest sedimentary level in the Willandra stratigraphic sequence and thought to be early MIS 4 or 5 in age. A [230]Th age of between 104 and 140 ka was obtained for WLH 52 in 2001 but that is not now accepted.

WLH 53, Site: WOC145, Lake Mungo Found: 1980 Sex; Undetermined

This individual consists of small fragments of white mineralised cranial and postcranial bone. The vault fragments have a maximum thickness of 8 mm, with some as little as thin as 3 mm. The outer cranial table is thick for an otherwise thin vault, measuring 3.4 mm in some places, while the inner table is about 1.5 mm.

WLH 55, Site: WOC 152, Lake Mungo Found: 1978 Sex: Undetermined

The bulk of these remains consist of highly fragmented cranial and postcranial remains bleached white from exposure. Some pieces are covered in a slight, reddish brown carbonate wash. Most of the cranial remains are from the vault, but two small pieces from the mandible body and gonial angle are present. The body of the right malar is delicately built, with a sharp lateral border. A small section of the left temporal at the root of the zygomatic process is also rather gracile in its construction. Maximum thickness on any of the cranial fragments is 7 mm, with 1-mm-thick inner and outer tables.

Left Humerus

This consists of the distal half of the diaphysis as far as the proximal surface of the olecranon. The bone is slim, with few distinctive markings and little development of the deltoid tuberosity. Its minimum circumference is 61 mm.

Right Tibia

This bone is resented by a medial section of the proximal diaphysis, with a strong popliteal line, is all that remains. Although I have left this individual unsexed, it is likely that it is female.

WLH 56, Site: WOC152, Lake Mungo Found: 1978 Sex: Male?

WLH 56 consists of some highly mineralised and eroded scraps of long bone and 11 pieces of cranial vault. The latter are pale yellow/white and indicate a very robust individual. Maximum vault thickness is 12.5 mm although it is 14 mm at asterion where the inner and outer bone tables are 2 and 4.6 mm thick, respectively. The few pieces with cranial sutures show advanced closure with the endo-cranial sutures obliterated and the ecto-cranial suture partially closed. Inferiorly, some of the mastoid air cells have spread extensively from the immediate area of the mastoid which is missing. The robust nature of these remains indicates a male individual of mature years.

WLH 57, Site: WOC 1, Lake Mungo Found: 1980 Sex: Undetermined

WLH 57 is represented by only five small pieces of cranial vault. Only the outer table remains on all of them. All pieces are highly mineralised.

WLH 58, Site: MA 1, Lake Mungo Found: 1979 Sex: Undetermined

This individual consists of only very small, poorly preserved scraps of cranial and postcranial bone. The former have a maximum thickness of 10 mm. All bones are covered in a yellowish-orange carbonate wash.

WLH 59, Site: WOC 1, Lake Mungo Found: 1980 Sex: Female

This individual consists of eroded and highly fragmented postcranial bone. Many are completely or partially covered in an extremely hard, thick, light grey calcium carbonate. Bone Identification was almost impossible because of the covering, but removal would have caused them to disintegrate because the bone was shattered within the hard setting. The only cranial remains consist of four tooth crowns and two very small pieces of vault. The latter have a maximum thickness of 7 mm. The crowns are from the upper right and lower left first and second molars. The lack or wear on the cusps could suggest a sub-adult probably in their middle-late teens. However, a section of the left ilium contains a part of the acetabulum. The maximum diameter is 39 mm most likely indicating a female person. A small section of the sciatic notch has a wide angle which indicates a female possibly a young adult. The right side of the anterior arch of the atlas includes both the superior and inferior articular facets.

WLH 63, Site: LW4, Lake Leaghur Found: 1981 Sex. Male?

This individual is represented by one large section of the cranial vault. Burning has given the bone a variable light to dark-grey colouration, with orange-yellow patches on the external surface. The piece is an antero-medial section of the frontal bone. It has a robust morphology, reflected particularly in its thickness, which is in excess of 12.5 mm. The average thickness of the inner and outer cranial tables is 3 and 5 mm, respectively. Maximum outer table thickness is 7.5 mm, which is most unusual for individuals in this series and indeed more recent human crania generally. On its internal surface there is a posterior section of the frontal crest. This was not sharp, but rather wide. Some pits in the inner cranial table on each side of the crest were probably caused by penetration by arachnoid villi. Such prominent arachnoid pits are usually associated with age, so it is likely that this individual was an older adult, perhaps over 55 years old. The robust vault suggests a male.

There are two shallow oval depressed fractures the largest lies at about mid-frontal, with the long axis in the sagittal plane. It measures 36 mm long and about 14 mm wide. The second fracture is situated to the right and slightly posterior of the first. This is orientated also in the sagittal plane and measures

21.3 mm long by 11.5 mm wide. Close inspection reveals that only the outer table has been affected and there are no signs of stellate fracturing that can occur in this type of trauma.

WLH 64, Site LW4, Lake Leaghur Found: 1981 Sex: Undetermined

WLH 64 consists of many highly fragmented and eroded cranial and postcranial pieces. The maximum thickness of vault pieces is 9 mm and the inner and outer cranial tables are very thin. The distal third of the diaphysis of the left humerus is the largest piece of postcranial remains and is small and delicate with a minimum circumference of **54 mm.**

WLH 65, Site: ML 3, Lake Mungo Found: 1977 Sex: Undetermined

WLH 65 is represented by one small, highly mineralised fragment from a femoral shaft. Cortical thickness is 8 mm.

WLH 66, Site: WCW 6, Willandra Channel Found: 1977 Sex: Undetermined

These extremely fragmented remains are covered in a heavy, dark brown sand or carbonate. Some kind of hardener has been applied to the surface of all bones.

WLH 67, Site: GL5, Lake Garnpung Found: 1982 Sex: Male

WLH 67 is represented by bone from most parts of the skeleton. There are major sections of the long bones, metacarpals and cranium but a lot of bone is fragmented and friable with a substantial covering of pale brown calcium carbonate.

The cranium is eroded, resulting in extensive loss of surface bone including parts of the inner cranial table. It is dolichocephalic, with marked sagittal keeling and a gracile morphology. The vault is very thin, measuring 9 mm maximum and the inner and outer tables are <1 and 1.5 mm thick, respectively. The superior nuchal line is well developed in the form of a crest or small torus. Below are moderately rugged muscle markings placing these features between male and female robust people. There is a distinct occipital 'bunning' covering almost the whole area superior to the nuchal crest. The superior occipital fossae are moderately deep, but the internal occipital protuberance and the transverse lateral and sagittal sulci are poorly developed. Almost all the left side of the vault is missing, together with a large section of the right. All sutures are fused and obliterated. This tends to indicate an individual >55 years.

Only a comparatively small portion of the right side of the frontal remains. The right supra-orbital region is very gracile, with little or no development of the zygomatic trigone or superciliary ridge. The superior orbital margin is sharp, but the bone here does not extend as far as the supra-orbital notch, so preventing assessment of medial structures. There is moderate postorbital constriction, but the temporal crest is very short and there is no trace of a temporal line.

There is little development of the malar crest and tuberosity and the body of the malar is rather gracile, with minimal thickening through the temporal surface to the orbital margin. Viewing the cranium in *norma lateralis*, two things are immediately apparent. There is little development of the supra-orbital region and no demarcation between it and the slope of the frontal. There is a distinct slope to and flattening of the frontal bone both in the sagittal and the coronal planes.

An inferior section of the left temporal and glenoid fossa shows extreme osteo-arthritic changes to the fossa and articular tubercle. The surfaces of both these features have been badly eroded altering their normal morphology. Such degenerative processes are usually caused by excessive stress on the temporo-mandibular joint over a long time, such as that derived from heavy chewing (see WLH 44).

The mandible fragment includes both coronoid processes, the right mandibular condyle and both gonial angles. The latter have very pronounced markings of muscle insertion medially and laterally, suggesting vigorous use of the medial pterygoid and masseter muscles. This supports the conclusion reached above concerning the condition of the left glenoid fossa and chewing stresses. Strong and consistent exercise of the medial pterygoid and masseter muscles are associated with extreme and long-term heavy chewing and other sorts of work. The stresses induced by these activities have probably been primarily responsible for the marked eversion of the inferior border of the gonial angle.

Right Femur

All bone superior to the gluteal tuberosity and inferior to the popliteal surface has been lost. Although a thin layer of carbonate covers about half of the diaphyseal surface, most of it is situated posteriorly and does not affect to any great extent the measurements taken. The gluteal tuberosity is deep and the *Linea aspera* is sharply defined.

Left Femur

This bone is in a similar condition to the right femur, but the diaphysis is shorter. The thickness of the cortical bone, medial to the gluteal tuberosity, is 9 mm.

Right Tibia

Both the proximal and distal ends of this bone are missing. It is robustly built, with a well-defined inter-osseous border. Carbonate covers about 70% of the surface, but there is no obvious sign of pathology.

Left Tibia

Only part of the proximal third and distal half of the diaphysis is present. No measurements are possible.

Right Humerus

This bone consists of a section of the diaphysis from the inferior point of the inter-tubercular sulcus to just above the olecranon fossa. There is marked development of the deltoid tuberosity, indicating heavy use of the right shoulder girdle and this, perhaps, gives an indication of handedness. A layer of brown calcium carbonate covers about 80% of the bone surface.

Left Humerus

The condition of this bone is similar to that of the right humerus, but it is slightly shorter. The deltoid tuberosity is not developed to the same extent, which supports the suggestion of right-handedness made above.

OTHER BONE

There seems to be a distortion of the diaphysis of a right fourth metacarpal. It appears to be a well-healed oblique fracture resulting in a slightly shortened bone and a 'sharpened' dorsal surface. The bone has not been X-rayed, which would help to identify the nature of the trauma. There are several phalanges from both hands among the remains.

Although the cranium is generally gracile, the postcranial remains are not. Nevertheless, the development of the nuchal crest, coupled with well-defined sites of muscular attachment on the long bones and mandible, cortical bone thickness and general robustness of the long bones, suggests that this is a male person. The height, if correctly estimated, is another factor which supports this conclusion.

Besides the longitudinal and transverse flattening of the frontal bone, there is a distinct pre-bregmatic bulge and an obvious superior angulation of the basi-occipital region on the cranium. In combination, these characteristics are those proposed by Brown (1982) as constituting evidence for head-binding in the Late Pleistocene Coobool Creek remains from the Murray River.

Metrical data (mm)

Cranial thickness

Frontal bosses	5
Vertex (near)	9
Right parietal boss (near)	5
Right asterion	7
Inion	14

Brow thickness

Middle	7
Lateral	9
Biasterion	110

Left femur

Actual length	299
Estimated length	504

Subtrochanteric diameter

Antero-posterior	28.2
Medio-lateral	36.0
Mid-shaft diameter	
Antero-posterior	2.0
Medio-lateral	30.0
Platymeric index	77.8
Right femur	
Actual length	360
Estimated length	504
Stature – if male	177 cm
Subtrochanteric diameter	
Antero-posterior	27.0
Medio-lateral	6.0
Mid-shaft diameter	
Antero-posterior	31.0
Medio-lateral	29.0
Shaft circumference	91.1
Platymeric index	75
Robusticity index	11.9
Right tibia	
Actual length	320
Estimated length	400
Stature – male	178 cm
Diameter at nutrient foramen	
Antero-posterior	42.0
Medio-lateral	23.2
Mid-shaft diameter	
Antero-posterior	36.0
Medio-lateral	24.0
Platycnemic index	54.8
Left humerus	
Actual length	215
Mid-shaft diameter	
Antero-posterior	20.0
Medio-lateral	18.0
Minimum circumference	56.6
Right humerus	
Actual length	241
Estimated length	361
Stature – if male	180 cm
Mid-shaft diameter	
Antero-posterior	23.0
Medio-lateral	20,0
Minimum circumference	62.8
Robusticity index	17.5

WLH 68, Site: GG25, Lake Gogolo Found: 1978 Sex: Female

This individual consists only of cranial pieces. The largest pieces are an anterior part of the frontal and a section of the right parietal. Other fragments are from various parts of the calvarium, mainly the temporal surface of the parietal and posterior part of the frontal. Bone colouration ranges from grey to blue-black and some parts of the outer cranial table have patches of calcination. Fragmentation and lengthy burning seem to be the only possible explanation for this bone charring. Because of its condition and its gracile morphology, this individual is very like WLH 1.

The bone is exceptionally gracile with an extremely thin vault and oval cranium giving the impression of a very small and delicate head. The large section of frontal extends from just posterior to but inclusive of the central section of the supra-orbital region, to a point in the middle of the right orbit, about 7.0 mm lateral of the supra-orbital notch, and to a point 110 mm lateral of the left supra-orbital notch, a total width of 68.0 mm. Both notches are very narrow, much narrower than usual in this skeletal series, and shallow. Missing bone from the superior orbital margin, however, gives the impression of shallower supra-orbital notches than may have been the case. Small portions of the right and left orbital plates of the frontal remain, extending posteriorly 9 and 10.5 mm, respectively, and are smooth, concave. Both orbital margins are damaged and the outer bone table removed. Nevertheless, it is easy to see that there was little, if any, supra-orbital development. In fact, there is no sign of any kind of brow ridge or structure to support superciliary ridges.

The surface table of the glabella has been lost revealing the cancellous bone beneath. It seems from the general gracile morphology of the glabella-nasal area that there was little or no development of this region. Inferiorly, there are two small openings in the roof of the ethmoidal sinuses. They lie either side and anterior of the frontal crest. An X-ray of the frontal shows no trace of a frontal sinus superior to the ethmoid sinuses and inter-orbital width is estimated at 24 mm. The medial margin of the left orbit is smooth and rounded. The frontal bone rises smoothly from the slender brow without any demarcation between the two. The maximum thickness of the frontal is 8 mm, with a range of 4–8 mm. There is inner and outer table on most of the frontal, but they are thin (about 1–2 mm thick). Internally, the frontal crest is prominent, long (32.7 mm) and sharp. The internal bone table is liberally marked with small sulci for the ascending veins of the sagittal sinus. The sagittal sinus sulcus is barely visible at the posterior end of the frontal crest. The vault is thickened at this point, with the bulk consisting of diploe. The right parietal consists of a medial section 82.0 mm long and 59.7 mm wide. The thin cranial vault, full frontal and smooth supra-orbital region are especially interesting. Grooves for the branches of the middle meningeal artery are prominent on the internal surface of the parietal and both the inner and outer bone tables are smooth and very thin (<1 mm). The maximum thickness of the right parietal is 6 mm and the inner and outer tables are similar in their morphology and size to those of the frontal.

There is little indication of advanced sagittal suture although there is some indication of preliminary fusion has begun along the medial border from a point just posterior to bregma to approximately obelion. Suture fusion usually suggests an adult person and originally, I believed that was the case perhaps an adult between 20 and 25 years. However, its general gracility is extraordinary for an adult and after reassessment I now believe it is a sub-adult in their teens. The gracile features of WLH 68 reflect those of WLH 1, but are even more gracile and both individuals were cremated. If the age is correct then this is possibly the oldest example of a sub-adult cremation anywhere in the world testifying that this type of ceremonial disposal being extended to sub-adults and possibly children.

Metrical data (mm)
Cranial thickness

Frontal range	4–7000
Bregma (near)	4
Vertex	5
Obelion	6
Right parietal boss(near)	4
Minimum thickness	3
Right tibia	
Actual length	190
Mid-shaft diameter	
Antero-posterior	42
Medio-lateral	23
Platycnemic Index	67.5

WLH 69, Site: LP1, Lake Leaghur Found: 1978 Sex: Male

A limited number of cranial and postcranial remains are associated with this individual. The former consist of two sections of cranial vault and some smaller fragments. The grey colour of the bone is similar to that of WLH 50. The general morphology of this individual is robust and in complete contrast to WLH 68. The robustness of development of the supra-orbital region is akin to that of WLH 50. The larger of the two vault sections comes from the right anterior frontal bone. It includes a large part of the right superior orbital margin from and including the right zygomatic trigone to a point just left of the supra-glabella fossa. The glabella is missing, but enough bone remains to indicate the presence of the shallow fossa above. There is an oblique break from this fossa to a point high on the medial wall of the orbit. From here the superciliary ridge arches in a supero-lateral direction, forming a large, thick, partial torus. It continues laterally for 37 mm, until it disappears at a point above the supra-orbital notch. The notch itself is shallow but wide and looks like a dent in the rounded superior orbital margin rather than a non-metric feature. A small section of the superior orbital margin is missing, but laterally there is a prominent zygomatic trigone accentuated by a flared zygomatic process. The sharp lateral extension of the latter highlights a marked postorbital constriction, which is emphasised by a sharply angled temporal surface that descends into the temporal fossa from a well-developed temporal crest. The angle at this point is 123° (WLH 50 is 126°). Emphasis of the demarcation between the frontal proper and the temporal fossa is made by the marked development of the temporal crest, which continues for a short distance as a strong temporal line.

The vault is evenly thickened, with a maximum thickness of 10 mm. The outer cranial table is thicker than that of many individuals (3.0 mm), but the inner table is only 1.0 mm. The frontal rises smoothly from the rear of the superciliary ridge and the lateral section of the supra-orbital region, with only minimal demarcation between them. No distinct ophryonic groove is present. The right frontal boss can be distinguished on the outer surface of the frontal bone. Apart from this feature, however, the frontal is rather smooth and flat, though maintaining a slight rounding of its surfaces when viewed in *norma frontalis* and *norma lateralis*. A triangular-shaped section of the left parietal reflects the basic morphology of the frontal bone. The only feature on this bone is part of the temporal line. Another small piece of vault has a section of suture along one side. Suture fusion is advanced and indicates a mature age.

Except for a few small fragments, the postcranial remains of WLH 69 consist only of the proximal half of the right tibial diaphysis, less all bone proximal to and including the tibial tuberosity, and the left first distal phalange. The tibia is robustly constructed, with a cortical bone thickness through its anterior border of 13 mm. The remains of a single Harris line can be seen lying transversely on the internal surface, at the proximal end of the posterior tibial wall. From the position of this line, it would seem that the metabolic stress event causing its formation occurred between 3 and 6 years of age.

The robust morphology of this undoubted male stands out in stark contrast to other individuals in the series, not only the females WLH 1, 11, 44 and 68, but also the males WLH 3,18, 22 and 101.

Metrical data (mm)
Cranial thickness

Frontal boss	10.0
Mid-frontal (at crest)	12.5
Brow thickness	
Medial	19.0
Middle	13.0
Lateral	12.0
Craniometrics	
Minimum post-orbital breadth	128
Minimum supra-orbital breadth	138

WLH 72, Site: ME1, Lake Mulurulu Found: 1974 Sex: Female

The exact site at which this individual was found cannot be accurately traced. It has been tentatively assigned to Mulurulu East 1 on such information as is available.

Most of the bone is highly fragmented and fragile and have are extensively eroded. A pale yellowish orange carbonate wash covers the remains and beneath there is extensive blue-black manganese staining. Cranial material includes some sections of the vault, the left temporal bone and most of the upper and lower dentition. There are large parts of the postcranial skeleton, including sections of the right and left femora, tibiae and humeri, as well as small parts of the right ulna, right radius and left fibula and some phalanges.

The lateral part of the frontal extends from the brow almost to the coronal suture. The vault itself is moderately thick, with a maximum thickness of 11.0 mm. Both inner and outer cranial tables are thin, measuring 1.0 and 2.0 mm, respectively. The superior orbital margin is sharp and there is little sign of a postorbital sulcus or groove. The anterior frontal bone rises smoothly and almost vertically from behind the superciliary ridge. The temporal crest is short and rapidly becoming a temporal line that is also short and postorbital constriction is minimal. On the occipital, the superior nuchal line not developed and the nuchal plane has few muscle markings. The right temporal has a deep glenoid fossa, with slight osteoarthritic pitting of the medial side of the articular tubercle. The supra-meatal pit lies posterior to the external auditory meatus. There is no supra-mastoid crest as such beyond the root of the zygomatic process. Mastoid air cells, which can be observed on the broken internal surface, extend to a position corresponding to that of the supra-mastoid crest and along the posterior wall of the auditory canal. The various features described here indicate a female.

Both the upper and lower dentitions show advanced, even attrition of all occlusal surfaces, exposing large areas of dentine and eliminating almost all surface enamel. No periodontal disease is visible, although much of the alveolar bone is missing.

The right femur lacks both proximal and distal ends. Thin cortical bone in the diaphysis, has contributed to the friable nature of the remains. Muscle markings are negligible. The left femur consists only of a distal section of the diaphysis and part of the lateral surface superiorly. Both tibiae are fragmentary and only represented by a small section of their respective diaphyses. They are thin and, as with the femora, have succumbed to the vagaries of interment.

Humeri consist only of the bone inferior to and including the deltoid tuberosity, minus the distal articular joint. They are slim, typically female, with thin cortices and little development of the deltoid tuberosities. The difference between the left and right shafts suggests a right-handed person. There are 11 phalanges and part of the shaft of the left fibula. Irregularity of the shaft and changes to its general morphology suggests it had undergone some form of trauma. The break was partial in the pattern of a greenstick fracture. Subsequent healing has been good, attesting to the general good health of the individual, but callus formation has tended to thicken the shaft. Irregular sub-periosteal bone apposition has formed areas of thickened cortex, giving the trauma an unusual appearance.

Metrical data (mm)
Cranial thickness

Right frontal boss	10
Right asterion (near)	11

Brow thickness

Medial	11
Middle	6
Right femur	
Actual length	340
Estimated length	450
Stature	162 cm
Sub-trochanteric diameter	
Antero-posterior	27
Medio-lateral	31
Mid-shaft diameter	
Antero-posterior	31
Medio-lateral	26
Platymeric index	87.1
Robusticity index	12.7
Right tibia	
Actual length	240
Mid-shaft diameter	
Antero-posterior	37
Medio-lateral	23
Platycnemic index	62.2
Left humerus	
Actual length	226
Mid-shaft diameter	
Antero-posterior	19
Medio-lateral	21
Minimum circumference	56.6
Right humerus	
Actual length	230
Estimated length	305
Stature	159 cm
Mid-shaft diameter	
Antero-posterior	23
Medio-lateral	21
Minimum circumference	56.6
Robusticity index	18.6

Dental data (mm) for WLH 72

	Left			Right		
	L-L	B-L	M-D	L-L	B-L	M-D
Maxillary						
I1	8.0	–	7.2	7.8	–	7.0
I2	7.0	–	6.0	7.7	–	6.0
C	9.2	–	6.1	7.4	–	7.0
Pm1	–	9.1	5.0	–	10.0	5.0
Pm2	–	10.0	5.0	–	9.3	–
M1	–	13.1	9.0	–	12.8	9.0
M2	–	12.1	10.8	–	13.1	9.4
M3	–	11.8	9.7	–	11.9	9.4
Mandibular						
I2	8.9	–	5.0	7.2	–	5.1
C	–	–	6.9	8.9	–	7.6

	Left			Right		
	L-L	B-L	M-D	L-L	B-L	M-D
Pm1	–	8.5	5.8	–	8.9	6.8
Pm2	–	8.9	7.0	–	10.3	8.0
M1	–	10.5	10.9	–	12.2	10.9
M2	–	10.0	10.5	–	12.1	11.9
M3	–	12.0	12.4	–	12.0	12.2

WLH 73, Site: ME2, Lake Mulurulu Found: 1974 Sex: Female

Most of these remains are highly fragmented, extremely eroded and weathered and display extensive blue-black manganese staining. Cranial and postcranial bone is represented, but the extent of fragmentation and the deeply etched surface abrasion render the postcranial remains of poor quality.

Cranial material consists of pieces from all major vault bones, the left maxilla and parts of the mandible. Most of the dentition is badly eroded and broken, so cannot be measured effectively. Almost all tooth crowns and sections of teeth are broken away to the cemento-enamel junction. Nevertheless, the occlusal surfaces of one or two teeth indicate that moderate attrition has taken place.

An antero-lateral piece of the left frontal bone includes part of the supra-orbital region. The zygomatic trigone is not well developed, but there is a prominent superciliary ridge. The left supra-orbital area extends just medial of the very wide and moderately deep supra-orbital notch, where a break has exposed part of the lateral wall of a small left frontal sinus. The superior orbital margin is not sharp, but neither is it thickened. A small anterior section of the right supra-orbital repeats the morphology of the left in all but the development of the superciliary ridge, which is somewhat smaller. A slight ophryonic groove or sulcus demarcates the supra-orbital region from the frontal bone. The temporal crest is almost non-existent, but erosion in this area may have reduced some original prominence. The temporal line is well developed and strongly demarcates the angulation (approx. 125°) between the frontal bone and the temporal surface.

The frontal bone rises smoothly, producing a rather rounded appearance when viewed in both *norma lateralis* and *norma frontalis*. The vault itself is uniformly thin with a maximum thickness of 8.0 mm and thin inner and outer cranial tables, 1.0 and 1.5 mm, respectively. The inner cranial table has been completely eroded away in several places. The little that remains of the frontal crest suggests that this feature was high, with a broad base and a sharp border. An estimate of the minimum postorbital breadth (approx. 94 0 mm) was established by measuring the distance between the temporal crest, just posterior to the zygomatic process, and the frontal crest and then doubling the resultant figure. However, because of inaccuracies inherent in this method, the measurement has not been included elsewhere.

Both mastoid regions show a confined spread of mastoid air cells. The left digastric fossa is deep and rises towards the occipital border. The glenoid fossae are deep, with transversely oriented, narrow, oval sockets for secure articulation of the mandibular condyle. A remaining section of maxilla indicates that the palate was shallow. A piece of the maxillary alveolar bone contains the roots of all the teeth from and inclusive of the lateral incisor to the first molar. A small mesial section of the mandibular body at the mental trigone contains the roots of all teeth from the right canine to and inclusive of the left lateral incisor with all crowns missing. The roots of all the left and right molars remain in two superior sections of the mandibular body.

The light structure of the supra-orbital region, the thin vault and the minimally developed zygomatic trigone suggest a female person. Molar attrition indicates she was fully adult, probably in her 40s.

Metrical data (mm)
Cranial thickness

Left frontal boss	7
Bregma	8
Obelion	8
Lambda	6
Parietal bosses (near)	6
Left asterion	8
Brow thickness	
Medial	14
Middle	7
Lateral	8
Mandible	
Symphyseal height	30
Symphyseal breadth	14
Body height	27
Body breadth	14

WLH 74, Site: ME 3, Lake Garnpung Found: 1977 Sex: Undetermined

A portion of the left mastoid area is all that remains of this individual. It was found with WLH 130. There is a part of the supra-mastoid crest and one large mastoid foramen situated posteriorly near the occipital border. The individual is probably fully adult.

WLH 75, Site GL25, Lake Garnpung Found: 1980 Sex: Undetermined

WLH 75 consists only of eroded fragments of cranium, probably recent in origin. None of the pieces is very thick, the thickest being about 9.0 mm. The individual is fully adult, with a medium-sized mastoid process and a moderately developed right superciliary ridge. On the small section of superior orbital margin that remains, there is a wide, but shallow, supra-orbital notch.

WLH 76–91

These individuals were not collected although their locations were recorded.

The following description of five individuals has been taken from the original report (Dowling et al., 1995) with minor editing. The bones were not photographed or collected.

WLH 92

These remains were found on the flood plain south-east of Lake Pan Ban (WCL2). They consisted of a section of the upper left mandible, three molars and the inferior one-third of the ascending ramus. Ten fragments of post cranial skeleton, probably from the lower limbs, were found scattered nearby. All bones were heavily weathered and extremely friable and tended to disintegrate with handling making close examination in the field difficult. All three molars had erupted and each showed extreme surface wear. These features point to a mature (over 40 years) person at time of death. The mandibular section was partly buried and more skeletal material may lie under the soil surface. No attempts were made to locate any further remains and after observation the exposed portion of the mandible was covered by scraping surrounding soil over it.

WLH 93

This individual was found on top of a residual mound in a gully reading into the south western shoreline of Lake Baymore (BW1). The remains consisted of several fragments (5 mm thick) from an undetermined area of the cranial vault and portions of the ulna and tibia. These fragments showed signs of having been burnt. It is thought that they may then be from a cremated individual with the bones having been smashed before or after being burnt. No reliable archaeological or stratigraphic feature was present to aid in dating these remains apart from a Lens of fresh water shell' located 50 m to the south. A minimum date from this lens could be expected to be Late Pleistocene, around 15–14,000 ka, during a period of final high water through the Willandra system. However, the association between the remains and the shell was not established.

WLH 94

This bone was located on the surface of a blowout on the north-western shore of Lake Baymore 1 km south of the Willandra Creek entrance (BW3) to the lake. Nine cranial fragments were identified. At least one of the fragments was from the sphenoid bone; the remainder were from the vault. Manganese staining was noticed on at least three of the vault fragments which would suggest some antiquity for this individual. Such staining has been noticed particularly in remains thought to be Late Pleistocene. The fragments appeared to have been eroded from the rim of a blowout and washed down-slope towards the shoreline. If so, then more skeletal material may be located in the more stable area above the rim of the blowout.

WLH 95

Approximately 50 skeletal fragments were found on non-stratified red sand deposits on the northern end of the Lake Garnpung lunette. None of the fragments were greater than 5–6 cm long making identification specific identification difficult. However, five tibial, three femoral and eight cranial fragments were identified. All the fragments were heavily mineralised and scattered in a circular fashion on the sandy surface of the lunette. The pattern suggested the individual was buried before the lake emptied and the lunette began to deflate.

WLH 96

These remains were also found on the Lake Garnpung lunette 200 m northwest of the WLH95 remains but 20 m higher on the lunette. At least 60–70 fragments were found representing the cranium, tibia and femur and other long bones, scattered on the surface. The condition and pattern of the fragments were identical to WLH95 suggesting similar deposition with exposure of the remains after deflation of the lunette.

WLH 97, Site LW 10

Not collected or studied.

WLH 98, Site: GL?, Lake Garnpung Found: ? Sex: Undetermined

Only a few very small mineralised cranial and postcranial fragments remain of WLH 98. They are cranial remains from the vault and have a maximum thickness of 8.0 mm. The inner and outer tables are each 2.0 mm thick. The postcranial remains cannot be identified to any particular bone.

WLH 99, Site: GL?, Lake Garnpung Found: ? Sex: Undetermined

The bone is highly mineralised and has a heavy covering of yellowish-brown carbonate. There is one piece of vault (9 0 mm thick), two very small vault fragments and the left body of the malar. Postcranial remains are mostly unidentified scraps. Some, however, originate from an anterior portion of the left tibial diaphysis and both femora. The largest piece is the proximal third of the diaphysis of the right femur. It shows a rather small bone with a 6.0 mm-thick cortex, a small gluteal tuberosity and low *linea aspera*.

Metrical data (mm)
Right femur
Subtrochanteric diameter
Antero-posterior	25.1
Medio-lateral	26.0

WLH 100, Site GG 16, Lake Gogolo Found: 1977 Sex: Female

WLH 100 consists of cranial remains only. They include the posterior half of the right parietal, the superior right half of the occipital, a portion of the right frontal bone, a superior portion of the right orbit, the right mandibular fossa and a small piece of the left supra-mastoid region. There is fairly strong and extensive manganese staining on all external surfaces, but almost none internally. Some small deposits of calcium carbonate can be seen, together with a silica sheen noted for others. All the remains of WLH 100 are completely mineralised.

Because of the complete obliteration of both the sagittal and lambdoid sutures, there is some difficulty in determining the boundaries and size of the right parietal. Tracing the route of the superior sagittal sulcus, however, reveals a small posteromedial section of the left temporal. The whole bone has a rather full, rounded appearance, with no indication of a temporal line. There is a slight, supero-medially directed temporal angle when the piece is viewed in *norma occipitalis*, which suggests that the greatest width of the cranium was low down on the temporal. A pre-lamdoid depression occurs above a slight bulge in the superior occipital plane, that slightly accentuates. The bulge is obvious when viewed in *norma lateralis*, but it does not constitute an occipital 'bun'. There is an angulation to the occipital plane inferiorly that tends to flatten the bone between the bulge and the small remaining section of the superior nuchal line which is neither torus nor crest. There is an interesting feature in the lateral section of the right lambdoid suture, just superior to asterion, which seems to have been caused by an overlap of the occipital bone by the parietal, perhaps as a continuation or consequence of the minimally developed angular torus that sits superiorly. As a result, a prominent fold now lies on top of and in the same direction as the suture. The maximum thickness of the occipital directly inferior to the fold is 15.7 mm. The maximum thickness on the parietal side, through the fold, is 14.9 mm. The maximum thickness of the parietal elsewhere is 11.0 mm.

Although remains of the superior nuchal line are confined to a small laterally placed section, measurement through this at the position of the lateral sulcus is 19.0 mm. Internally, the superior occipital fossae are very shallow, particularly when compared to WLH 19 and other individuals in the series. The superior sagittal sulcus is broad, but less well defined than in others. A 10.5-mm-diameter hole sits in the right superior occipital fossa, but there is nothing to suggest a pathological origin for it.

The section of frontal lacks most of the temporal surface and medial portions of the bone. The vault is thick, with a maximum thickness of 13.5 mm, and so too are the inner and outer cranial tables (2 0 and 5.0 mm, respectively). A frontal boss can be distinguished, which gives a slightly rounded appearance to the bone. Laterally, the temporal crest is prominent and demarcates the sharp angle (132°) between the surface of the frontal and the temporal fossa. Postorbital constriction is marked.

The right supra-orbital area includes most of the orbital plate of the frontal. It extends 44 mm medially from the fronto-zygomatic suture to a point medial to the supra-orbital foramen. The latter is, in fact, a double fossa, with a thin bony bridge placed medio-laterally slightly inside the entrance to the foramen. It exits as one opening on the orbital plate. The zygomatic trigone is small, but there is some development of the superciliary ridge but only half of this structure remains. There is no apparent supra-orbital notch. The superior orbital margin is rounded but lacks any significant robust development, in keeping with the general character of the brow. The lateral wall of a small sinus is present on the medial edge of the orbital margin, indicating a centrally located frontal sinus.

The minimal development of the supra-orbital region is consistent with this individual being female, while the robust construction of the vault is in keeping with a male. I propose that we are dealing with a robust female, similar to WLH 45.

WLH 101, Site GG 16, Lake Gogolo Found: 1977 Sex: Male

This individual is represented by a large section of left frontal bone and parts of both temporals around the glenoid fossae. All bones are highly mineralised with a dark manganese stain covering internal surfaces while outer surfaces are almost white and shiny. The frontal section extends from the supra-orbital border to the coronal suture. Most of the right side is missing, together with the left temporal surface, the nasal area and the right orbit. The frontal bone is rather full and rounded with marked sagittal keeling. The vault is thick (13.0 mm) with inner and outer cranial tables of 2 and 5 mm thick, respectively. The supra-orbital area is moderately developed with a prominent left superciliary ridge and zygomatic trigone suggesting a male individual. A small depression lies between the supra-orbital and frontal. Remnant lateral and posterior walls of the frontal sinus indicate this feature was small. The superior orbital margin is rounded and a large supra-orbital foramen lies lateral of the medial corner of the orbit.

Temporal bones are represented by portions above and including the glenoid fossae. The fossae are broad, deep and roughly triangular, although not to the extent found in WLH 100. Reduction in fossa depth has been caused by an advanced, bilateral osteoarthritic condition, which has deposited new bone across the articular surfaces of both fossae and then was eroded probably by the same repetitive mechanical stresses that caused the initial condition. Osteoarthritic pitting is present on the articular tubercle. The anterior half of the fossa has been substantially altered following wear of the articular surface. The left articular tubercle has been almost destroyed by loss of articular cartilage and degenerative wear reducing its height and flattening it almost to the level of the fossa floor. There is an impression of the mandibular condyle on the tubercle suggesting it had repositioned itself anteriorly. There is pitting in the glenoid fossa, together with minor osteophytosis, although this has been worn away by interment erosion. Osteoarthritic degeneration involves the root of the left zygomatic process caused by lateral excursion of the mandibular condyle following loss of TMJ integrity. In turn, that may have emerged from the raising of the glenoid fossa surface and flattening of the articular tubercle, allowing lateral movement of the condyle within the resulting wider plane of articulation. The wear pattern suggests long-term bilateral stress on the jaw possibly because of manufacturing vegetable string using the anterior tooth complex as seen in WLH 3.

Behind the posterior wall of the glenoid fossa lies the roof of the external auditory meatus. The notch of Rivinus (*incisura tympanica*) is positioned posteriorly. A well-defined supra-mastoid crest lies above the ear, almost as a continuation of the superior crest of the zygomatic arch, running posteriorly and slightly superiorly to the occipital border, close to the parietal notch. This crest is not so clear on the right but there the bone is missing. On the left, there is an extensive area of large mastoid air cells adjacent to the mastoid. Smaller cells extend posterior to the mastoid, near the occipital border, and run anteriorly as far as the external auditory meatus.

On the right side of the cerebral surface of the cranium is the lateral half of the *foramen spinosum* medial to the glenoid fossa and placed in a small section of the greater wing of sphenoid. Although the bone on each side of the foramen is eroded, it is likely that it was not fully enclosed. The progress of the middle meningeal artery, conveyed by the *foramen spinosum,* can also be traced on the cerebral surface. After curving forwards, the groove for this vessel runs antero-laterally on the anterior part of the squamous temporal. A posterior or parietal branch can be traced from close to the foramen. Both grooves end at the broken margins of the temporal bone.

WLH 102, Site GG16, Lake Gogolo Found: 1977 Sex: Male

All bone is evenly covered in a brown-orange carbonate wash and is poorly preserved. Remains consist of teeth, portions of the mandible, a right malar, two parts of the petrous portion of the temporal and small pieces of vault. Only two small pieces of cranial vault remain. The thickest of these is 12 mm and originates on the posterior temporal surface of the left parietal. There are two small pieces of vault and two sections of the petrous portion of the right temporal that show the internal auditory meatus and the superior petrosal crest with the cochlear canal running inferiorly. The superior petrosal sulcus is shallow and wide.

The malar is robust, but much of the inferior border and all the temporal process are missing. The malar tuberosity is large and the body of the malar is thick through the temporal surface to the orbital margin. Much of the bone surface is weathered and eroded pitting the bone surface. The orbital margin is smooth and well rounded. The marginal process is prominent and similar to those of other individuals with rugged malars. The few diagnostic features present suggest a male person.

All teeth are in a poor condition and heavily worn. Occlusal surfaces are worn with some teeth having lost their crowns, others have lost some or all of their roots. There are 42 teeth, of which 31 have been measured. There is a fragment of the right maxilla, with a first molar in situ. The dentition represents at least two adults and there could be four.

A portion of the right mandibular body extends from the genial spines to the sub-maxillary fossa and has the first and second molars in situ. The superior body surface is missing, including all teeth and the alveolar borders. A large mental foramen lies on the lateral surface, together with a prominent mental tubercle. The posterior marginal tubercle is prominent and above that is the antero-inferior part of the external oblique line. Posteriorly, all bone has been lost. On the internal or medial surface the digastric and sublingual fossae and the anterior section of the sub-maxillary fossa are well defined. The area of the mental spines is developed to the stage of a tubercle with a lateral ridge extending to below the sublingual fossa. External surfaces are stained with patchy brownish orange carbonate and manganese spotting.

Dental Measurements (mm)

	Left			Right		
	L-L	B-L	M-D	L-L	B-L	M-D
Maxillary						
I1	10.0	–	8.8	–	–	–
I2	9.1	–	7.0	7.1	–	8.4
C	8.6	–	6.1	9.9	–	6.2
Pm1	–	10.0	6.0	–	10.7	6.2
Pm2	–	–	–	–	9.5	5.6
M1	–	12.4	11.4	–	13.3	10.6
M2	–	12.8	12.1	–	–	–
M3	–	–	–	–	13.6	9.4
Mandibular						
I1	–	–	–	–	–	–
I2	6.3	–	5.0	–	–	–
C	8.9	–	7.4	8.2	–	6.9
Pm1	–	–	–	–	9.1	7.7
Pm2	–	8.2	6.8	–	9.2	6.0
M1	–	11.3	12.5	–	12.8	12.0
M2	–	11.7	11.3	–	11.7	11.4
M3	–	–	–	–	12.3	13.5

WLH 103, Site: GG16, Lake Gogolo Found: 1977 Sex: Male

WLH 103 consists of only two pieces of mandible in a similar condition to that of WLH 102. A small section of the left maxilla has the first molar in situ, but only a very small part of the alveolar bone and socket remain. The tip of the bucco-mesial root of the second molar is embedded in the bone. The tooth has been badly eroded and subject to heavy attrition. On the anterior section of the body there is a prominent mental tubercle extending posteriorly into an anterior marginal tubercle. Above is a large, oval-shaped, mental foramen with its anterior edge missing. A posterior marginal tubercle is present and the internal surface has lost a large section of the tubercle for the genial spines. This tubercle extends inferiorly as a *spina interdigastrica*, dividing the deep digastric fossa into left and right sections although most of the right fossa is missing. The only other feature on this very eroded and fragmentary bone is the antero-inferior edge of the shallow sub-maxillary fossa. A piece of the right mandibular body has the first and second molars in situ and the inferior border missing. Both teeth have extensive weathering, causing cracking and enamel loss. The teeth have marked attrition that has removed all cusp features and exposed large areas of dentine.

WLH 104, Site: GG16, Lake Gogolo Found: 1977 Sex: Undetermined

WLH 104 consists of a restricted area of the right side of the mandibular body, at the position of the second and third molars. Major features include a deep *sulcus extramolaris*; the second and third molars in situ; the distal root of the first molar still embedded in the tooth socket; and a wide breadth to the body.

Both teeth are large and have lost some enamel from their buccal surfaces. There is no obvious sign of pathology, but attrition has removed all cusps from the two teeth. This pattern indicates a mature adult who, because of the size of the teeth and the body breadth, was probably male, although gender cannot be confidently assigned. A small section of the right acromial process accompanies the mandible.

Metrical data (mm)
Mandible
Body breadth   17.5
Right mandibular dental measurements (mm)

	BL	**MD**
M2	13.8	14.0
M1	12.8	13.9

WLH 105, Site: GG16, Lake Gogolo Found: 1977 Sex: Undetermined

These remains consist of only a few heavily eroded and fragmented deciduous teeth. The occlusal surfaces are very worn, which indicates a child probably over 5 years of age. It is interesting to note that the attrition so prevalent on adult dentitions is just as severe on deciduous teeth. The diet of children obviously contained similar abrasives to those in the adult diet. While this observation may appear obvious, the diet and nutritional requirements of the human infant and child have yet to be discussed in any comprehensive manner in the context of the Australian Late Pleistocene.

WLH 106, Site: GG16, Lake Gogolo Found: 1977 Sex: Undetermine

Right Femur

The surface of this bone is badly eroded in the mid-shaft area. Heavy brown calcium carbonate covers the surface in many places. Underneath this cover is a brownish-yellow wash and some blue-black manganese staining. The bone itself extends from the superior end of the intertrochanteric line to the popliteal surface. The gluteal tuberosity is deep and well-marked. The spiral line and *linea aspera* are prominent also. Cortical bone is thick (10.2 mm). The general appearance, size and robust nature of the bone indicates a male individual.

Left Femur

The condition of this bone is similar to the right. Only the middle third of the diaphysis remains and this is without any notable features, except for a prominent *linea aspera.*

Right Tibia

This bone has a badly eroded surface, particularly on the lateral side, as well as patches of solid, thickly deposited brown calcium carbonate similar to that seen on the right femur. Much of the antero-proximal portion is missing, together with the proximal and distal ends. The bone extends from the proximal end of the soleal line to a point which I have termed the supra-malleol fossa.

Right Humerus

With the exception of a small part of the trochlear, all the bone below and including the deltoid tuberosity is present. All bones are covered in a brownish-yellow carbonate wash, with small nodules of carbonate adhering to the outer surfaces in some places. Manganese staining is visible below the wash. Articular surfaces at the elbow joint have various osteoarthritic features: erosion of the mid-trochlear ridge and osteophytosis around the lateral. The deltoid tuberosity is well developed, together with the lateral supra-condylar ridge. A marked medial bowing of the diaphysis is prominent. This is not believed to have been caused by artificial or taphonomic processes, either during interment or since exposure. It is, moreover, a feature that can be observed in low frequencies among more recent Aboriginal humeri in larger collections. Bone deformities can be caused by nutritional factors, by way of such diseases as rickets or scurvy. These almost always affect the skeleton bilaterally, which is impossible to establish in this case, because of the lack of sufficient remains of the left humerus. However, the robust nature of the right humerus, including its thick cortex, suggests that dietary factors were not involved.

Left Humerus

Only a 74-mm section of the mid-shaft of this bone remains, which has been affected in the same way as the right humerus by post-depositional processes. Because of the small piece, it is impossible to say whether it displays the same pathological condition as the right humerus.

Right Ulna

This bone consists of a section of the trochlear notch and a large portion of the diaphysis, from just below the ulna tuberosity to just proximal of the narrowest part of the shaft. The inter-osseous, posterior and anterior borders are well defined. The supinator crest is prominent also and becomes a vertical line which runs about a third of the way down the posterior side of the diaphysis. The bone is well built and rather robust. A small patch of osteoarthritic erosion is present on the lateral articular surface of the coronoid process. The latter corresponds to a similar condition on the medial surface of the mid-trochlear ridge of the right humerus, described above.

Right Radius

The right radius extends from the neck to just proximal of the quadratus surface.

Left Radius

Only the medial third of the diaphysis remains.

Left Ulna

Only a small section of the diaphysis remains.

FIBULAE

There are four fragments of the left and right fibulae. A mid-shaft portion of the left fibula has received a complete fracture from which it has not healed. This has left jagged edges on both pieces of bone at their broken margins. There are only minimal signs of healing on the medial side, where new bone has been laid down across the medullary cavity. Because it is kept steadier by the attachment of the inter-osseous border, the medial side of the bone would have more chance to mend. The fracture pattern suggests that the trauma occurred near or close to the time of death.

Metrical data (mm)
Right femur

Actual length	327
Estimated length	485
Stature	173 cm
Subtrochanteric diameter	
Antero-posterior	29.2
Medio-lateral	36.0
Mid-shaft diameter	
Antero-posterior	29.5
Medio-lateral	32.5
Platymeric index	80.6
Robusticity index	13.6

Left femur

Actual length	174
Mid-shaft diameter	
Antero-posterior	32.2
Medio-lateral	31.2
Right tibia	
Actual length	338
Estimated length	388
Stature	171 cm
Diameter at nutrient foramen	
Antero-posterior	37.0
Medio-lateral	23.0
Platycnemic index	62.2
Right humerus	
Actual length	268
Estimated length	340
Stature	173 cm
Mid-shaft diameter	
Antero-posterior	25.5
Medio-lateral	22.8
Actual length	121
Minimum circumference	65.0
Robusticity index	19.1
Right ulna	
Actual length	185
Estimated length	260
Stature	164 cm
Left ulna	
Actual length	75
Right radius	
Actual length	180

WLH 107, Site: GG16, Lake Gogolo Found: 1977 Sex: Male

The remains consist of postcranial bones only. The general condition of the bone is almost exactly like that of WLH 106. Patches of erosion have removed some surface features and there is a brown carbonate wash over most surfaces which is thick in places.

Right Femur

The most striking aspect of both femora is their robustness and, in particular, their cortical bone which ranges from 7.0 mm to 16.0 mm in thickness. The right femur extends from the inferior part of the neck to well down onto the popliteal surface. Both the gluteal tuberosity and the *linea aspera* are prominent, the latter particularly so. Its width is about 8 mm on average, with a maximum of 10.2 mm. The line is very rugose and displays some minor exostoses. These are typical of a strong and well-muscled limb and use of the vasti and adductor muscle groups.

Left Femur

Although this bone extends from just below the lesser trochanter to near the end of the lateral supra-condylar line, little bone remains below a point about two-thirds of the way down the diaphysis. The *linea aspera* is prominent (6–11 mm wide) but inferiorly, just above the division of the lateral and medial supra-condylar lines, it gives the appearance of being pathologically swollen. Bone definition seems to have been lost in this area because of this condition. The maximum thickness of the bone at this point is about 16 mm.

Left Tibia

The development of the cortical bone on the left tibia is like that on the femora. The extraordinary thickening reaches a maximum of 17.0 mm midway down the anterior crest. The right tibia is missing.

Right Ulna

Only the proximal half of the right ulna, without the olecranon and coronoid processes, is present. The radial notch and the articular surface of the trochlear notch have been badly damaged by erosion. The robustness seen in the larger long bones is also reflected in the right ulna. The marked robustness of the femora and the strongly developed areas of muscle attachment suggests a male person.

Metrical data (mm)

Right femur

Actual length	353
Estimated length	473
Stature	170 cm
Sub-trochanteric diameter	
Antero-posterior	29.3
Medio-lateral	33.0
Mid-shaft diameter	
Antero-posterior	32.0
Medio-lateral	29.0
Platymeric index	87.9
Robusticity index	12.9

Left femur

Actual length	346
Sub-trochanteric diameter	
Antero-posterior	29.0
Medio-lateral	34.0
Mid-shaft diameter	
Antero-posterior	34.0
Medio-lateral	29.0
Platymeric index	85.3

Left tibia

Actual length	190

WLH 108, Site: GG16, Lake Gogolo Found: 1977 Sex: Undetermined

The mid-shaft sections of two femora have severe erosion of their external surfaces. In some places this erosion has completely removed the periosteum and sub-periosteal bone. The left femur consists only of the middle third of its diaphysis. The *linea aspera* is well developed, but the extent to which this has taken place is masked by the general poor condition of the bone. The thickness of the cortex through the *linea aspera* is 12.0 mm, which seems too developed for a female. The paucity of the remains, however, prevents a definite assignment of gender. The tibiae are represented only by large fragments of the diaphysis. Surface features have been removed by erosion.

Metrical data (mm)
Right femur
Actual length   300
Mid-shaft diameter
Antero-posterior   35.0
Medio-lateral   25.0
Left femur
Mid-shaft diameter
Antero-posterior   34.0
Medio-lateral   25.0

WLH 109, Site: GG16, Lake Gogolo Found: 1977 Sex: Undetermined

These postcranial remains are so badly eroded that they are some of the poorest in the series.

Femora

There are two heavily eroded pieces of femoral shaft. One is a middle section of the right femur, 240 mm long, the other, 88 mm long, unidentified to exact position.

Tibiae

The sections of left and right tibial diaphyses are in a very poor condition also. There is 225 mm of the right mid-shaft, but only 80 mm of the left.

Fibulae

There are small fragments of the shafts of the left and right fibulae. The condition and limited amount of bone precludes further description.

Humeri

The proximal and distal ends of the right humerus are missing. The remaining bone extends from a point inferior to the bi-occipital groove to the flair of the medial supra-condylar surface. Surface features have been eroded, but the deltoid tuberosity seems to have been long and prominent. The bone is narrow with a thickened cortex reducing the medullary space to a narrow passage approximately 5.0 mm in diameter. The remains of the left humerus are shorter and even more weathered than those of the right but it reflects the same morphology: slim, long, thick cortical bone, narrow medullary cavity.

Right Radius

The right radius has both the head and distal end missing. The inter-osseous border, although sharply defined, is not as sharp as that seen in other examples in the series.

Metrical data (mm)
Right femur
Actual length   240
Left femur
Actual length   88
right tibia
Actual length   184
Mid-shaft diam.
Antero-posterior   35
Medio-lateral   24
Platycnemic index   69
left tibia
Actual length   80
right humerus
Actual length   258
Estimated length   328
Stature   170 cm
Mid-shaft diam.
Antero-posterior   21
Medio-lateral   24
Minimum circumference 67
left humerus
Actual length   184
right radius
Actual length   225
Estimated length   270
Stature   174 cm

WLH 110, Site: GG16, Lake Gogolo Found: 1977 Sex: Male

This individual is represented by humeri only that are in a much better condition than those of WLH 108 and 109. They are heavily mineralised, have almost no trace of carbonate but are covered in manganese spotting over most surfaces.

Right Humerus

The right humerus has both proximal and distal ends missing. The remaining section of shaft extends from the extremely prominent deltoid tuberosity to the lateral and medial supra-condylar ridges and proximal surface of the coronoid fossa. The bone has a robust construction, thick cortices (7–8 mm), a narrow medullary cavity (7 mm), a broad distal end, highlighted by a buttressed flaring of the lateral supra-condylar ridge, and a large, heavily developed trabecular lattice formation within the distal medullary cavity. The bone is in two sections that articulate perfectly. The solid character of the bone strongly indicates a male.

Left Humerus

The left humerus consists of almost the same section of diaphysis as the right. It is somewhat smaller and slimmer with a less rounded distal anterior border, less lateral buttressing and smaller deltoid tuberosity. Those characteristics suggest this person was right handed. Cortical bone thickness ranges from 5 to 6 mm and the medullary cavity diameter averages about 6 mm.

There is a marked contrast in size between left and right humeri. Handedness may account for overall size difference between them. The absence of any obvious pathology suggests only differential use has resulted in the marked size difference.

Metrical data (mm)
Right humerus

Actual length	192
Estimated length	312
Stature	164 cm
Mid-shaft diameter	
Antero-posterior	20.7
Medio-lateral	23.6
Minimum circumference	69
Robusticity index	20.4
Left humerus	
Actual length	188
Mid-shaft diameter	
Antero-posterior	20.5
Medio-lateral	19.0
Minimum circumference	60

WLH 111, Site: GG16, Lake Gogolo Found: 1977 Sex: Undetermined

WLH 111 consists of some small pieces of humeral shaft and an anterior section of the mid-shaft of the (right?) tibia. All are extremely eroded and nothing more can be said about them.

WLH 112, Site: GG16, Lake Gogolo Found: 1977 Sex: Undetermined

These remains have a heavy blue-black manganese staining all over them. They are fragmentary, very friable and extremely eroded on all surfaces. Only long bones are represented, but none is remotely complete. All the pieces seem to be from a delicate, rather gracile individual. Cortical bone is much thinner (3–4 mm) than others described in this series.

WLH 113, Site: GG16, Lake Gogolo Found: 1977 Sex: Undetermined

WLH 113 is represented by parts of both tali. The right talus has part of the middle calcanian, posterior calcanian and trochlear articular surfaces, and a deep tali sulcus. Erosion has removed many parts of the bone. The left talus is in a similar condition, but is less complete.

WLH 114, Site: GG16, Lake Gogolo Found: 1977 Sex: Undetermined

Only the superior or dorsal aspect of the right talus is present for this individual.

WLH 115, Site: GG16, Lake Gogolo Found: 1977 Sex: Undetermined

These highly mineralised remains are thoroughly broken and burnt. Fracture patterns are typical of those resulting from the smashing of wet bone, with transverse, crescentic breaks and cracks. Colour variation of charred fragments suggests differential burning from being in different parts of a fire at different intensities of temperature. It is a cremation with bone smashing.

Unfortunately, due to the scrappy nature of the fragments, individual bones cannot be reconstructed or identified. There is, however, a very small piece of cranial vault amongst the pieces, which appears not to have been burnt, although it is difficult to know whether this belongs to the same individual. It is thin, with a maximum thickness of 2.8 mm. Both cranial tables are <1.0 mm.

WLH 116, Site: GG16, Lake Gogolo Found: 1977 Sex: Undetermined

These remains consist of a few highly mineralised, eroded, white, post-cranial fragments only.

WLH 117, Site: GL13, Lake Garnpung Found: 1977 Sex: Undetermined

All the pieces are covered to some extent with a rough, brown carbonate incrustation. Some manganese spotting can be seen on some of the bones. There are no cranial remains.

Femora

A major portion of the moderately robust left diaphysis remains, but the proximal end above and including the lesser trochanter, and distal end including the popliteal surface, are missing. The whole medial side of the right femur is missing, together with the proximal and distal ends. The cortex is 9.5 mm thick at the lateral mid-shaft position.

Tibiae

Both tibiae lack proximal and distal ends and all areas where measurements are usually are missing. New bone has been deposited on the popliteal line of the left tibia taking the form of a periostitis. This type of infection usually originates from a wound to overlying soft tissues but not affecting the cortical bone beneath.

Humeri

The left humerus consists of the proximal half of the diaphysis extending from the bicipital groove to the distal edge of the deltoid tuberosity. The latter is well developed, as are crests for the insertion of *pectoralis major* and the origin of the lateral head of the triceps. Almost exactly the same section of diaphysis represents the right humerus. The deltoid tuberosity, however, is not as well developed and left handedness may be the reason.

Ulnae and Radii

Only the middle two-thirds of the diaphyses of the right ulna and radius remain. They are fused in their natural anatomical position by a thick brown coating of calcium carbonate. The length of the radius is 129 mm and the ulna 158 mm. The latter has suffered a closed comminuted fracture about 75 mm from the distal end that did not fuse. The fracture was held in place by soft tissue and carbonate deposition within the fracture shows the trauma took place close to the time of death. Swelling of the bone close to the broken ends suggests post-traumatic tissue response and initiation of callus took place before death. The break is in the typical position as a parrying fracture. These occur more frequently in the left arm due to predominant handedness, with the left arm normally used for parrying blows while the right wields the weapon. The circumstances here afford further support for the individual being left-handed. The distal two-thirds of diaphysis of the left ulna and radius have been fused by carbonate. Any measurements taken on these bones would not reflect the correct morphology because of carbonate deposition and little time allowed for study and preparation.

CLAVICLES

The proximal two-thirds of the right clavicle and the diaphysis of the left have manganese spotting beneath carbonate incrustation. Neither is robust, although there is some development of the coronoid tubercle on the right clavicle.

Other Bones

There is a third proximal left phalange (hand), proximal two-thirds of a left fourth metatarsal, less the articular head, and another small piece of metatarsal bone, probably from the proximal end.

Metrical data (in mm)

Left femur	
Actual length	310
Estimated length	465
Stature – if male	169 cm
Stature – if female	166 cm
Sub-trochanteric diameter	
Antero-posterior	29.0
Medio-lateral	32.0
Mid-shaft diameter	
Antero-posterior	29.0
Medio-lateral,	30.0
Platymeric index	90.6
Robusticity index	12.7
Left humerus	
Actual length	121[a]
Mid-shaft diameter	
Antero-posterior	22.0
Medio-lateral	22.0
Right humerus	
Actual length	125
Mid-shaft diameter	
Antero-posterior	26.0
Medio-lateral	20.0[a]

[a]*Estimated measurement.*

WLH 118, Site: GL13, Lake Garnpung Found: 1978 Sex: Undetermined

This individual is represented by a distolateral portion of the left humerus only. It is mineralised, has a greyish white colour and a well developed supra-condylar ridge.

WLH 119, Site: GL13, Lake Garnpung Found: 1978 Sex: Undetermined

WLH 119 consists only of the proximal third of the left clavicle. It has a strong manganese stain and is completely free of carbonate incrustation.

WLH 120, Site: GL13, Lake Garnpung Found: 1978 Sex: Undetermined

Only a small section of the cranial vault and two other scraps of cranium are present. They have a light ash-coloured appearance, due to burning. The maximum thickness of the vault section is 7 mm and both cranial tables are thin.

WLH 121, Site: GL13, Lake Garnpung Found: 1978 Sex: Undetermined

This is a small triangular piece of femoral shaft, with part of an underdeveloped *linea aspera*. The cortical bone is very thin (3 mm) and looks almost too thin to have come from a femur. The bone is grey in colour but shows no evidence of having been burnt. There is brown carbonate incrustation, below which is a blue-black manganese stain.

WLH 122, Site: GL13, Lake Garnpung Found: 1978 Sex: Undetermined

WLH 122 is represented by a few small cranial-vault fragments showing thorough burning probably by cremation. Colouration varies from grey to blue-black, in the manner of other cremated individuals in the collection. All bones are heavily mineralised and produces a metallic ring when tapped indicating fossilisation. The vault is thin and delicate in its construction, with a maximum thickness of 6 mm and a minimum of 3 mm. Both inner and outer cranial tables are very thin (<1 mm) and some fragments have very little diploeic bone. One piece shows a very small section of cranial suture. From this it can be assumed that suture fusion was not complete but probably in its early stages. Bone gracility indicates female and suture fusion suggests a young adult, in her early 20s, all reminiscent of WLH 1 and WLH 68.

WLH 123, Site: GL24, Lake Garnpung Found: 1978 Sex: Male?

These remains are mineralised, highly fragmented and completely burnt. Colour varies from a chalky white (calcination) to deep blue-black typical of cremation and is designated a cremation with bone smashing. Vault fragments are in poor condition but they show the cranium was very thin, with a maximum thickness of 7 mm (asterion?).

Postcranial remains consist of small fragments and none can be attributed to any particular bone. Although cortical bone thickness of some pieces is as much as 8 mm (femur?), these pieces have a gracile appearance, like those of the cranium. Although the gracility suggests a female, the remains are too poor to confirm the sex of this individual but cortical bone thickness may suggest male or a mix of two individuals.

WLH 124, Site: GL24, Lake Garnpung Found:1978 Sex: Undetermined

WLH 124 consists of cranial and postcranial bone most covered in a pink-orange carbonate wash. The largest section of cranial bone is a right anterior section of frontal bone. It extends from the glabella, just left of nasion, laterally almost to the zygomatic process, although the process itself is missing. Posteriorly, the bone extends only to a point near metopion, where its broken edge then proceeds antero-laterally to an anterior part of the temporal line. Proximal parts of both nasal bones are fused to the nasal portion of the frontal, forming a deep infra-glabella depression. The bony nasal septum can be seen inferiorly, together with the roofs of two ethmoid sinuses. Large spaces for the frontal sinus exists each side of the glabella. The right frontal sinus extends as a large hollow over the right orbit.

Although the surface of the right superciliary ridge has been removed by erosion, it was probably not large. The glabella is small and the slope of the forehead is more accentuated than on most others in the series. The frontal bone rises smoothly from the supra-orbital region without a supra-orbital sulcus. The internal surface has been badly etched by erosion and is covered in an orange-brown carbonate layer. Few features are present except a long (41 mm) and sharply frontal crest. The vault is medium thickness (8 mm) but the bone tables are comparatively thick. The orbit is large and rounded but erosion has damaged almost the whole superior orbital margin as far as the supra-orbital notch which is shallow and wide. The remaining bone around the orbital margin suggests it was rounded.

The left temporal consists only of bone around the glenoid fossa which is not deep and has a transverse oval 'sub-fossa' or gutter posteriorly. The shape fits into one of four basic fossa forms in the collection. The other three are the triangular form (WLH 100), the shallow, circular or oval fossa (WLH 44 and 101) and a deep oval form (WLH 72). Minor pitting laterally on the articular tubercle could be osteoarthritis or erosion. A section of the roof of the external auditory meatus is present, together with the anterior border of the tympanic plate. Some remnant mastoid air cells on the posterior edge of the temporal section suggest they extended out from the mastoid. The groove for the middle meningeal artery lies on the internal surface and anterior of the glenoid fossa.

The right temporal consists only of an inferior section of the mastoid. The thickness through the suture near the occipital border is 13.5 mm. Inferior to this is a deep and long digastric fossa, which starts well above the position of the mastoid crest. The depth of this fossa highlights the medial para-mastoid crest and a deep and sharply defined posterior mastoid crest. Mastoid air cells have been exposed by removal of the inner bone table. They extend, superiorly, almost to the occipital border and, posteriorly, into the para-mastoid crest.

Only two small sections of the left parietal remain. They have accentuated ridging on the temporal surface emphasising a strong articulation with the temporal bone. This may indicate something of the stress acting upon the area from a well-developed *temporalis* muscle that overlay the area. There is also a small section of the mastoid angle.

A small piece of the right maxilla extends from the sub-nasal area at the canine to the position of the second molar. The bone above the canine shows the inferior half of a small canine fossa. Scraps of tooth root, of varying proportions, are still in position within the alveoli, but no crowns remain.

There is a large section of the left mandibular body and part of the ramus, less the gonial angle. The coronoid process and condyle are missing, together with much of the alveolar bone and the anterior section of the body. The body is not robust, but neither is it gracile. Other features include a mylohyoid groove and ridge; deep sub-maxillary and sublingual fossae; a large mandibular foramen; a long *sulcus colli*; a well-developed posterior marginal tubercle; and a wide mental foramen.

There are a number of cranial-vault pieces, which have a maximum thickness of 11.5 mm. Some of them have sections of cranial suture which have begun to fuse indicating an adult person over 30 years old. The light build of the cranial skeleton, including a minimally developed supra-orbital region and moderately thin vault, indicates a female individual.

Left Humerus

Only the distal third of the diaphysis of this bone remains. Much of its surface has been removed by erosion, but it is slim, with thick cortices (7 mm).

Right Femur

The right femur is represented only by the distal quarter of the diaphysis above the popliteal surface. The bone is very eroded and highly mineralised. There are only fragments of the left femur, with no outstanding features.

Right Tibia

A small section of the mid-shaft is all that remains of this bone.

Other Bone

Other bone includes: part of the neck of the left acromion; the proximal third of the right radial shaft, a small section of the left ulna; and the fifth right proximal phalange.

WLH 125, Site: GL24, Lake Garnpung Found: 1977 Sex: Undetermined

This individual consists of a few highly mineralised and heavily eroded cranial vault fragments with a maximum thickness of 8 mm.

WLH 126, Site: GG16, Lake Gogolo Found: 1977 Sex: Undetermined

This individual is represented by a single, very small piece of cranial vault, 5 mm thick. The bone is burnt to a greyish blue colour.

WLH 127, Site: WOC152, Lake Mungo Found: 1978 Sex: Undetermined

WLH 127 consists of mineralised cranial and postcranial fragments. The cranial material is made up of one large (left parietal?) and one small piece of vault. Their maximum thickness is 10 mm, mainly comprised of diploeic bone. The inner and outer cranial tables are thin, 1 and 2 mm, respectively.

A mid-diaphyseal section of the left ulna has a well-healed fracture. A large callus has formed around the trauma, indicating a complete healing of the bone during continued use. The fracture position indicates a parrying fracture received from using the left forearm to block a blow. This kind of fracture is common in more recent Aboriginal male skeletons, but it is also seen in some females.

WLH 128, Site: GL24, Lake Garnpung Found: 1977 Sex: Undetermined

This individual is represented by one very small piece of thin cranial vault, with a light brown carbonate incrustation. Maximum thickness is 6 mm with thin cranial tables.

WLH 129, Site: GL24, Lake Garnpung Found: 1977 Sex: Undetermined

This individual consists of five small pieces of white, mineralised, very fragmented cranial vault. Maximum thickness is 10 mm at a point near asterion and 8 mm elsewhere.

WLH 130, Site: GL24, Lake Garnpung Found: 1978 Sex: Female

WLH 130 consists only of highly mineralised cranial remains. The main section comprises the posterior two- thirds of the calvarium, extending from the coronal suture, across the occipital, to the inferior nuchal line. Only the petrous portions of the temporals remain and the left mastoid. A section of the frontal bone, over the right orbit, is also present. The calvarium is neither gracile nor robust. In general appearance it is very similar to many modem Aboriginal crania. A posterior view shows no sagittal keeling or para-sagittal depressions but presents a rounded morphology with slight sagittal ridging along the suture. Both parietals curve uniform towards the temporal surface. The vault has a fairly uniform thickness up to 9 mm. The inner cranial table is 1–1.5 mm thick and the outer table 1.5–2.0 mm. Diploeic bone is not a prominent part of vault thickness as in some individuals.

There is a small parietal foramen in the left parietal near obelion. A small supernumerary Wormian bone lies within the left lambdoid suture but a much larger one was present laterally indicated by a large curve set in the lambdoid suture that extends into both the left parietal and occipital bones. There is also a 6 mm diameter hole in the lambdoid suture, medial of the small wormian. This may, however, be artificial and have nothing to do with a missing supernumerary, which seems the only other likely reason for its origin. Sutures are open internally or externally.

There are two 19 mm narrow, shallow grooves etched into the external surface of the left parietal. They are old but do not conform to cranial trauma, such as a depressed fracture. They may be post-mortem made by an implement with a thin, sharp edge. The individual was not excavated so is unlikely to have contracted these scars from such activities.

The internal surfaces of both parietals show sulci to accommodate the anterior branches of the middle meningeal artery. The posterior branches are not as well defined on the posterior section of the left parietal. The superior and inferior occipital fossae are deep and the internal occipital protuberance and crest are prominent. It is through the internal occipital protuberance that the greatest measurement for cranial thickness occurs at 17 mm.

The posterior suture of the occipital border can be traced superiorly, then anteriorly to the parietal notch. Here, a small part of the mastoid angle from the left parietal is partially fused. Inferior to this is a superior portion of a well-developed supra-mastoid crest. Anterior to the occipital border are two large mastoid foramina. They lie 6 mm above the extreme supero-posterior point of the digastric fossa which is deeply etched and large, flanked medially by a paramastoid crest and laterally by a small crest on the posterior surface of the mastoid. The mastoid itself is long and rather narrow. Pneumatisation occurs from the tip to the supra-mastoid crest and posteriorly almost to the lambdoid suture. Air cells are large up to 6 mm in diameter. A small section of the petrous portion of the left temporal has an internal auditory aperture. The opening for the aqueduct of the vestibule and part of the superior petrosal sinus are visible.

Only a section of the left orbit remains of the fronto-facial region. It extends medially from the fronto-zygomatic suture to an almost closed, moderately deep, supra-orbital notch. The zygomatic trigone and the superciliary ridge are not developed and the height of the superciliary ridge corresponds to the space for the frontal sinus which lies beneath. The superior orbital margin is sharp and there is no sign of a postorbital sulcus. The frontal rises smoothly from the minimally developed supra-orbital region. The very small area representing the frontal vault has a thin external table. All the above characteristics suggest a female person.

A distal section of the right mandibular body extends posteriorly from the gonial angle to just below the distal roots of the second molar. Externally, there is a well-marked posterior marginal tubercle and a mylohyoid groove is visible on the internal surface. Bone deposition in and around the third molar socket indicates its pre-mortem loss. This suggests that the individual was older than cranial-suture fusion might indicate, perhaps >40 years.

Metrical data (in mm)
Cranial thickness

Bregma	9
Vertex	9
Obelion	7
Left parietal boss (near)	9
Lambda	9
Inion	17

Craniometrics

Lambda-bregma	122
Lambda-inion	69
Lambda-asterion (est.)	82
Biasterion (est.)	112
Bistephanion	91
Bregma-inion	161
Inion-asterion (est.)	68

Mastoid

Length	*23*
Width	13
Breadth	16
Module (female)	47.8

WLH 131, Site: GG25, Lake Gogolo Found: 1977 Sex: Female

There are one very small piece of cranial vault and nine postcranial fragments. All are eroded, mineralised and covered in a yellowish-brown carbonate wash. The inner cranial table is missing from the vault section, so that no measurement of thickness can be taken.

WLH 132, Site: GG16, Lake Gogolo Found: 1977 Sex: Female

Only postcranial remains are represented. Most of them consist of small fragments of bone, covered by a brown calcium carbonate. Almost all have been completely burnt. They have been included in the category of cremation with bone smashing, described in Chapter 2.

WLH 133, Site: GL13, Lake Garnpung Found: Unknown Sex: Female

These remains consist of a few mandibular fragments and one tooth, all with some manganese staining and a light carbonate wash. A left gonial angle has strongly developed muscle markings at the medial pterygoid. There is, however, no eversion of the gonial edge. A superior portion of the body located at the infero-lateral border of the ramus has a shallow intra-molar sulcus. This runs in a mesial direction, showing the buccal root impressions of the second and third molars on its internal surface. All bones below the alveolar border and mesial to the second molar are missing.

A left lower central incisor displays some moderate attrition of its occlusal surface. A hypoplasic line is situated 1.7 mm from the cemento-enamel junction.

Dental measurements (in mm)
Lower left central incisor

Labio-lingual	7.0
Bucco-lingual	5.3

WLH 134, Site: GL26, Lake Garnpung Found: 1977 Sex: Female

WLH 134 is only four mineralised cranial fragments. The largest is a section of the right superior orbital margin, minus the zygomatic trigone. The orbital margin is sharp and there is only a hint of a supra-orbital notch. The superciliary ridge is moderately developed but the lateral brow is smooth and devoid of any rugosity. The morphology suggests female and although the zygomatic trigone is missing, it is likely from the construction of the adjacent area that it was small. Laterally, a small segment of the temporal is negligible and more like a line. Infero-medially there is a superior section of a large frontal sinus. Except for the superciliary ridge, there is a smooth transition from the supra-orbital region to the frontal bone. Maximum thickness of other small vault pieces is 6 mm. The inner and outer cranial tables are very thin.

Metrical characteristics (in mm)
Orbital thickness

Medial	14.4
Middle	5.0
Lateral (eat.)	6.0

WLH 135, Site WOC 1, Lake Mungo Found: 1986 Sex: Female

WLH 135 was found early in 1986 by a party of visitors on the southern end of the Lake Mungo lunette and was reported to NSW Parks and Wildlife rangers. Its position was approximately 200 m north of WLH 1 possibly on the same stratigraphic level as WLH 1 and WLH 3 that was found another half a kilometre further north (Figs. 10.26–10.62).

I was asked to visit the site and inspect the individual by Traditional Owners in the area and did so with a colleague from the Australian National University. A park ranger and community members accompanied us to the site and the ranger uncovered the find. This was a difficult time for the community members because a moratorium on skeletal research was in place but they were curious about this individual because of its proximity to WLH1 and position between that individual and WLH 3. When uncovered only the top of a cranium was visible. Because of the moratorium only a certain amount of the cranium was allowed to be revealed under the guidance of a ranger and community members. Sand was removed from around the cranium by hand exposing the top and side. The position of the head suggested it was tilted forward facing down into the sediments (Fig. 10.63).

The vault structure was in generally good condition and was covered in a calcium carbonate wash. However, it was cracked across the parietal bones in an antero-posterior direction forming several larger fragments as well as some smaller pieces that were loose. The larger fragments remained in place but some of the smaller sections had separated. Two of the large cracks met where a circular section of bone was missing roughly at the position of inion (Fig. 10.64). The occipital bone seemed complete although the sub-occipital region was below the sand. The occipital was slightly separated from the parietal along the lambdoid suture. Carbonate encrustation along the open suture indicated it was open before the carbonate event had occurred. Medium-sized fragments from the right parietal were loose. Examination of their thickness showed a very thin

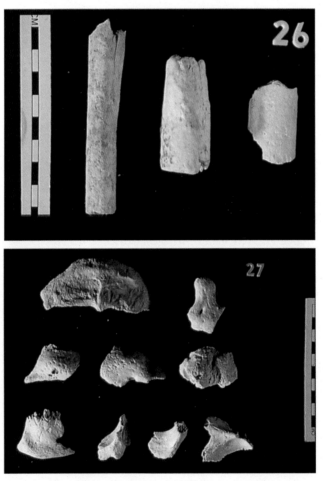

FIGURE 10.26 Robust cranial fragments of WLH27 (above) with rugged nuchal area (top left) and left malar (top right) and a left anterior section of mandible (bottom left).

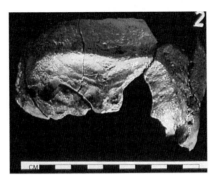

FIGURE 10.27   Sub-occipital view of WLH 28 (top) showing robust occipital base with prominent muscle insertion markings and nuchal crest that curves bi-laterally towards the mastoids. Internal architecture of the occipital (below). Note varying shades of grey from burning/ cremation.

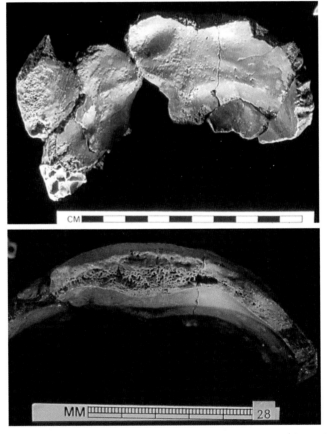

FIGURE 10.28   Superior view of occipital just above the torus of WLH 28 showing varied grey/blue colouration from burning/cremation.

FIGURE 10.29   External surface of vault sections of WLH 28 showing burning.

FIGURE 10.30   Close-up inverted view of WLH 28 vault showing thorough burning of dip-loeic bone through full vault thickness that is 12 mm at this point (*line*). Note very thin inner (top) cranial table.

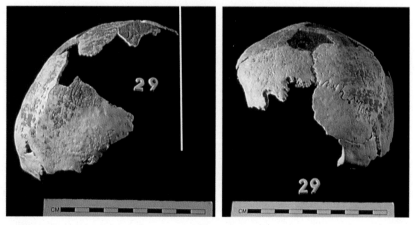

FIGURE 10.31   Right lateral view of WLH 29 cranium (left) and posterior (right) view show-ing parietal keeling.

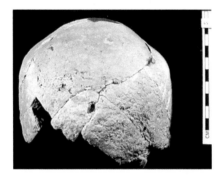

FIGURE 10.32   Section of cranial table from WLH 42. Note the percentage of diploeic bone.

FIGURE 10.33   Posterior view of WLH 45 showing parietal keeling.

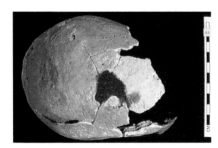

FIGURE 10.34   WLH 45 right lateral view.

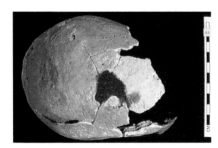

FIGURE 10.35   Superior view of WLH 45.

FIGURE 10.36   Long bone fragments from WLH 46 note brown-orange carbonate wash.

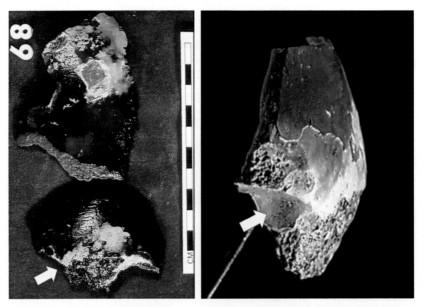

FIGURE 10.37   Large pieces of frontal and right parietal (top, left) WLH 68 and (top, right) right lateral view of the frontal. The colour of the bone is the same as the cremated remains of WLH 1. High temperature burning is shown by the white patches of calcination. Bone morphology also shows the same gracile morphology of WLH 1.

FIGURE 10.38   This view of the right parietal along the sagittal suture indicates the delicate nature of the cranial vault with very thin walls like WLH 1. The sectional view of the sagittal suture shows only minimal suture fusion indicating a young adult person.

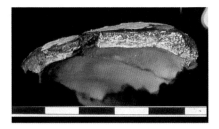

FIGURE 10.39   Two views of the frontal bone of WLH 68 before the removal of calcium carbonate.

vault with little diploeic space between thin inner and outer tables reflecting those of WLH 1 and others in the series described as 'gracile' and the general appearance and consistency of bone indicated complete fossilisation.

Because complete excavation or further uncovering of these remains was not permitted an opportunity was lost to take the investigation further. It was more than likely this individual was another example of the very gracile individuals in the collection and possibly an individual contemporary with the oldest individuals yet found. It was decided by community members to rebury WLH 135. It would be stabilised using sand, tree branches and tin sheets. During the following years that work was shown to be futile with wind and rain producing the forces of erosion that are constant on the Mungo lunette and across the region. The structure was, therefore, slowly destroyed together with the remains of the individual and a tremendously important fossil was lost. Like other individuals it had a story to tell that would not now be heard. As I suggested in the beginning of this description of WLH 135, the face seemed to be looking down into the sand. That suggested to me that is could have been a crouch burial, that

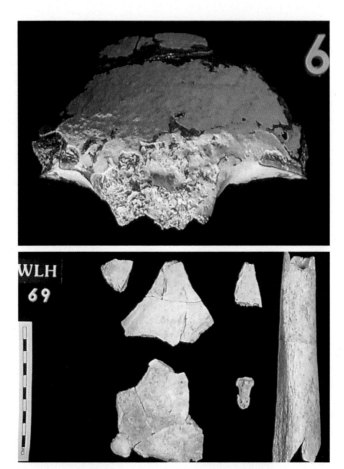

FIGURE 10.40 Cranial fragments of WLH 69 (top) together with a proximal section of the right tibia.

FIGURE 10.41 A section of the right frontal displays a prominent brow ridge and zygomatic trigone and temporal crest. The trigone accentuates a marked post-orbital constriction.

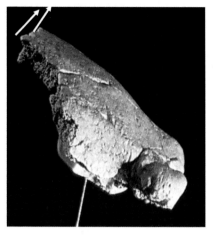

FIGURE 10.42   Right lateral view of right frontal section of WLH 69 shown in previous picture.

FIGURE 10.43   Cranial pieces from WLH 72.

may have had its lower limbs flexed and placed below the head in the pit. If that was the case it would have been the oldest burial of that kind to be recorded in Australia. The above pictures show its bone was not burnt and so it was not cremated, nor was it covered in red ochre as WLH 3. It might be drawing a long bow but this individual may have shown us that not just two types of burial were being carried out on the side of Lake Mungo but a third may have occurred. We will never know.

FIGURE 10.44 Sections of long-bone from WLH 72.

FIGURE 10.45 Various cranial fragments of WLH 73.

WLH COLLECTION NOTE

Individuals WLH 136–151 were not collected, their locations were not recorded and they were not studied or photographed as per agreement with the Aboriginal community.

WLH 152, 153 and 154, Site GG16, Lake Garnpung Found: 2002, Sex: Various

From time to time during the moratorium on research, the Aboriginal community decided to relax the halt on research to rescue burials that were usually found by accident. The WLH 135 description above was one of those, although

FIGURE 10.46   WLH 73 left lateral view of zygomatic region of the frontal bone with prominent brow.

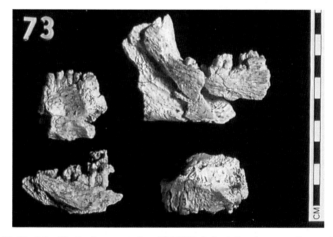

FIGURE 10.47   Mandibular fragments from WLH 73 (below) showing extensive erosion of surface features as well as fragmentation and splitting of the few remaining teeth with loss of crowns.

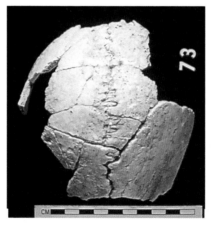

FIGURE 10.48   A superior portion of the cranial vault of WLH 73 consisting of fragmented medial portions of both parietals.

FIGURE 10.49   Posterior portion of WLH 100 cranium showing complete fusion of the sagittal and lambdoid sutures. The bone is highly fossilised and has robust buttressing around the base of the occipital and mastoid regions.

FIGURE 10.50   Left frontal region from WLH 101, note the brow and prominent zygomatic trigone top left.

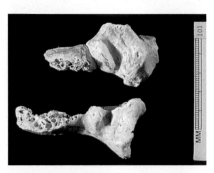

FIGURE 10.51   The glenoid fossae from WLH 101.

FIGURE 10.52   Tooth fragments and the left body section of the mandible from WLH 103.

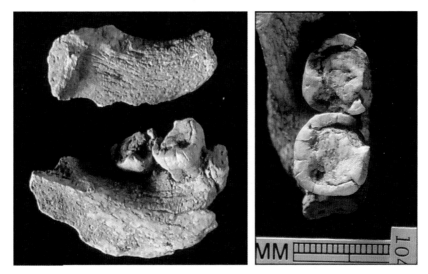

FIGURE 10.53   Mandibular molar teeth from WLH 104 showing heavy attrition of surfaces.

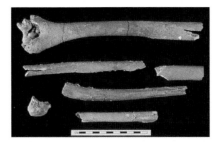

FIGURE 10.54   Upper limb bones from WLH 106 covered with thick red carbonate indicating burial before high lake levels that were then precipitated by carbonate enriched groundwater rise covering burials.

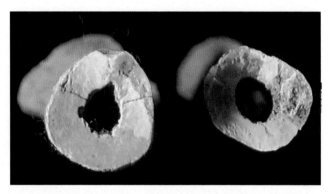

FIGURE 10.55   These mid-humeral sections from WLH 110 show very thick cortical bone reminiscent of archaic human cortices. They align with archaic cranial features displayed in the collection.

FIGURE 10.56   Cremated cranial fragments of WLH 122.

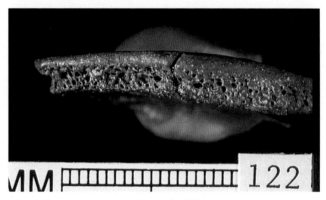

FIGURE 10.57   Thin cranial vault structure of WLH 122.

FIGURE 10.58　Thin, cremated cranial fragments of WLH 123.

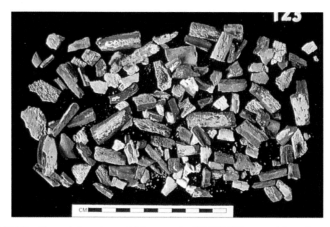

FIGURE 10.59　Cremated post-cranial fragments of WLH 123 some with carbonate wash.

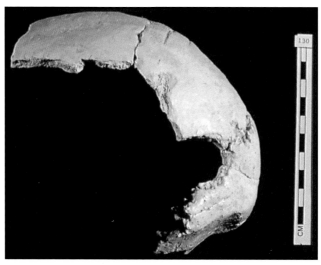

FIGURE 10.60　Left lateral view of WLH 130 showing a large hole from a lost supernumerary bone in the lambdoid suture.

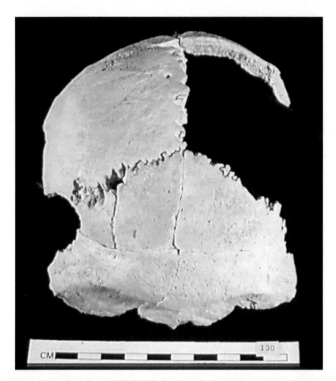

FIGURE 10.61   Posterior view of WLH130 showing the hole where a very large supernumerary bone had been and a second much smaller one in the left lambdoid suture.

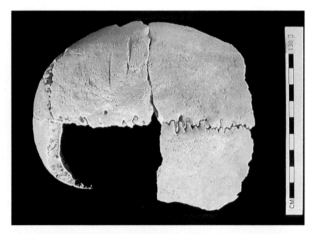

FIGURE 10.62   Superior view of WLH 130 (below) with cut marks in the left parietal bone.

FIGURE 10.63   The WLH 135 cranium (above) as it appeared when uncovered for the first and only time. The face was buried in the sediments revealing the occipital to the right demarcated by an open lambdoid suture. Large parietal fragments were separating from the cranial body and other smaller fragments were loose.

FIGURE 10.64   A superior view of WLH 135 showing the occipital and posterior sections of both parietals. A 20-mm-diameter area around lambda may have occurred pre-mortem.

that was not rescued. Another opportunity came to rescue a burial in May 2002. The burial was named WLH 152, this time it was excavated after clearance from Members of the Elders Committee of the Three Traditional Tribal Groups. It had actually been known about 7 years before. The skeleton was generally well preserved although fragmented. At that time, a photographic record was made followed by stabilisation of the burial with branches to gather wind-blown sand. A second burial (WLH 153) was also discovered in 2001 4 m west of WLH 152. Both burials were located within a large dune blowout exposing Late Pleistocene dune core sediments containing indurated sands. They included light coloured loose sands containing regular, dark orange-red coloured stratified, inter-bedded carbonate sands. The burials were reported to the Elders Committee and discussions regarding future management and study of the two burials were held. In February of 2002 the author was invited to carry out a biological investigation of both individuals. All works were carried out on site at the wishes of the Elders Committee. However, weather conditions forced collection of the bone and its transport to the Ranger's Office of the Visitor's Centre at Lake Mungo for study there. After examination, the remains were deposited in the Mungo Keeping Place, a cellar in the Rangers Office, where WLH1 is stored as well as other remains found in the region, some that have never been studied (Figs. 10.65 and 10.66).

FIGURE 10.65 A portion of the right temporal of WLH 135 showing impressions of meningeal arteries.

FIGURE 10.66 Thin cranial vault bone from WLH 135 covered in calcium carbonate.

Few individuals in the Willandra Lakes collection, except WLH3 and these remains, have been excavated using a standard set of procedures. Even then, excavation has only consisted of the removal of loose sand or in some instances bone from only a few centimetres down that was revealed during initial removal of already exposed bone (Fig. 10.67). In the case of these individuals all loose surface bone was collected then a shallow excavation into the centre of the burial area was made to ascertain its extent and depth and recover all fragments. There was little depth to the burial which had been partially exposed during the 7 years. Both WLH 152 and 153 had some surface bone deterioration and were scattered. The bone is light in colour and although fragmented they is well preserved and lacks friability often occurring on bone exposed in the arid environment but was also due in part to robust bone construction.

Bone pieces covered an area 2 m in diameter and associated skeletal elements had become widely separated. Some pieces were in close anatomical position not moving far from each other since exposure. It was obvious that WLH 152 had been highly disturbed with fresh breaks apparent on several bones, something not unexpected after a seven-year exposure.

Wlh 152

Parts of the vertebral column, ribs, cranial vault, face, pelvis, scapulae, the dentition and a large section of the mandible were missing. It is uncertain whether this was due to taphonomic processes following lengthy exposure, or possible burial of a partial skeleton. No burned bone was present. Besides eroded surfaces, the bone is in good condition particularly when compared to many in the WLH series. The few pieces of cranial vault show a uniform thickness of almost 14 mm although that increases to 24 mm through inion. Other prominent cranial features include a pronounced brow ridge and large

FIGURE 10.67   **GG16 site with two burials positioned closely together in dark brown inter-bedded calcified sands.**

mastoid suggesting a male individual with a very robust morphology typical of a number of individuals in the collection. Long bone measurements indicate a personal height of around 179 cm. A posterior section of the right mandible shows a particular robustness emphasised by a very broad ramus. Pronounced muscular marking on the medial side of the mandibular angle probably originates from strenuous use of the medial pterygoid muscle during life, possibly bighting down during implement manufacture or similar activity. Attrition of the mandibular dentition is marked but even, particularly on the molars. Substantial attrition of the third molar suggests an older adult person of perhaps 45–55 years.

Epiphyses are fused and thick carbonate encrustation has fused the right ulna and radius in anatomical position along their diaphyses as well as encasing the tarsals of the right foot. Although separate, a similarly encased group of proximal phalanges suggests they too were part of this block at one time. Many of the bones of the hands and feet are present but the restricted time available for examination did not allow full release of these bones from their casing and reconstruction.

WLH 152 was a generally healthy and well-nourished person. There is nothing to suggest any form of nutritional inadequacy, systemic disease, congenital malformation or chronic illness. There are, however, various traumatic pathologies including:

- well healed distal fractures on both ulnae;
- a well healed fracture of the left clavicle with large, smooth callus;
- a small infection of the left parietal, possibly as a result of trauma;
- a large penetration in the outer cranial table on the left parietal;
- there is exposed diploeic bone above the left mastoid which could be from trauma but it could equally have originated post mortem from interment processes although that seems unlikely in the light of other trauma;
- a distal fracture of the left third proximal phalange resulted in its lateral displacement which has set irregularly causing arthritic osteophytosis or exostosis around the injury site,
- several small, depressed fractures occur on the left frontal immediately behind the orbits (Fig. 10.68).

The above trauma pattern suggests blows received from hand to hand fighting. The lower arm fractures are typical of 'parrying fractures' derived from a glancing blow hitting the lower arm, or taking the full force of a blow on the raised arm (Fig. 10.69). An attack by a right-handed person could have caused the broken clavicle, although so could a stumble onto an outstretched hand which could also have caused the distal fracture of the left ulna. The distal end of the third phalange is the most prominent knuckle so it could have fractured from a thrown punch.

Wounds on the left side of the frontal bone and left parietal also strongly suggest blows from an instrument wielded by a right-handed person as in violent

FIGURE 10.68   Burial WLH 152. Note the tight clustering of bone in the central region of the site.

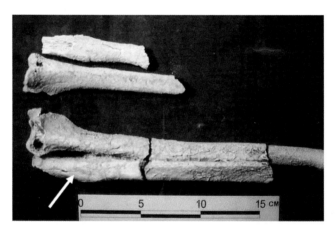

FIGURE 10.69   The left ulna and radius fused by calcium carbonate. The distal ends of the right ulna and radius are above with well healed parrying fractures on the distal one third of both ulnae (arrows).

confrontation (Fig. 10.69). Osteophytosis on the finger joint suggests trauma occurred some years pre-mortem. It also points to age in the onset of such bony changes, further support for the suggested age range of this individual. Good callous formation at fracture sites on the wrist and clavicle indicate good physical health and an adequate nutritional intake.

Thick cortical bone observed on sections of major long bones has been noted previously among Willandran skeletal remains. This, together with thick cranial vault structure is a robust feature previously noted and associated with some Late Pleistocene Willandrans. The prominent supra-orbital development of WLH 152 reflects that of other individuals in the series and supports the general robustness of this male person.

Wlh 153

The appearance of WLH 153 reflects that of WLH 152. The pattern of bone scatter, its discrete placement, bone fragmentation, colouration and fossilisation, close association and stratigraphic position strongly suggest a shared age of interment. The burial rested on a carbonised layer as though the supra-orbital had been placed on a small hearth or burnt area of some kind. The lower bones were covered in a fine damp black ash that turned to grey dust when dry, but none of the bone has been burned. The burial was put onto a cold ash layer of an old fireplace. Following removal of the bone the hearth was excavated which revealed its elongated shape, 50 cm across. A cross-section showed it to be lens shaped typical of a hearth or fireplace but little more can be added to the reasons for the placement of the bones or body on it.

Elements of the skeleton were missing but as pieces were collected it became apparent that two adults were represented. The long bones are robust, the left mastoid is very well developed together with a rugged sub-occipital region, all characteristic male features. A section of left brow, however, does not fit this pattern. It lacks a superciliary ridge and a zygomatic trigone and has a thin rather sharp upper brow margin typical of a female person. This second individual was given the series designation WLH 154 (see below).

Two sections of the right mandibular body emphasise the double burial with one having larger proportions than the other. The largest skeletal elements were with WLH 153. The anterior portion of the WLH 153 mandible has prominent lingual tubercles and is very thick through the mental symphysis (13.5 mm). Both these features support it being a male person. The second section includes a piece of the right mandibular body extending from the third molar to the posterior edge of the socket of the right first premolar. The bottom of the mandibular body is missing together with the crowns of all teeth. This mandibular section does not articulate with the mental symphysis of the first, it is small in its general appearance and has been included with WLH 154.

Some sections of the WLH 153 cranial vault are thin for a robust male and have been put with the female WLH 154. Nevertheless, other features point to most bone in this burial being from a male. They include the pathology noted below, the angle of the sciatic notch, robust cranial features and the mandibular mental tubercles. Personal height was assessed using arm bone lengths. The humerus and radius individually produced different results for height but that difference indicated that the humerus was from a female (WLH 154) and the radius from the male WLH 153. This conclusion was based on the lack of close approximation or overlap between personal height determined from the radius and humerus. Therefore, given that it is unlikely the female was taller than the male, it is suggested that the 163 cm height estimated using the humerus is from a female and the radius indicating a height of 173 cm is male. The latter indicates that the WLH 153 male was somewhat shorter than WLH 152 with a 179-cm stature (Figs. 10.70 and 10.71).

All articular bone surfaces comprising the right elbow of WLH 153 are affected by chronic osteoarthritis. That of the distal humerus has been almost destroyed by arthritic wear with severe pitting. It also displays severe osteophytosis that follows the line of the articular cuff or capsular ligaments both on the anterior and posterior sides of the bone (Fig. 10.72). The proximal ends of the ulna and radius are also damaged by chronic arthritic degeneration with an accompanying osteophytosis and erosion of the surface of the radius (Fig. 10.73). This individual must have suffered severe pain associated with this advanced arthritic condition. The WLH 153 pattern of arthritic wear reflects that on WLH 3 but is substantially worse in terms of general articular destruction of the elbow joint. Increasing reduction in joint flexibility would have occurred for WLH 153 as it did in WLH 3 and it is suggested that this condition has the same aetiology as that of WLH 3: long-term use of the spear thrower. That condition would have chronic eventually preventing this man from using a spear thrower or spear and disabled him from many activities where the use of this elbow was required.

Two right mandibular condyles were found with WLH 153. The smallest (20.2 mm across) articulates with a section of right glenoid fossa which has manganese staining similar to the associated mandibular condyle. The smaller condyle articulates with the right body of the female mandible found with WLH 152. The second condyle is larger than the first measuring 28.5 mm across and is considered part of the male WLH 153. A section of

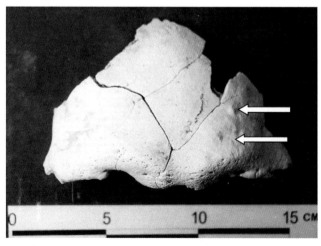

FIGURE 10.70   Frontal bone of WLH 152 showing prominent brow ridge and small depressed fractures/indentations on the left side of the bone (*arrows*).

FIGURE 10.71 The GG16 hearth structure after all bone was removed.

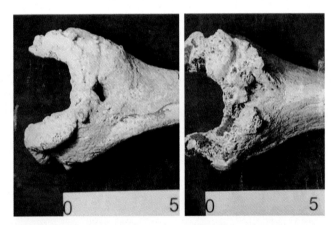

FIGURE 10.72 Posterior (left) and anterior (right) views of distal right humerus of WLH 153. Severe damage of the distal humerus is shown with breakdown of the entire articular surface. This has resulted from continued use of the joint after arthritic onset. Excessive proliferative osteophytic bone growth also traces the line of attachment for the capsular ligaments. The condition would have been very painful and debilitating and is reminiscent of that seen on WLH 3.

left gonial angle has prominent muscle markings and was part of the male mandible (Fig. 10.74).

Wlh 154

The section of mandibular body extends posteriorly from the second premolar to the ramus. The second molar is moderately worn but retains a basic cusp pattern. In contrast, the third molar has little if any attrition in complete contrast to the third molar of WLH 152. The pattern indicates a young adult possibly

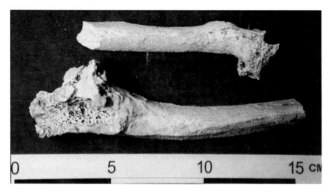

FIGURE 10.73   The associated proximal ends of the right ulna and radius of WLH 153 showing both proliferative bone growth and erosion of their articular surfaces reflecting the osteoarthritic damage observed on the distal humerus (see above) of WLH 3.

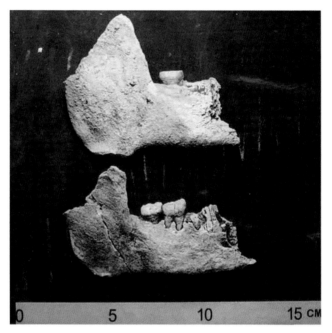

FIGURE 10.74   Two right sections of mandible from WLH 152 (top) and WLH 154 (bottom) showing the difference between their respective rami widths possibly indicating sexual dimorphism.

20–25 years old. The bone is significantly smaller than the WLH 152 mandible and can only underpin the likelihood of this individual being female. As mentioned above the bone has now been included with those fragments determined as female found with the WLH 153 burial.

The close placement of WLH 152 and WLH 153 suggests they possibly lived at a similar time and received a similar disposal. That is suggested by their respective mode of burial, grave shape, time of exposure and post mortem bone scatters together with bone condition, degree of fossilisation and colouration. Bone distribution was confined to a specific area, denser in the middle, circular in its spread and the prominent position of the cranium and/or cranial sections in the centre. The loss of major sections of their respective skeletons suggests that they may have been partial bundle burials occupying a moderately narrow pit covering a small, discrete area. However, it is difficult to explain how elements of both skeletons remained in anatomical approximation if they were just burials of dry bones. Other taphonomic evidence for whole body disposal includes the carbonate encrusted lower right arm and right foot of WLH 152 that is maintained integrity which is hard to explain given the lack of carbonate encrustation on the other bones. Moreover, the loose but closely associated feet and hand bones of WLH 153 together with the head of the right femur still held within the right acetabulum of the pelvis are also hard to explain unless they retained some soft tissue such as ligament or sinew that held them together at burial retaining their natural anatomical position. Such burials may indicate they were the last stage of a complex burial procedure or ceremony involving laying out the body for partial decomposition before final interment. This might also explain the loss of some parts of the skeletons perhaps by animal or bird scavenging during the laying out phase.

The two male burials have an obvious robust morphology. The WLH 152 male has a brow ridge like but not as prominent as that of WLH 50. Similarly, its maximum cranial thickness (13.7 mm) is one of the thickest cranial vaults in the Willandra series except for WLH 50. These include those of WLH 19 (15.0 mm), WLH 27 (14.0 mm), WLH 45 (14.0 mm), WLH 18 (13.5 mm) and WLH 101 (13.5 mm). Moreover, with the few cranial fragments found it is likely that the maximum thickness of the WLH 152 vault may be greater elsewhere. Cortical bone thickness is not just a reflection of nutritional standard but is a standard morphological characteristic associated with robustly built individuals from the region. Both males have thick cortices but appear to have somewhat contrasting stature indicating stature may not be a factor in cortical bone thickness but rather a genetic trait of the population.

The WLH 153 burial contains the remains of both a male and a female person. The female was short (163 cm) but stocky with some moderately robust features suggesting this woman was from the same group of people as the robust males. The reason for her inclusion with WLH 153 is unknown although she may have been related either through marriage or family. She was younger than the man and may have died earlier. Custom may have required that he retained parts of her skeleton until he died and both were then buried together. The stature of the individuals fall within the range attributed to people leaving their foot prints around an Ice Age wetland area not far from this burial site which is dated to 20 ka (Webb et al., 2006; Webb, 2007).

Metrical data WLH 152

Right femur

Length (estimated)	505		
Mid-shaft diameter			
Antero-posterior	33		
Medio-lateral	32		
Head diameter (est.)	52		
right ulna			
Length	310		
Femoral head	Thickness	32	
	Width	29	
right radius			
Estimated length	290		
Right humerus:			
Mid-shaft diameter at humeral tuberosity			
Antero-posterior	22		
Medio-lateral	21		
right tibia			
Mid-shaft diameter			
Antero-posterior	45		
Medio-lateral	28		
right patella			
Width	44		
Length	51		
Thickness	23		
left patella			
Width (est.)	43		
Length	51		
Thickness	23		

Personal height (cm)

Using Ulna length	180
Using Radius length	179
Using Femoral length	176–178 (177)
Average of all three bones	179

Cranial measurements (mm)

Bi-frontale	135
Bi-maxillofrontale	33.7
Mastoid height	44.1
Mastoid width	19.6

Brow ridge	**Lateral**	**Middle**	**Medial**
Right	12.1	15.2	17.6
Left	–	18.1	20.0

Cranial thickness:	Maximum	13.7
	At inion	24.0

Right glenoid fossa depth	5.3
Left glenoid fossa depth	6.6

Right malar

Height of body   30.7
Overall height   47.1
Thickness, lower edge 9.5
Thickness through   12.2
body
Width   39.5
Width of zygomatic   16.2
Mandibular measurements (mm)
Right mandibular body
Acromial width   26.3
Ramus minimum   51.0
breadth
Body height   37.5
Body breadth (esti-   18.0
mated)
Gonial angle 105°
Left mandibular body
Breadth   18.2
Height   37.4
Dental measurements (mm)

	MD	BL	EH
Upper Right Pm^2	13.0	9.5	–
Lower Right M1	13.1	12.7	4.7
Lower Right M2	13.2	13.6	5.0
Lower Right M3	13.0	12.4	5.5

Metrical data WLH 153
right radius
Length   267
Right humerus
Length   320
Mid-shaft diameter
Antero-posterior   24.6
Medio-lateral   19.6
Cortical bone thick- 6.6
ness
Left femur
Mid-shaft thickness
Antero-posterior   33.5
Medio-lateral   27.1
Cortical bone thick- 8.5
ness

Stature (cm)	Male	Female
Using radius	173	170
Using humerus	166	163

Cranial measurements (mm)
Mastoid length   30.2
Inion thickness   14.9
Max cranial thickness 11.0
Left sciatic notch angle 40^0
Dental measurements (mm)

	MD	BL	EH
Lower left canine	7.1	9.8	9.2

Mandibular measurements (mm)
Symphysial thickness   13.5
Right condyle   28.5
Right gonial angle   104⁰

Metrical data WLH 154

Brow ridge	Lateral	Middle	Medial
thickness: Left	8.1	6.6	12.5

Mandibular measurements (mm)

Gonial angle	108°		
Minimum ramus breadth	35.5		
Body height	33		
Body breadth	14.1		
Acromial width	23.3		
Right Condyle	20.5		

Dental measurements (mm)

	MD	BL	EH
Lower right M2	12.5	10.9	5.0
Lower right M3	11.4	10.8	5.2

A PIONEER IN UNDERSTANDING THE FIRST AUSTRALIANS

A Personal Recollection

Peter Clark 1955–2004

Peter Clark has been mentioned several times in this book and some of his work and data have been included in it. Therefore, it is appropriate that I say a little more about this generous, hardworking and well-meaning man that deeply cared for and loved the Willandra and played a fundamental role in its history and development in its early stages.

I knew Peter Clark for over 25 years until his premature death on 1 July 2004 from a brain tumour. As an undergraduate in Prehistory at the Australian National University he had worked on various archaeological sites including Lake Mungo. He knew about archaeological methods, was a capable photographer, had skills in making and casting stone tools and bone and had an innate ability to read the landscape and match it with the archaeology. In March 1979, he took a position as resident Archaeologist at Lake Mungo National Park and his interests in megafauna, zoology, geomorphic weathering and erosion processes of archaeological surfaces, stone sourcing, park management and conservation, won him the position. The job was daunting but Peter's resourceful nature and innate bush skills were up to the challenge, there was nobody more suitable. It was required he live on site, alone in the old Mungo Station homestead, the isolation would not have suited many. While there, researchers came and went, but he was always immersed in the Willandra's special landscape and he made the most of it applying his talents. His life required a good deal of self-sufficiency as well as electrical and mechanical competence and Peter was nothing if not self-sufficient and mechanically savvy.

Peter grew up on the family sheep property outside Boorowa, New South Wales and he made and fixed things from a very young. Living at Mungo he had to maintain the facilities and electrical and mechanical equipment, maintain roads and tracks and continue his archaeological responsibilities. Moreover, he maintained a base for those who wanted to work out there because it was 140 km of rough dirt tracks to the nearest town.

Peter was a true 'multipologist' able to understand and converse in the realms of geology, palaeontology, ecology, geochronology, geomorphology, anthropology. He understood and could link up interdisciplinary relationships in all those areas rather than treating archaeology as a discipline in isolation. Fieldwork was his forte and he was at his best within and moving across the landscape. Visiting Lake Mungo it was a revelation because Peter had always discovered something new and I was immediately immersed in his infectious enthusiasm about what it was. After driving kilometres and then walking over a sand dune or two, across some bare scalds and through some scrub there would be what he wanted you to see. You were lost, but there *it* was. He appreciated people, landscape, climate and environmental connections and visited communities in the Simpson Desert where Wangkangurru Elders taught him these skills. He realised why those people were a very special source of knowledge for interpreting archaeology.

He actively encouraged others, without having agendas, jealousies or claims to knowledge all of which he shared freely. Peter was saddened by the comparative lack of interest in the area as he was at those who would put bureaucratic obstacles in the way of research, something that was increasingly becoming a problem. He was only interested in the wider story of archaeological debris and finding the disparate clues that he could join up to unravel that story. His animated gestures, first pointing to the archaeological debris then out to the surrounding landscape, brought the two together for the onlooker and described the way things were likely to have happened as he drew listeners onto the ancient stage upon which he stood. His innate bush skills, good nature and personal enthusiasm for his work underpinned his highly infectious enthusiasm.

Peter's site survey and excavation data produced a very large, rich tapestry of Willandra's prehistory. He soon realised the whole region was just one big archaeological site; the biggest in Australia. It had to be correctly managed from the beginning and a detailed site inventory was the logical place to start. There was obviously a great deal of information out there and, while he had not scratched the surface, what he could see about him he knew required a system of recording that would last, be adaptable and user-friendly. It had to be developed as quickly as possible because he could literally watch valuable archaeological remains emerge, become weathered, fragment and blow away in weeks or months. In response he developed the Site Designation Areas (SDA) system that could be overlayed on the five major lakes. I have described the SDA system in Chapter 9, and it is all we have to trace the original sites recorded mostly by Peter in the late 1970s and 1980s. He recorded site and artefact type, faunal content, distribution of burials and, where he could, the age of the site, either deduced stratigraphically or occasionally radiocarbon dated. In this way, he recorded over 200 sites including over 100 places where skeletal remains were emerging. Unfortunately, most of these places have now been lost through erosional forces.

Much of Peter's data remained unpublished but during 1986 and 1987 but he assembled several large reports for the NSW Department of Environment and Planning in Sydney as well as the Western Lands Commission of NSW. By this time the Willandra had been made a World Heritage Area and Peter's extensive knowledge of the region, his years of work and expertise, not only in its archaeological and geomorphological heritage but also in its general ecology and history, drew him into many issues regarding future management of the region. As an expert, he naturally became the focus of attention for major government organisations who were now involved in management of the new World Heritage Area.

A particularly important aspect of his work was the close association and friendship that he built with local Aboriginal communities over many years. His innate honesty, forthright manner and the way he spoke plainly about Willandra's heritage as well as his obvious feeling for the land and understanding of Aboriginal wishes and aspirations regarding the region, helped build a special trust between him and those communities. Aboriginal people in the area saw Peter not only as a friend but a special person that they could talk to about the future and confide in and trust with confidential cultural and personal information.

Peter left Mungo in 1984, just about the time it was put forward for World Heritage listing. His legacy was much more than a local archaeologist in residence. His ability to explain things simply and clearly also helped him network with local landholders. His quiet but confident nature, heartfelt attachment to the land, knowledge of the pastoral industry and its problems and balance that with the archaeology and his good nature warmed him to a wide variety of agencies, organisations, pastoral property owners and individuals. Peter was never someone to promote himself; indeed, his modesty was a very valuable asset.

He drew together research, land care agencies, indigenous people, government organisations, pastoralists, academics, other scientists, bureaucrats and friends all of which grew even when he moved on to rise to various managerial positions in State Government land and water departments. But he never lost his interest in archaeology, the wellbeing of the Willandra and his mates.

Peter was never afraid to 'arm wave' about certain issues and ideas in prehistory, so that was another reason he and I always got on well. His philosophy like mine was that we should put everything on the table for discussion, even ostensibly crazy ideas. Discussion should not be undertaken with a limited outlook bound by the present evidence alone which sometimes was miniscule. He believed, as I do, that you cannot move forward or come up with strategies for progress if you wait for definitive truths and the complete data to appear before you. He and I agreed that theoretical discussion was life blood to keep the engine of research, thought and interpretation going in archaeology and palaeoanthropology as well as present a vision of possibilities even, as I say, if they seemed crazy.

I arrived in Broome with my family on the evening of the 31 May 2004. At 4.00 the next morning I received a call telling me of Peter's passing. I did not realise how strong the bond between us had become over the years. Peter had a compartmentalised life that reached into many disparate areas. But he was always an archaeologist, palaeontologist, environmentalist and Professor of GSCS (Good Sound Common Sense).

In many ways, I regarded him as a brother. Peter was a one-off and I was extremely proud to know him, call him my friend and have him as a brother colleague. I remember sitting with him on clean, ancient water-washed sand under a wide blue sky and shady tree at Lake Mungo with a lunchtime 'billy' boiling in the days when a small lunch time fire was allowed, he would look around, turn to me, grin and use his favourite saying: 'you wouldn't be dead for quids…!!'

REFERENCES

Brown, P. (1982). Coobool Creek, a prehistoric Australian hominin population (Unpublished PhD Thesis). Australian National University, Canberra.

Dowling, P., Hamm, G., Klaver, J., Littleton, J., Sanderson, N., & Webb, S.G. (1985). Middle Willandra Creek archaeological site survey (Unpublished report to the New South Wales National Parks and Wildlife Service). Canberra, Australian National University.

Shipman, P., Walker, A., & Bichell, D. (1985). *The Human Skeleton*. Cambridge (Mass.): Harvard University Press.

Webb, S. G. (1995). *Palaeopathology of Australian Aborigines*. Cambridge: Cambridge University Press.

Webb, S. G. (2007). Further research of the Willandra Lakes fossil footprint site, southeastern Australia. *Journal of Human Evolution, 52*, 711–715.

Webb, S. G., Cupper, M., & Robbins, R. (2006). Pleistocene human footprints from the Willandra Lakes, southeastern Australia. *Journal of Human Evolution, 50*, 405–413.

Index